Oxidative Stress: Human Health and Medicine

Oxidative Stress: Human Health and Medicine

Editor: Everette Wheaton

FOSTER
ACADEMICS
www.fosteracademics.com

www.fosteracademics.com

FOSTER ACADEMICS

Cataloging-in-Publication Data

Oxidative stress : human health and medicine / edited by Everette Wheaton.
p. cm.
Includes bibliographical references and index.
ISBN 978-1-64646-563-7
1. Oxidative stress. 2. Oxidative stress--Physiological effect. 3. Cardiovascular system--Pathophysiology.
4. Oxidation, Physiological. I. Wheaton, Everette.
RB170 .O95 2023
616.07--dc23

Foster Academics,
118-35 Queens Blvd., Suite 400,
Forest Hills, NY 11375, USA

ISBN 978-1-64646-563-7 (Hardback)

Contents

Preface

This book has been an outcome of determined endeavour from a group of educationists in the field. The primary objective was to involve a broad spectrum of professionals from diverse cultural background involved in the field for developing new researches. The book not only targets students but also scholars pursuing higher research for further enhancement of the theoretical and practical applications of the subject.

The cells of the body generate free radicals throughout the metabolic processes. These cells also generate antioxidants, which neutralize the free radicals. Generally, the body is capable of balancing free radicals and antioxidants. Oxidative stress refers to a condition characterized by an imbalance of antioxidants and free radicals in the body that can cause tissue and cell damage. It occurs naturally and contributes to the process of aging. Long-term oxidative stress plays a role in the development of a variety of chronic diseases, such as heart disease, cancer and diabetes. Various factors that increase the risk of oxidative stress include obesity, alcohol consumption, pollution, diets having high components of sugar and fat, and consumption of processed foods. Some dietary and lifestyle changes may aid in the reduction of oxidative stress. It requires consumption of a well-balanced nutritious diet that is rich in vegetables and fruits, maintaining a healthy body weight, and exercising regularly. This book explores all the important aspects related to the impact of oxidative stress on human health. It is an essential guide for both academicians and those who wish to study this topic further.

It was an honour to edit such a profound book and also a challenging task to compile and examine all the relevant data for accuracy and originality. I wish to acknowledge the efforts of the contributors for submitting such brilliant and diverse chapters in the field and for endlessly working for the completion of the book. Last, but not the least; I thank my family for being a constant source of support in all my research endeavours.

Editor

1

Oxidative Stress and Maxi Calcium-Activated Potassium (BK) Channels

Anton Hermann [1,*], Guzel F. Sitdikova [2] and Thomas M. Weiger [1]

[1] Department of Cell Biology, Division of Cellular and Molecular Neurobiology, University of Salzburg, Salzburg 5020, Austria; E-Mail: thomas.weiger@sbg.ac.at

[2] Department of Physiology of Man and Animals, Kazan Federal University, Kazan 420008, Russia; E-Mail: sitdikovaguzel@gmail.com

* Author to whom correspondence should be addressed; E-Mail: anton.hermann@sbg.ac.at;

Academic Editors: Michael Breitenbach and Peter Eckl

Abstract: All cells contain ion channels in their outer (plasma) and inner (organelle) membranes. Ion channels, similar to other proteins, are targets of oxidative impact, which modulates ion fluxes across membranes. Subsequently, these ion currents affect electrical excitability, such as action potential discharge (in neurons, muscle, and receptor cells), alteration of the membrane resting potential, synaptic transmission, hormone secretion, muscle contraction or coordination of the cell cycle. In this chapter we summarize effects of oxidative stress and redox mechanisms on some ion channels, in particular on maxi calcium-activated potassium (BK) channels which play an outstanding role in a plethora of physiological and pathophysiological functions in almost all cells and tissues. We first elaborate on some general features of ion channel structure and function and then summarize effects of oxidative alterations of ion channels and their functional consequences.

Keywords: Maxi calcium-activated potassium ion channels; BK channels; KCNMA1; redox mechanisms; oxidation; hypoxia; ischemia

---

1. Introduction

Ion channels play a pivotal role for the functioning of any cell in the animal as well as in the plant kingdom. An important class of ion channels is the family of potassium (K^+) channels, they are not only in charge of the membrane resting potential or the repolarization of the action potentials, but also control cell proliferation or transmitter/hormone release, to name a few. A subgroup of K^+ channels are the so called calcium (Ca^{2+}) activated K^+ channels which need either an increase of Ca^{2+} at their intracellular face to open or a combination of Ca^{2+} and voltage to function properly. Maxi Ca^{2+} activated K^+ channels, also named BK channels which constitute a subgroup of Ca^{2+} activated K^+ channels, are in the focus of our review. These channels modulate a number of physiological events, like blood pressure, smooth muscle relaxation or electrical tuning of hair cells in the cochlea and have a leading role in many pathophysiological conditions such as epilepsy or the behavioral response to alcohol, to give only a few examples. Oxidative stress on the other side is a physiological byproduct of any aerobic metabolic process and as such common for cells to deal with. Oxidation modulates many pathways in the cell including activity of ion channels like BK channels. The modulatory actions of oxidative stress on BK channels will be in the focal point of this paper. First, in Section 2, we will address general properties of ion channels and in particular the qualities of BK channels. In Section 3, we will specify what we mean by the term oxidative stress, while Section 4 will deal with the impact of redox modulation of BK channels. At the end, we will discuss these findings in the light of their clinical relevance and conclude with perspectives and vistas.

2. Ca^{2+}-Activated K^+ Channels (K_{Ca})

2.1. General Introduction to Ion Channels

The task of cell membranes (plasma/organelles) is to separate cellular compartments, such as extra- and intracellular or extra- and intra-organelle areas to allow for proper performance of biochemical processes. Since the lipid bilayer membranes are almost impermeable for ions, tunnel proteins are inserted to allow for communication between compartments for electrical and chemical signaling. In addition membranes themselves are sites for processing biochemical reactions. Readers familiar with ion channels may omit this section on general features of ion channels.

Since cellular ion channels form minute pores across membranes they are suitable for the passage of ions, in particular monovalent sodium (Na^+), K^+, protons (H^+), the divalent cation Ca^{2+} or the anion chloride (Cl^-). These ions are differently concentrated across cell membranes and various mechanisms are involved in the separation of ions, *i.e.*, (a) active transport systems, which directly consume energy (adenosine triphosphate, ATP) to transport ions against a concentration gradient; (b) transport systems which do not directly consume ATP but use the concentration gradient across the membrane as driving force; or (c) the redistribution of ions due to fixed negative charges provided mostly by intracellular proteins (Donnan equilibrium). In essence, this creates an electro-chemical gradient for each ion which provides for a driving force since for example at the membrane resting potential, endowed in all living cells, none of the ions is at equilibrium [1].

Most ion channels are gated, which means they contain an intrinsic mechanism which allows for either a closed or open configuration. Once the pore is open ions can flow according to their electro-chemical gradient and as moving charges produce a current which provides a potential difference across the membrane. Hence, in difference to our technical current which is brought about by a flow of electrons, living cells use ions as charge carriers.

An enormous amount of knowledge about the structure and function of ion channels has been gathered over the last decades. This development was based on the seminal work by Hodgkin and Huxley (1952) [2]. They provided the basis for proper measurement of membrane potentials and they measured the ionic current flow across cell membranes by application of the "voltage clamp" technique which inspired generations of scientists. Only about 30 years later Neher and Sackmann (1981) [3], by installing the "patch clamp" method, succeeded to measure the flow of ions through single channels in native cell membranes. This technique, in fact allows us to visualize a protein at work. Again, 20 years later, MacKinnon and his group (1998) [4] succeeded to combine molecular-, electrophysiological techniques and structural analysis to develop a 3-dimensional scheme of an ion channel and its functions. It is clear now that there are many more types of ion channels than previously imagined. Although not all details concerning ion channels have been resolved to date, however, the knowledge gathered by scientists on structure and function over the last decades is overwhelming.

In general, ion channel proteins consist of hydrophobic amino acid α-helices inserted into the membrane lipid bilayer connected by hydrophilic amino acid linkers. Cation channels usually contain four domains with two, four, or seven α-helix segments. In some channels amino acids built one continuous fibrous structure (*i.e.*, Na^+-, Ca^{2+} channels), while others consist of four separate domains which form a tetrameric channel within the membrane (*i.e.*, K^+ channels). In most channels "α-subunits" form the pore for the passage of ions. Attached to the α-subunits are in most cases a variety of other proteins—called auxiliary subunits (β, γ, ε, δ) [5], or special channel specific proteins, and/or enzymes, such as kinases, phosphatases or heme [5,6]. To classify ion channels by their mechanism of activation (gating) has been considered as a proper means.

The structure and function of ion channels has been summarized in several reviews [7–11]. The today's extensive realm of ion channel structure and function is brought about by the separation into many special topics, such as for Na^+, Ca^{2+}, K^+, Cl^- [12–16] channels. The field of ion channels developed immensely and we therefore limit our brief introduction to maxi calcium activated K^+ or BK channels which are the main focus of this review. K^+ channels can be devoid of auxiliary subunits but usually have various types of β-subunits, *i.e.*, BK channels consist of four types of β subunits (β1–β4), and a set of γ-subunits (Figure 1). These subunits are of particular interest because of their potential to diversify the transport properties and distribution of channels in various cells and tissues. Attached to the α-subunits these auxiliary subunits cause a variety of different effects on channel gating, on current kinetics or conductance, on trafficking the α-subunit to the cell membrane, link to intracellular cytoskeleton and extracellular matrix proteins, modulation of channel expression or they provide for binding of drugs.

In general, the activity of ion channels which composes the ion current and hence the potential across the membrane is governed by three factors: the number of active channels (N), the channels in the open state (Po = open probability) and the single channel current (i). This can be summarized by the relationship: $I_{tot} = N\ Po\ i$, where I_{tot} is the total current, N is the number of functional channels and i is the single channel current. In most cases Po is modulated by ligands or drugs.

2.2. Types of K_{Ca} Channels

K_{Ca} membrane permeability was first described from experiments using erythrocytes where the presence of internal Ca^{2+} increased the permeability of the cell plasma membrane to K^+ [17]. It took more than a decade to transfer this idea to other cells and to further investigate this phenomenon in more detail. To date a variety of K_{Ca} channels have been described in a great variety of excitable and non-excitable cells of many species and are mainly defined by their biophysical and pharmacological properties.

The K_{Ca} channel family contains eight members according to the sequence homology of transmembrane hydrophobic segments [18] with four subfamilies: **BK** (or K_{Ca}1.1, Slo1, Maxi-K, *KCNMA1*), **SK** (K_{Ca}2.1 (SK1, *KCNN1*), K_{Ca}2.2 (SK2, *KCNN2*), K_{Ca}2.3 (SK3, *KCNN3*), **IK** (K_{Ca}3.1 (IK$_{Ca}$1, SK4, *KCNN4*) and **other subfamilies** (K_{Ca}4.1 (*KCNT1*), K_{Ca}4.2 (SLICK, Slo2.1, *KCNT2*), K_{Ca}5.1 (Slo3, *KCNU1*). These channels generally consist of four α-subunits with six membrane-spanning α-helices (segments) per domain (S1–S6) containing the K^+ pore. Both, N- and C-terminal are usually at the internal, cytosolic side of cells. In addition, Slo 1 (BK) and Slo 3 channels have one more transmembrane helix (S0) which locates the N-terminus to the external side (see Figure 1 and section BK channels). Not all of these channels are responsive to Ca^{2+}—since some are activated by Na^+ (Slo 2) or Cl^- (Slo 3).

2.3. Maxi Calcium-Activated K^+ Channels (BK)

BK channels are expressed by a single gene. The conductance of single BK channels is up to 300 pS and hence comprises the largest single-channel conductance of all K^+ channels. To date a vast amount of information has been gathered regarding their biophysical, structural and functional, physiological, pathophysiological and pharmacological properties (recently reviewed in [19–37]). BK channels are usually synergistically activated by both—Ca^{2+}/Mg^{2+}—metal ions and by membrane voltage. The channels can be also activated in the absence of Ca^{2+} but then require an extremely large depolarization of 100–200 mV which is not physiologically relevant but important for the understanding of some biophysical properties of these channels. BK channel α-subunits consist of a total of seven transmembrane segments with a S0 segment preceding the six transmembrane segments (S1–S6). This renders the N-terminus (amino terminal) at the extracellular side of the membrane (Figure 1). The transmembrane segments (S1–S4) consist of various charged amino acid residues which confer voltage sensitivity to the channels. The S0 segment appears mainly required for interaction of α-subunits with auxiliary β-subunits as well as for targeting the channels to the plasma membrane. As in other voltage dependent K^+ channels, four pore-forming loops between the segments S5-S6 of each α-subunit configure the ion selectivity filter. The large C-terminus (carboxyl terminal) comprises about two thirds of the α-subunit protein and contains various binding sites for kinases, phosphatases and negatively charged Ca^{2+} binding sites as well as two so called RCK-domains (regulatory domain of K^+ conductance) (Figure 1).

Multiple splice variants of the α-subunit have been identified resulting in a great variety of channel properties in various cell types [38]. Through alternative splicing the pore forming α-subunit contains at its C-terminus a cysteine-rich 59-amino-acid insert between RCK domains called stress-axis regulated exon (STREX). STREX exon expression is under physiological conditions initiated by the stress-axis adrenocorticotropic hormone [39]. STREX inserts cause BK channels to activate at more negative potentials, enhance activation and decrease deactivation which leads to increased repetitive firing of action potentials. STREX inserts can also be artificially produced by growing cells in phenol red which causes a significant increase in channel sensitivity to inhibition by oxidation [40]. Phenol red, which is routinely used as a pH indicator in tissue culture media, bears structural resemblance to some nonsteroidal estrogens and has significant estrogenic activity [41]. Further findings indicate that maxi-K channel transcripts are differentially spliced by 17β-estradiol, which may contribute to changes in isoform expression [42] and may also lead to STREX expression in GH3 cells which carry estrogen receptors [43].

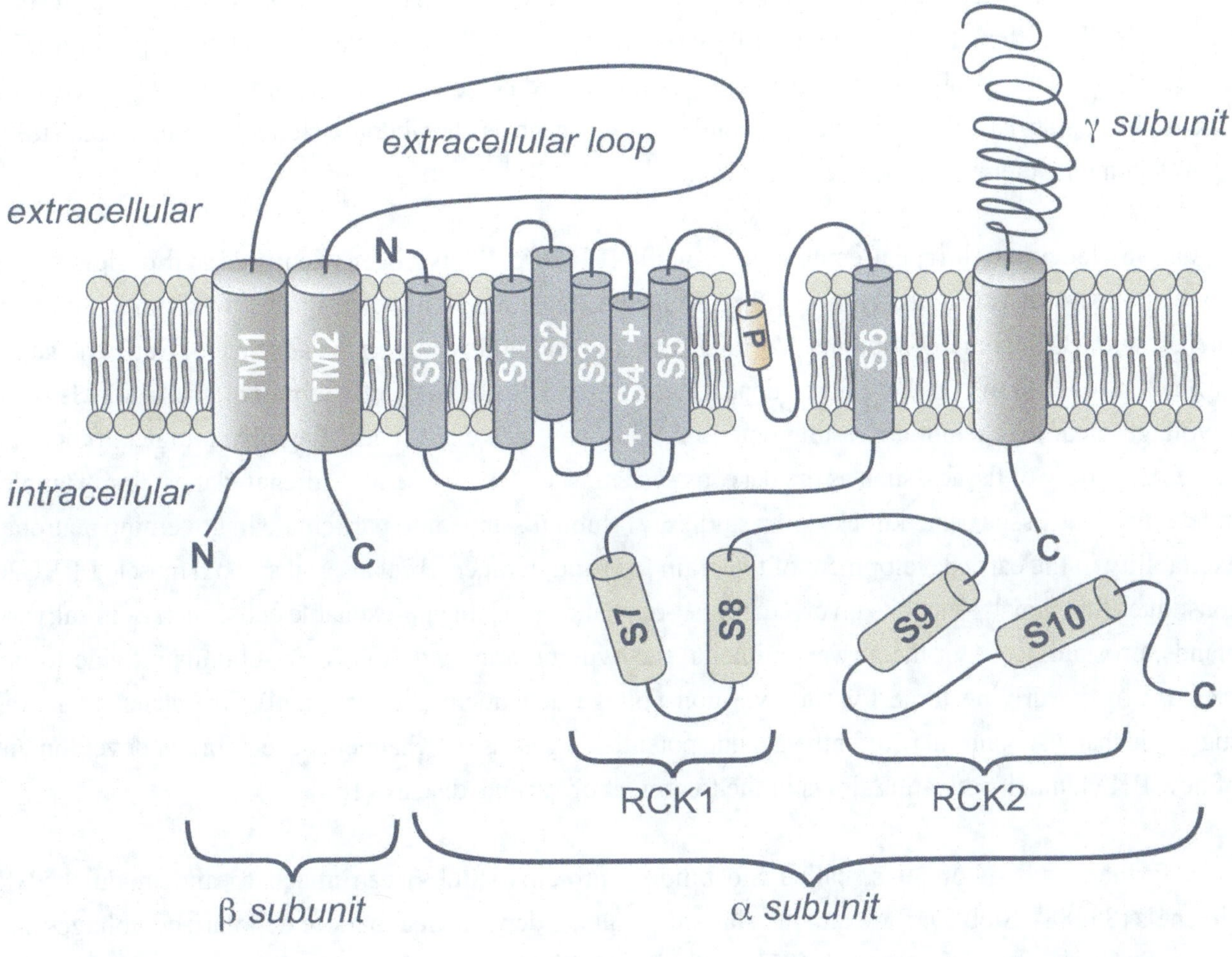

Figure 1. Schematic structure and membrane topology of maxi calcium-activated potassium (BK) channel α, β and γ subunits (adapted from [44]). See text for a discussion of the structure.

Specific blockers of the channels are tetraethylammonium (TEA) at submillimolar concentrations [45,46], iberiotoxin (from scorpion) or paxilline (mycotoxin from *Penicillium paxilli*) [47,48]). BK channels can be expressed together with other types of K_{Ca} channels (SK or IK) [49] or in particular with Ca^{2+} channels [50,51] which makes them important modulators of electrical discharge activity and synaptic transmission. Ca^{2+} as a major player in cellular signaling is involved in a large variety of cell functions and links cell excitation to cell metabolism and gene expression.

BK channels are associated and modulated by a wide variety of intra- and extracellular factors, such as auxiliary subunits (β, γ), Slobs (slo binding protein), phosphorylation, gasotransmitter action (nitrosylation, carboxylation, sulfhydration) or by redox mechanisms [11,44,52–67]. The BK α-subunits assemble 1:1 with four different auxiliary types of **β-subunits** (β1, β2, β3 or β4) (Figure 1). β-subunits are members of the protein superfamily of oxidoreductases [68,69] which contain 191–235 amino acids [70–73]. BK β-subunits consist of two transmembrane segments (TM1, TM2) with both, the N-terminal and the C-terminal located intracellularly. In different tissues four different genes can be expressed, as in smooth muscle, adrenal chromaffin cells or in neurons. β-subunits are involved in modifying voltage sensitivity, current kinetics and/or pharmacological properties of BK channels [65,70,74]. β-subunits are also responsible for tissue specificity, they can alter channel activity by activation of protein kinases, confer hormone (estradiol) activation, and alter toxin binding to the channels [75]. They are involved in current inactivation by a flexible N-terminal chain and ball structure which blocks the channel's pore [76,77]. In the brain, beta 4-subunits for example, inhibit BK channel activation and slow down channel kinetics [65,78–80], confer resistance to peptide blockers such as charybdotoxin and iberiotoxin [65,81,82] or protect against temporal lobe seizures [79].

Various leucine-rich repeat containing proteins (LRRCxx), have been identified and designated as auxiliary **γ-subunits** (Figure 1) [66,83,84]. These proteins, which are clearly distinct from β-subunits, interact with BK channel α-subunits. Various types of γ-subunits cause in the absence of Ca^{2+} a negative shift of up to ~140 mV in the voltage dependence of the BK channel activation, *i.e.*, the channels open at voltages near the membrane resting potential. This way they exert interesting physiological functions in a great variety of tissues, such as the nervous system, skeletal muscle, and adrenal glands. For example, in fetal nervous tissue using knock-down studies, γ1-subunits appear to participate in governing neuronal excitability in the early development of the brain [83] and in rat cerebral arterial smooth muscle LRRC26 constitutes functional γ-subunits involved in vaso-regulation [84]. In non-excitable cells, such as in salivary glands, prostate, testes or the airway epithelia, the hyperpolarizing actions of γ-subunits provide for an increase of the driving force for Ca^{2+} via non-voltage dependent (ligand gated) Ca^{2+} channels. It was suggested that γ-subunits may offer therapeutic potential targets as BK channel openers for the development of new BK channel–activating drugs in the treatment of various diseases [66].

A further class of proteins, called Slo binding proteins (**Slobs**) can attach to and modulate BK channels [85–88]. Slob 57, for example, shifts the voltage dependence to more depolarized voltages and causes faster closure of the channels [87]. Furthermore, Slob exerts a diurnal cycle *in vivo* [89] indicating that BK channel activity changes as a function of day time imparting circadian rhythmicity to neurons. Other Slobs like Slob71 or Slob 53, shift the voltage dependence to less depolarized voltages but have no effect on channel kinetics [88].

Some BK channels can be activated by stretch or pressure. These stretch-activated BK channels (SAKCaC) [90,91] are expressed in a variety of tissues such as in myocytes or neurons and modulate vascular smooth muscle tone and endocrine cell secretion. A STREX insert between RCK1 and RCK2 domains at the channel's C-terminal α-subunit is indicated to confer stretch sensitivity to the channels [90,92]. However, other BK channels lacking the STREX insert still remain sensitive to membrane stretch suggesting that additional structures of the channel may be responsible for mechanical coupling to the cell membrane [93].

2.4. Function of BK Channels in Norm and Pathology

BK channels contribute to various functions, such as controlling electrical discharge activity of nerve and muscle cells. Since the opening of K^+ channels drives the membrane potential towards the K^+ equilibrium potential, this will result in hyperpolarization of the membrane resting potential which will lead a nerve cell away from excitation. On the other hand, it may speed up the repolarization of action potentials, which makes their duration shorter and more action potentials per time can be generated. At the synapse the duration of the excitation, *i.e.*, duration of the action potential is translated into the amount of opening of Ca^{2+} channels, influx of Ca^{2+} into the cell and an increase of the internal Ca^{2+} concentration which directly relates to the quantity of transmitter release or hormone secretion [94]. Similar mechanisms play a role in vaso-regulation, auditory tuning of the hair cells, in erectile processes and participate in mediating drug actions such as ethanol or acetaldehyde. In circadian rhythm generation BK channels are expressed in neurons of the supra chiasmatic nucleus during night and are removed during the day causing silencing or excitation, respectively [95]. Targeting of BK channels to appropriate membranes is important for the proper functioning of cells and organelles. Trafficking to and expression of BK channels in the plasma membrane has been found to be regulated by distinct splicing motifs located within the intracellular C-terminal RCK domains [96]. In particular a splice variant that excluded these motifs prevented cell surface expression of BK channels and suggests that such a mechanism impacts physiology and pathophysiology.

BK channels are not only present in the plasma membrane of cells but are also located in the membranes of cellular organelles such as mitochondria, nucleus, endoplasmic reticulum or the Golgi apparatus (reviewed in [97]). Information on the function and regulation of BK channel in organelles is at its infancy. A recent report indicates that nuclear BK channels (nBK) regulate gene expression of hippocampal neurons in a synaptic activity-evoked, Ca^{2+} dependent manner, suggesting that nBK channels play an important role as modulator and molecular linker of neuronal activity dependent functions and nuclear Ca^{2+} [98]. Since most information is available from mitochondria we will briefly discuss these channels in a later section.

Channelopathies are caused by mutations at one or more amino acids of the channel protein which may lead to dysfunction of cells and organs. These mutations can occur at any site of the channel protein, the pore, the voltage sensor, or the inactivation structures, including auxiliary subunits. Channelopathies affect all kinds of channels (voltage-, ligand-, mechano-gated, *etc.*), and hence are involved in a multitude of disorders, such as epilepsy, stroke, paroxysmal movements, cerebellar ataxia, hearing loss, autism, asthma mental deficiency, myotonia, heart diseases, hypertension, cystic fibrosis, bladder or gastric hypermotility, erectile dysfunctions, *etc.* (for recent reviews see [25,32,99–102]). BK channel mRNA expression is lower in the prefrontal cortex of schizophrenic, autistic, and mentally retarded persons [103,104]. Mutation at the α-subunit which is associated with idiopathic generalized epilepsy and paroxysmal dyskinesia [105] appear to result from augmented Ca^{2+} sensitivity at the RCK1 binding site together with mutations at the brain specific β4-subunit. Mutant BK channels being more sensitive to Ca^{2+} were found to increase excitability in humans by causing a more rapid repolarization of action potentials which in turn limits the amount of Ca^{2+} flowing into the cell. An enhanced repolarization favors a faster removal of inactivation from Na^+ channels and thus lets neurons fire at a higher frequency. A nonfunctional β4-subunit which under wild type conditions broadens the actions potential on the other hand also leads to faster action potential repolarization. Eventually this results in an increased discharge of action potentials which may lead to epilepsy, paroxysmal movement disorders, or alcohol dependent initiation of dyskinesias [79,106]. Although depletion of BK channels in the brain appears not to be lethal it produces a variety of deficiencies as indicated above.

This introduction to the realm of ion channels as well as BK channels in detail was intended to provide some background for (a) getting to know and summarizing the structure and function of these proteins and for (b) providing a basis for the following elaboration of their modulation by oxidative stress.

3. Oxidative Stress

Oxidative stress in general has been introduced by other authors of this special issue, defined, interpreted and generally covered and therefore will not be further discussed. In this section we intend to summarize some aspects of redox effects and oxidative stress on BK channels. Reactive oxygen species (ROS) are involved in a magnitude of oxidative stress modulations but also in the physiological regulation of a great variety of proteins and cell functions as outlined in various reviews [34,61,97,107–117]. The sulfur containing **cysteine** (Cys or C) and/or **methionine** (Met or M) residues (Figure 2) within proteins are preferred targets of redox modulation, *i.e.*, reversible oxidation/reduction. Cysteines can be oxidized at their thiol groups where two cysteine molecules link to form disulfide bonds internally within the protein or with other proteins. For experiments, cysteine specific reagents 5'5-dithio-bis(2-nitrobenzoic acid) (DTNB), methanethiosulfonate ethylammonium (MTSEA) or p-chloromercuribenzoic acid (PCMB) have been used. Biochemically, disulfide bonds are readily reduced, *i.e.*, by dithiothretiol (DTT) or mercaptoethanol (β-ME) to convert them back to thiol cysteines. *In vivo* reversibility of cysteines is brought about by antioxidants such as the tripeptide glutathione (GSH) or the small protein thioredoxine (TRX) which is present in all organisms [118]. Methionine can be preferentially oxidized by agents such as chloramine-T (Ch-T) under physiological conditions or by *N*-chlorosuccinimid (NCS) [119]. Methionine residues have been hypothesized to function as endogenous antioxidants in proteins [119]. Oxidation leads to polar methionine sulfoxide (Met-O) which can be reversed by methionine sulfoxide reductases (MSRA). Some oxidizing/reducing (ROS) agents are listed in Table 1.

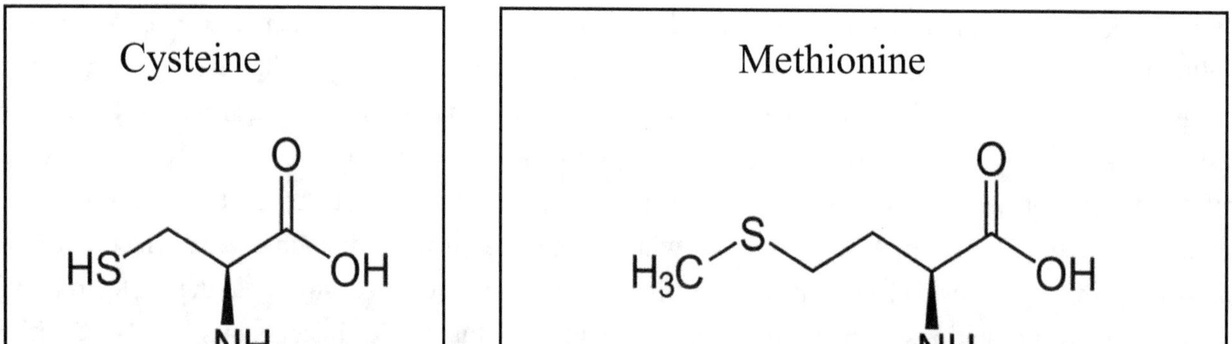

Figure 2. Structure of the sulfur containing amino acids cysteine and methionine.

Table 1. Reactive oxygen species (ROS) agents, reducing agents and gasotransmitters that are produced by cells and/or are experimentally used chemicals for modulating BK channels.

Oxidizing/Reducing agents at BK channels			
Chemical nomenclature	Abbrev./ Formula	Company	Notes/literature
Oxidizing agents			
Hydrogen peroxide	$\mathbf{H_2O_2}$	Fisher Scientific; Sigma	commonly used oxidizing agent. H_2O_2 is naturally produced by the cell metabolism, is membrane permeable and relatively stable [120,121].
Superoxide anion	$\mathbf{O_2^{-\bullet}}$	Sigma	radical; range of 10^{-11} to 10^{-10} M; [122] antagonist: superoxide dismutase.
Glutathione disulfide	**GSSG**	Sigma	physiological oxidized form of GSH [123].
Chloramine-T	**Ch-T**	Sigma	oxidizes methionine [119]; oxidation leads to polar methionine sulfoxide (Met-O), reversed by methionine sulfoxide reductases [109].
N-chlorosuccinimide	**NCS**	Sigma	oxidizes methionine [119].
4,4'-dithiodipyridine	**4,4DTDP**	Sigma	[124]
2,2'-dithiodipyridine	**2,2DTDP**	Sigma	[125,126]
5'5-dithio-bis (2-nitrobenzoic acid	**DTNB**	Sigma	cysteine-specific reagent [109,123,127].
(2-Aminoethyl) methan-ethiosulfonate hydrochloide	**MTSEA**$^+$	Toronto Research	cysteine-specific, covalently modifies free thiol groups [125].
2-(trimethylammonium) ethyl methanethiosulfonate, bromide	**MTSET**	Toronto Research	[128]
p-chloromercuribenzoic acid	**pCMB**	Sigma	cysteine-specific [109].
Sodium (2-sulfanatoethyl) methanethiosulfonate	**MTSES**	Toronto Research	[128]
Thimerosal		Sigma	sulfhydryl reagent; its oxidizing ability is related to the presence of mercury [124,129].
Diamide		Sigma	[125]
Hydroxyl radicals	$\mathbf{OH^{-\bullet}}$		[130]
Peroxynitrite	$\mathbf{ONOO^-}$		formed at a low rate by the reaction of $NO^\bullet$ with ($O_2^{-\bullet}$) in a 1:1 stoichiometry for synthesis cf. [131].
Rose bengal	**4,5,6,7-tetrachloro-2',4',5',7'-tetraiodo fluorescein**	Sigma	Stain, clinical trials in some cancer therapies; generates singlet oxygen from triplet oxygen [132].
Normoxia			test solutions equilibrated with room air [126,133,134].

Table 1. *Cont.*

Oxidizing/Reducing agents at BK channels			
Reducing agents			
Chemical nomenclature	Abbrev./ Formula	Company	notes/literature
Dithiothreitol	**DTT**	Sigma	reduction of disulfide links [124,126,129,135].
β-mercaptoethanol	**β-ME**	Sigma	[125]
Glutathione	**GSH**		Physiological reduced form [136].
Nicotinamide adenine dinucleotide hydrate	**NADH**	Sigma	reduced form of physiological oxidized form NAD^+ [137].
Thioredoxine	**TRX**	Sigma	in reduced form act as oxidoreductases [118].
Methionine sulfoxide reductase	**MSRA**	Sigma	converts Met-O to Met [109].
Hypoxia			bubbling of experimental solutions with nitrogen gas [126,133,134].
Gasotransmitters			
Nitric oxide, gas—donors Diethylamine NONOate Sodium nitroprusside S-Nitroso-*N*-Acetyl-D,L-Penicillamine, *etc.*	**NO** **DEA-NO** **SNP** **SNAP**	Alexis Corp. Cayman Sigma	short lived radical (seconds) acts directly or via the sGC—cGMP—PKG signaling pathway [138]. NO range of (10^{-9} to 10^{-7} M) [124,132].
Hydrogen sulfide, gas—donors Sodium hydrogen sulfide Sodium sulfide, *etc.*	**H_2S** **NaHS** **Na_2S**	Sigma	[60,129,139]
Carbon monoxide, gas—donors Carbon monoxide releasing molecules	**CO** **CORM-A1** **CORM-2** **CORM-3**	Sigma	[140] [141,142] [143] [144]
N-ethylmaleimide	**NEM**	Sigma	alkylating agent [128]

Since these amino acids are particularly sensitive to redox reactions ROS may act on these targets in an autocrine or paracrine manner, *i.e.*, on the cell itself or in its near vicinity. Other amino acids, such as arginine, lysine, proline, histidine, tryptophan and tyrosine may also provide targets for oxidation. The reactions may proceed extremely fast as with hydroxyl radicals (•OH) with a life-time of a few nanoseconds and a diffusion radius of a few nanometers [130], whereas the weaker radical nitric oxide (NO) has a life span of several seconds and a diffusion radius of several hundred micrometers [145]. Redox mechanisms linked to cell metabolism appear extremely important in the modulation of cellular signaling pathways and for the regulation of ion channels and electrical activity.

Human BK (Slo1) channels have 29 cysteines and 30 methionines per α-subunit which adds to 116 and 120 residues for each tetrameric complex [113]. Due to conformational properties many of these amino acids are buried within the 3D-structure of the protein and hence are not readily accessible to modifications. Functional consequences of oxidation of cysteine and methionine residues have been reviewed by Sahoo *et al.* (2014) [113].

4. Ion Channels and Oxidative Stress

Ion channels are modulated and regulated by a vast array of mechanisms, drugs and ligands. One of the first recognized and more detailed investigated posttranslational modifications of ion channels was their modulation by phosphorylation. In the human genome more than 500 putative kinase genes have been identified [146]. The attachment of phosphate groups to amino acids by protein kinases and the antagonistic action of phosphatases paved the way in our conceptualization of posttranslational modification of ion channel functioning [52,53,59,147–149]. Thereafter the field was open for more modulatory factors from signaling pathways, such as by G-proteins or second messengers, by redox mechanisms, by the action of gasotransmitters, such as nitric oxide (NO), carbon monoxide (CO), or hydrogen sulfide (H_2S) [1,44,57,59,86,150] or more recently by S-acylation (the post-translational attachment of fatty acids to cysteine residues) [151]. Modulation of ion channels is involved in physiological processes such as transmitter release, hormone secretion or muscle contraction to name just a few [11,52,152–155]. Alterations of channels proteins by oxidative stress and its functional consequences constitutes one further way of channel modulation in a great variety of cells and tissues which will be covered in the following sections.

4.1. BK Channels and Redox Modification/Regulation

The brain, although it only accounts for 2% of the body weight, consumes ~20% of O_2 and hence during metabolic activity produces large amounts ROS [156]. High amounts of O_2 appear to be needed for the production of ATP which is used to maintain ion homeostasis in the brain in order to run active ion transport to establish transmembrane ion concentration gradients, intracellular neuronal transport systems or the synaptic transmission/neurosecretion machinery.

From brain-derived human BK channels (hslo) expressed in HEK cells (Human Embryonic Kidney cells) DiChiara and Reinhart (1997) [135] provided first evidence that the reducing agent dithiothreitol (DTT) shifted the voltage activation of the channels to more negative potentials and increased current activation as well as channel open probability. These modulations cause the channels to open earlier, *i.e.*, during action potential repolarization and shorten the action potential duration which speeds up repetitive firing. Oxidation, by using hydrogen peroxide (H_2O_2), had the opposite effect. In contrast, BK channels cloned from *Drosophila* (dslo) were not modulated by DTT. It was suggested that in hslo proteins disulfide bonds were formed between cysteines—predominantly at the channels' large interior C-terminus region. Soh *et al.* 2001 [136], reported that in neonatal rat hippocampal neurons the reducing reagent glutathione (GSH) increased BK channel activity whereas its oxidized form (GSSH) had the opposite effect indicating a redox modulatory mechanism when applied to the intracellular side of the cell membrane. On the other hand, after intracellular application of the oxidizing agent DTNB, Gong *et al.* (2000) [127] and Gao and Fung (2002) [157] reported an increase of open probability and decrease of closing times of BK channels from adult native hippocampal CA1 pyramidal neurons, while the reducing GSH had no apparent effect on the channel activity.

As natural reducing agent, the sulfhydryl specific reagent GSH is present in millimolar concentrations within cells. Depending on the cell type the intracellular concentration of GSH ranges between 1 mM and 10 mM whereof 98% exist in reduced form [158–160]. GSH rapidly (within 1 minute) in a concentration dependent manner increased BK channel activity in excised patches (a membrane patch which is removed from the cell like an inside out or outside out patch is called excised) of the cell when applied to the intracellular face of the membrane (of non-identified rat hippocampal cells) and this effect was rapidly reversible after wash-out [136]. In its oxidized form glutathione (GSSG) inhibited single channel activity which again was readily reversible. Hence, under physiological conditions GSH and GSSG form a redox pair where the ratio of GSH/GSH+GSSG appears responsible for the modulation of channel activity. Evidence suggests that GSH varies between brain regions and is augmented in glia cells compared to neurons [160–162] which makes neurons more susceptible to oxidative stress such as in Parkinson's disease where GSH levels in neurons are decreased [163,164].

Mammalian neurons are highly vulnerable to oxygen (O_2) deprivation. It has been hypothesized that the resulting depolarization of the membrane resting potential of these cells by hypoxia maybe in part mediated by inhibition of K^+-currents. In neocortical neurons of mice hypoxia inhibited BK channel open probability in cell-attached patches but not in the excised inside-out mode [165]. Glutathione and application of DTT increased BK channel open probability. The experimental results suggested that O_2 deprivation modulates BK channel activity by some cytosolic process(es) altering Ca^{2+} sensitivity resulting from intracellular pH changes or phosphorylation. These channels were, however, iberiotoxin and charybdotoxin insensitive which makes it questionable if these channels are genuine BK channels or if these channels were associated with β4 subunits which causes insensitivity to these toxins [65].

Lewis *et al.* (2002) [134] reported that the open probability of recombinant BK channel α-subunits from human brain, when co-expressed with β1 subunits in HEK cells in excised inside out patches, is oxygen-sensitive and reversibly suppressed by hypoxia. The experiments further indicated that the inhibition of channels by hypoxia was voltage-independent, reduced their Ca^{2+} sensitivity and did not require soluble intracellular factors.

Wyatt and Peers, (1995) [166] described O_2-sensitive K^+ currents in carotid body chemoreceptors cells of neonatal rats. In single channel and perforated patch whole cell recordings the authors reported on charybdotoxin sensitive channels which were inhibited by lowered PO_2 and required cytosolic factors for normal functioning as well as O_2 sensing. Hypoxia, anoxia or charybdotoxin depolarized the cells suggesting that closure of these channels leads to cell depolarization which is sufficient to activate voltage-gated Ca^{2+} channels and hence increased transmitter release. Jiang and Haddad (1994) [133] reported Ca^{2+} and voltage-dependent K^+ channels of large conductance from rat central neurons in cell free, excised membrane patches, which were blocked by ATP. These channels were reversibly inhibited by hypoxia - but independent of cytosolic factors. Although ATP inhibition of BK channels has been reported [167] their identity as genuine BK channels remained unclear. Furthermore, hypoxia has also been reported to inhibit BK channel activity of rat carotid body type I cells in both whole cell and in excised single channel recording [126]. However, reduction of the channels using DTT increased, whereas oxidation using DTDP decreased channel open probability. From their experiments the authors concluded that hypoxic inhibition is not related to channel reduction and modulation of channels does not require cytoplasmic mediators. Interestingly, CO reverts hypoxic inhibition of the BK channels suggesting that CO binds to the channels or some hemoprotein sensor linked to the channels. In fact, such hemo-proteins have been also postulated for other preparations [168]. Oxidative modulation of K^+ channels in the nervous system has been recently comprehensively summarized [113,117].

Perturbation of ion homeostasis is fundamentally involved in causing cell death after ischemia. Ca^{2+} excitotoxicity, which leads to the release of ROS, generates degrading enzymes or apoptotic factors [111,169]. One of the counteracting measures initiated by intracellular Ca^{2+} accumulation is the activation of BK channels. In excitable cells opening of BK channels causing hyperpolarization of cells provides a negative feedback for Ca^{2+} influx through voltage dependent Ca^{2+} channels and hence is involved in the regulation of action potential duration, neurotransmitter release or modulation of excitotoxicity [48,170–172]. Organotypic hippocampal slice cultures *in vitro* exposed to oxygen and glucose deprivation initiates cell death of CA1, and less of CA3 neurons. Treatment with the BK channel blockers, paxilline or iberiotoxin, increased this effect suggesting that BK channels act as a kind of "emergency brake" in ischemia [173]. Interestingly these treatments also increased the vulnerability of granula cells which are normally resistant to ischemic treatment. This raises the question as to the possible differences in the BK channel setting of these neurons. Accordingly, BK channels are considered as potential molecular targets for neuro-protective therapy in stroke. Using *in vivo* experimentation the issue was further investigated [174]. Focal ischemia (by cerebral artery occlusion) produced neuronal death in $BK^{-/-}$ knock-out mice which was significantly increased compared to wild type animals. Organotypic hippocampal slice cultures if exposed to ischemia-like conditions experienced neuronal death in BK knock-out animals which was significantly increased compared to wild type cultures indicating that neuronal BK channels are important for protection against ischemic brain damage [174]. On the other hand, Kulawiak and Szewczyk (2012) [175] reported that inhibition of BK channels by the specific toxin paxilline dose dependently protected hippocampal neurons against glutamate induced cell death. However, iberiotoxin and charybdotoxin were not cytoprotective. From this and further data the authors concluded that the cytoprotective effect of paxilline was not dependent on BK channel inhibition. Although the experiments are not directly comparable to the studies of Liao's group [174], since neuronal death was initiated by different procedures (low oxygen vs high glutamate) the results cause concern about the general mechanism.

Investigation of field excitatory postsynaptic potentials (fEPSPs) revealed that fEPSPs were depressed in an Alzheimer's disease (AD) mouse model compared to age-matched controls [176]. BK channel blockers (charybdotoxin, paxilline) enhanced the fEPSP-potentials giving rise to the notion that impaired Ca^{2+} homeostasis and/or ROS generation may be considered as the underlying mechanisms.

Preconditioning, a phenomenon in which a non-harmful stress stimulus renders cells tolerant to a following otherwise damaging stimulus (see also below), has been tested in rat cortical neuronal cultures [177,178]. The BK channel opener NS1619, a potent inducer of delayed or immediate neuronal preconditioning, dose-dependently protected cells against toxic insults (oxygen/glucose deprivation, H_2O_2, or glutamate excitotoxicity) but protection was not blocked by BK channel inhibitors. Since NS1619 increases ROS generation, activates the phosphoinositide 3-kinase pathway and inhibits caspase activation it was proposed that it acts cytoprotective via these mechanisms rather than via BK channel activation. In immediate preconditioning modulation of NMDA receptors by ROS and up-regulation of superoxide dismutase activity followed by decreased Ca^{2+} influx, a reduction in oxidative stress upon glutamate exposure independent of BK channel activation was proposed [177].

Vertebrobasilar insufficiency (VBI) represents transient clinical symptoms including vertigo and hearing loss caused by decreased blood supply to the brain. Using whole cell patch clamp experiments at medial vestibular nucleus (MVN) neurons in brain slices Xie *et al.* (2014) [179] found that brief hypoxia elicits depolarization of the resting membrane potential and an increase of action potential discharge frequency. Furthermore, hypoxia decreased BK-mRNA levels as well as BK channel activity, the latter being alleviated by application of the BK channel opener NS1619. These experiments suggest a neuro-protective role of BK channels and a potential target for treatments in ischemia or stroke.

Hemin, an oxidation product of heme is released from decomposed erythrocytes and appears to play a role during hemorrhagic stroke or brain trauma (indicating intracerebral hemorrhage) [180]. Hemin inhibits BK channels of plasma and mitochondrial membranes [180]. The BK channel opener NS1619 attenuates ROS generation of brain mitochondria under conditions that allow for reverse electron flow [181], which may provide the reason for the different action of NS1619 on mitochondria compared to cells. The addition of hemin inhibits this effect by about 30% which is comparable to the specific BK channel blocker iberiotoxin [180]. Hence, the inhibition of mtBK channels by hemin may constitute a novel mechanism of neurotoxicity contributing to intracerebral hemorrhage. The authors suggest that an early anti-hemin therapy could help to prevent or diminish cytotoxicity [182].

There are no consistent results on the effects of oxidizing *vs.* reducing agents on BK channel activity. Although there is only one gene for BK channel expression many other factors can influence channel behavior, such as auxiliary subunits being present or absent, or differential splicing, as outlined in the previous section. Furthermore, different recording techniques using either whole cell recordings, where signaling pathways may interfere, *vs.* excised patch recordings, where channel activity is more directly accessed, may play a role. Also, the natures of the oxidizing/reducing compounds and their accessibility to the target site have to be considered.

4.2. BK Channels and Oxidative Stress at Muscle/Endothelial Cells

BK channel activity from excised inside-out tracheal myocytes was modified by sulfhydryl redox agents [125]. The reducing agent DTT augmented, whereas the oxidizing agent thimerosal inhibited channel open probability which persisted following wash-out of the drugs but were reversed by counteracting reagents. Alkylation of channel proteins to remove free thiols prevented the action of sulfhydryl altering agents. The experiments suggested that the inhibition by oxidizing compounds is caused by covalent modification of cytosolic channel thiol groups, likely cysteine residues. Glutathione (GSH) at low concentrations also significantly augmented BK channel activity indicating that alterations by the reducing GSH are of physiological relevance [125].

These results were in contrast to findings by other authors. In inside-out patches from isolated smooth muscle cells of rabbit pulmonary arteries, BK channel activity was increased by oxidizing agents such as nicotinamide adenine dinucleotide (NAD), 5'5-dithiobis(2-nitrobenzoic acid) (DTNB) or the oxidized form of glutathione (GSSG), whereas the reducing agents such as dithiothreitol (DTT), 2-hydroxy-1-ethanethiol (β-mercaptoethanol, (BME)), the reduced form of nicotinamide adenine dinucleotide (NADH) or GSH decreased channel activity [123]. Extracellular application of hydrogen peroxide (H_2O_2) relaxes porcine coronary arteries by increasing BK channel activity. The effect was observed in the cell attached mode and after excision of the patch to the inside-out mode and was mimicked by arachidonic acid suggesting the involvement of the lipoxygenase signaling pathway [120]. At the same preparation, Hayabuchi *et al.* (1998) [183] reported that the H_2O_2 initiated BK channel dependent vascular relaxation was mediated in part by direct action on the channels and in part by activation of the cGMP signaling pathway. From the experimental results, it appears feasible that both direct redox modulation at BK channel residues and/or indirect modulation occurs via some signaling pathway, which is initiated by redox processes. Explanations for these different findings may be the specificity of the preparations used, the expression of different auxiliary subunits with different redox sensitivity, the effect on the Ca^{2+} sensitivity directly or via auxiliary subunits *per se* or the preconditioned status of the channels [184].

Peroxynitrite ($ONOO^-$) in biology can be produced from the reaction of the superoxide anion radical ($O_2^{-\bullet}$) with the nitric oxide radical ($NO^\bullet$). In rat cerebral artery smooth muscle cells, $ONOO^-$ has been found in whole cell BK current recordings as well as in inside out patches to decrease BK currents, BK channel open probability and channel mean open times, respectively [131]. The effects could be reversed by reduced GSH which also protected BK currents/channels from oxidation. The experimental results are consistent with $ONOO^-$ being a contractile agonist of cerebral arteries and myocytes implicating a physiological mechanism in the modulation of vasoconstriction.

Infarct size in heart is profoundly reduced by ischemic preconditioning (IP)—a technique in which brief periods of ischemia precede sustained ischemia and provide tolerance to subsequent damaging insults. BK channels appear to be involved in this mechanism since pharmacological BK channel openers were found to be cardio-protective which was blunted by BK channel blockers, such as iberiotoxin or charybdotoxin [185–187] or after knock-out of the *KCNMA1* gene of neurons involved in the regulation of the heart beat [188].

ROS in addition to many other functions also appears involved in altering cellular ion homeostasis [189]. In heart ROS via oxidation of kinases (PKA, PKC, Calcium/Calmodulin Kinase II (CaMKII)) together with dysfunction of the sarcoplasmic reticulum can lead to perturbation of the Ca-homeostasis (which will indirectly affect BK channel activity) and finally cause heart failure.

4.3. Effects of Oxidative Stress on BK Channels of Epithelial Cells

In cultured alveolar epithelial A549 cells acute changes of oxygen tension (PO_2) increased BK channel mean open time whereas chronic changes in PO_2 did not affect expression, recruitment or function of BK channels or Na^+ channel activity [190]. The authors concluded that BK channels serve as oxygen sensors. The mechanism of oxygen sensing was further elaborated by demonstrating that hemoxygenase-2 (HO-2), which is associated to BK channels enhances channel activity in normoxia [191]. It was found that carbon monoxide (CO), which is produced from O_2 via HO-2, is a mediator of this function, indicating that HO-2 is an oxygen sensor.

In cultured human epithelial pigment cells, the oxidants, t-butyl hydroperoxide (t-BHP) or thimerosal suppressed BK outward currents recorded in whole cell configuration as well as BK channel open probability in the cell attached mode [192]. Internal application of ceramide prevented the oxidizing effect suggesting that BK channel inhibition may involve the intracellular generation of ceramide. In fact, ceramide can be produced in cells via the sphingomyelin pathway or by *de novo* synthesis [189,190]. H_2O_2 and other stressors have been reported to generate ceramide in various cells [193,194] and ceramide *per se* suppresses BK channel activity [195]. More information on this issue appears an interesting research endeavor.

It remained ambiguous for some time why oxidation in some cases increased, however, in others decreased channel activity. In their publication, Tang *et al.* (2001) [109] provided an explanation for the oxidative modulation of hSlo (human BK) channels considering the reversible oxidation of the amino acids cysteine and methionine. They showed that the agent chloramine-T (Ch-T), which oxidizes preferentially methionine to methionine sulfoxide (met(O)) or to met(O_2), increased hSlo channel open probability. In contrast cysteine specific reagents, such as DNTB and other oxidizing agents decreased channel activity. Oxidation of these amino acids may also provide an explanation for the run-down of channels after patch excision (loss of channel activity with time), which exposes the channel inside to a more oxidized environment compared to the more reduced, normal intracellular milieu, and may also explain some earlier conflicting results. The authors were able to further provide evidence where at the channels the functional modification may take place. Internal channel blockers tetraethylammonium (TEA) or 1-methyl-4phenyl-1,2,3,6-tetrahydropyridine (MPTP) protected the channels from oxidation by Ch-T, indicating that residues oxidized by Ch-T are located near or within the channel pore and β-subunits are unlikely to contribute to this process. As outlined by the authors, oxidation of critical amino acids may be clinically relevant. In reperfusion after ischemic conditions the formation of free radicals may affect vascular BK channels. Methionine oxidation causing activation of BK channels may limit Ca^{2+} entry into cells and serve as neuronal protectant.

Tang *et al.* (2004) [61] further provided evidence that ROS to a large extent inhibited BK channels (α + β subunits expressed in HEK cells) by targeting cysteine residues near the intracellular Ca^{2+} binding site (Ca^{2+} bowl). In particular, a single oxidized cysteine residue, Cys911, prevented Ca^{2+} sensitivity. Further evidence that the α-subunit was the target for oxidation was obtained from experiments where its sole expression was sufficient for the suppression of currents. The authors line out that oxidative stress causing inhibition of BK channel activity and hence impairment of vascular relaxation and blood pressure may have crucial implications in disease and aging.

Auxiliary β-subunits also contain various amounts of cysteines (4–7) and methionines (3–11). Modifications at these amino acids are therefore amenable to affect channel activity. Zeng *et al.* (2003) [184] observed that reduction of extracellular disulfide linkages of the β3-subunit abolished current rectification and improved charybdotoxin blockade of the channels. The results indicate that the β-subunit appears to be close to the ion permeation pathway of the α-subunit and may regulate the gating mechanism and access of blocking molecules. Physiological consequences may be the dynamic regulation of channel inactivation by oxidative modification of SH groups.

The bovine β-subunit contains five cysteine residues which are conserved among various species such as rat, dog and humans [196]. Mutagenesis of each of the four cysteines present in the extracellular loop caused a profound reduction of charybdotoxin binding suggesting the generation of disulfide bridges between these residues. Methionines are only present at the intracellular N-terminus (3), and one is present at the end of the second transmembrane domain [197]. Oxidation of methionine in BK channel α-subunits (hSlo) by chloramine-T induced a leftward shift at the voltage axis to more negative values [109], *i.e.*, the channels are activated already at more negative potentials closer to the membrane resting potential. In the presence of β1-subunits and under low intracellular Ca^{2+} conditions the hyperpolarizing shift was largely augmented, channel open probability increased and deactivation slowed [198]. However, this shift was independent of oxidation of methionine or cysteine residues of β1-subunits indicating that this auxiliary subunit has no direct effect but appeared to prime or to amplify the oxidation of the pore forming α-subunit residues. As a physiological implication it appeared that in electrically quiet cells at low baseline Ca^{2+} concentrations BK channels may have an increased impact on cellular electrical activity and hence influence vascular tone. In a following publication the authors demonstrated that the mutation of three conserved methionines located within the RCK domains (regulators of K^+ conductance) at the cytoplasmic C-terminus eliminate oxidative sensitivity of the BK channels [199]. Oxidation of at least one of these key methionine residues was sufficient to increase the open probability of the channels and slowing of inactivation at low internal Ca^{2+}. From theoretical considerations the authors proposed a mechanism of conformational changes in the gating ring structure which are transmitted to the voltage sensor domain. The functional effects caused by oxidation may serve as a protective mechanism by driving the cells into a more hyperpolarized, resting state and hence may be advantageous in pathological conditions such as after post-ischemic reperfusion or neurodegenerative disease to prevent neurotoxicity.

4.4. Oxidative Stress, BK Channels and Vascular Regulation

BK channels in the vascular system are modulated by agents naturally produced in the body, such as angiotensin II (Ang II), high glucose or arachidonic acid (AA) [200] which is modulated in diabetes by oxidative stress (ROS) [131,200,201]. In coronary smooth muscle cells BK channels are activated by arachidonic acid metabolites, in particular prostacyclin 2 (PGI2) [200]. In Zucker diabetic fatty rats channel activation is impaired caused by reduced production of prostacyclin 2 (PGI2) due to reduced PGI2 synthase. As a possible mechanism, ROS formation appeared to be involved. Lu *et al.* (2006) [201] reported that high glucose reduced BK channel density and channel kinetics by increasing ROS generation, in particular via H_2O_2-dependent oxidation of cysteine 911. In a further study it was shown that caveolae (small 50–100 nanometer invaginations of plasma membrane micro domains containing signaling components in close approximation) play an important role in mediating inhibition of BK channels by ROS [202]. In diabetic rat aortas, cav-1 expression (indicating caveolae) is upregulated and accompanied by an improved physical link between BK channels and the activated angiotensin-1 receptors (AT_1R) signaling complex. Angiotensin II (Ang II)-activated AT1R receptors in diabetic rats stimulate enhanced non-phagocytic NAD(P)H oxidase and (NOX1) expression. This leads to an increased ROS-induced redox inhibition of BK channels and also to phosphorylation and nitration of tyrosine residues of the channels. The absence of caveolae (initiated by caveolin-1 (cav-1) knock-down or gene ablation) caused disaggregation of the Ang II-BK channel signaling cascade and preserved BK channel function in diabetes. The molecular mechanisms delineated so far indicate that the composition and spatial arrangement of the signaling system is important for proper functioning of coronary blood flow and maybe an interesting target for further pharmacological developments in the treatment of hypertension or atherosclerosis [202].

Another mechanism of BK channel activation of vascular tissue by H_2O_2 was presented by Burgoyne *et al.* (2007) [203] and Zhang *et al.* (2012) [204]. H_2O_2 which serves as an endothelium-derived hyperpolarizing factor (EDHF) causes dilation of human coronary arterioles by activation of BK channels. Cell-attached single channel recordings as well as inside out channel recordings from coronary smooth muscle BK channels revealed strong evidence that H_2O_2 increased channel open probability (Po) which requires intracellular signaling by a mechanism involving disulfide dimerization of the redox sensitive cGMP-dependent protein kinase (PKG-Iα). Hence the H_2O_2 mechanism of action appears different to the NO mediated sGC-stimulated cGMP-PKG pathway but acts downstream directly at the PKG-Iα level. Release of H_2O_2 from endothelial cells appears to play a prominent role as endogenous regulator under normal conditions and in diseased states [204].

There is also evidence that anoxia may act via protein kinase C (PKC). In Western painted turtle cortical pyramidal neurons anoxia reversibly reduced the open probability of BK channels in the cell-attached mode but not in excised patches [205]. The PKC inhibitor chelerythrine prevented the anoxic effect whereas the PKC activating phorbol ester PMA (phorbol-12-myristate-13-acetate) decreased Po during normoxia. It was concluded that the inhibition of BK channels activity prevents K^+ efflux, preserves K^+ homeostasis and hence reduces cellular ATP usage and promotes survival of neurons which enables the animals to adapt to low oxygen levels. Similarly, an anoxia dependent inhibition of Na^+ currents recorded from rat hippocampal neurons via a PKC dependent mechanism has been reported previously [206].

In cat cerebral arterial muscle cells, hypoxia superfusion produced a transient increase in mean open time of BK channels in excised inside-out patches and this effect was independent of changes of the internal Ca^{2+} concentration or pHi [207]. Furthermore, reduced PO_2 caused dilation of cerebral arterial segments which was attenuated by tetraethylammonium (TEA). In a following publication Gembremedhin *et al.* (2008) [208] reported that in rat cerebral arterial muscle cells hypoxia reversibly enhanced BK channel open-state probability in cell-attached patches but this effect was absent in either excised inside-out or outside-out patches. The results further indicated that hypoxia induced the generation of superoxide which caused a reduction in endogenous level of 20-hydroxyeicosatetraenoic acid (20-HETE) that may account for the hypoxia-induced activation of arterial BK channel currents and cerebral vasodilatation.

BK channels, among other K channels, are associated with decreased uterine vascular tone and an increase of uterine blood flow during pregnancy in order to optimize adequate nutritional supply and tissue oxygenation for the fetus. Chronic hypoxia during gestation by suppression of BK channel function increases uterine vascular tone and decreases uterine blood supply which increases the risk of preeclampsia and retarded fetal growth [209–211]. In addition, expression of BK channels during pregnancy is upregulated but down-regulated during chronic hypoxia. At the same time, protein kinase C (PKC) activity is increased during hypoxia which inhibits BK channels, whereas under normoxia these effects are reversed [212].

Sex steroid hormones (estrogens) regulate uterine vascular tone and uterine blood flow by increasing BK channel density (in particular β1-subunits) and activity, whereby also PKC signaling appears an important modulatory mechanism. In contrast, progesterone appears to inhibit BK channels [213] but its contribution to regulating uterine vascular tone needs further investigation. Exposure to hypoxia during pregnancy attenuates the effects of sex steroid hormones/receptors, leading to enhanced PKC activation resulting in the inhibition of BK channel activity and increased pressure-dependent myogenic tone in pregnant uterine arteries [211]. Furthermore, hypoxia-mediated ROS activation during gestation inhibits steroid hormone mediated up-regulation of BK channel activity which may contribute to malfunctions during gestational hypoxia [214]. The evidence suggests that modulation of K^+ channel activity during pregnancy conveys an important mechanism underlying hypoxia induced uterine vascular dysfunction [215].

4.5. Mitochondrial BK Channels (mtBK) and Oxygen

Mitochondria are a major source of ROS generation targeting BK channels [122,216–218]. The inner membrane of mitochondria contains BK channels (mtBK) [219–221] which appear essential in the production of ROS. mtBK channels appear to be inserted into the mitochondrial membrane with the toxin binding sites for charybdotoxin and iberiotoxin exposed to the mitochondrial intermembrane space (accessed by using outside-out patch configuration of the inner mitochondrial membrane). Consequently the C-terminal tail domain including the Ca^{2+} binding site is localized to the mitochondrial matrix. mtBK channels of cardiac tissue have recently been specified as being encoded by the *KCNMA1* gene and hence appear to be identified as genuine BK channels [222].

Opening of BK channels allows K^+, which is present in a high concentration in the cytosol, to flow into the negatively charged mitochondrial matrix and depolarize the organelle (reviewed in [216]). This reduces the driving force for Ca^{2+} and hence Ca^{2+} influx which reduces Ca^{2+} overload of mitochondria. In fact, mtBK channel activation and K^+ uptake was reported to confer cytoprotection to heart infarction [221]. Further studies revealed that preconditioned activation of mtBK channels by the channel opener NS-1619 or NS11021 reduces superoxide production and reduces Ca^{2+} overload which improved Ca^{2+} homeostasis and redox state after ischemia/reperfusion in isolated guinea pig hearts [122,223–225]. Another mtBK channel opener 12,14-dichlorodehydroabietic acid (diCl-DHAA) was also reported to reduce ischemic injury in rat cardiac myocytes [226]. The steroid, 17β-estradiol, enhances the activity of cardiac mtBK channels, but only in the presence of the auxiliary β1-subunit and increased survival of myocytes under simulated ischemia [227]. In addition the β1-subunit appeared to interact with the cytochrome c oxidase subunit I. These findings may help to improve ischemic diseases such as heart attack in postmenopausal women by applying an estrogen-induced cardio-protective treatment. Furthermore, opening of mtBK channels of brain-derived mitochondria appears to inhibit ROS production suggesting that this may also contribute to the beneficial effects of BK channel openers on neuronal survival [175].

In the heart, BK channels appear absent from the sarcolemma, but the channels are present in mitochondrial membranes (reviewed in [97,228]). The opening of BK channels by hypoxia exerts a cytoprotective effect, which has been attributed to the mechanism mentioned above. In addition β1-subunits are highly expressed together with mtBK α-subunits that contain two *N*-glycosylation sites at their C-terminus which on enzymatic deglycosylation cause activation of BK channels [229,230]. Borchert *et al.* (2013) [231] reported that the BK channel opener NS11021 or H_2O_2 increased survival of cardiomyocytes in simulated ischemia/reperfusion experiments. The cytoprotective effect was abolished by the specific BK channel inhibitor paxilline or tempol, a specific antioxidant. The study indicates that the mechanism for this protection requires ROS signaling suggesting that activation of mtBK channels protect the cells against injury. Experiments using chronic hypoxic rats exposed to brief intermittent re-oxygenation, attenuated cardio protection possibly by a mechanism involving oxidative stress and suppression of mtBK activity [232].

Mitochondrial BK channels (mtBK) derived from streptozotocin-induced diabetic rat brain incorporated into lipid bilayer membranes were found to exhibit a decreased open probability and conductance [233]. In addition, both BK α- and β4 subunits expression was down-regulated. This evidence and an increased production of ROS during diabetic conditions, as proposed by others [201], were suggested to account for the abnormalities in channel gating. In the studies on molecular level, Lu *et al.* (2006) have shown that C911 is a major molecular target at the channel protein for the redox modulation by high glucose [201]. This body of acquired knowledge may be useful for the development of improved treatment of diabetics.

4.6. Gasotransmitters and ROS Effects on BK Channels

In this review, we only briefly outline some aspects of the interaction of BK channels and gasotransmitters. Recent information on this issue has been summarized by [63,117,234]. Nitric oxide (NO) produced by endothelial NO synthase (eNOS) appears to prevent atherosclerosis by interaction with the vascular NO generating system [235]. NADPH oxidases are a major source of ROS leading to various types of vascular pathophysiology. In atherosclerosis bioactivity of NO is decreased by reduced NO synthesis due to an increased NO inactivation.

Endothelial cells by appropriate stimulation (humoral or hemodynamic), release vasodilatory factors, such as NO, whereas vasoconstrictory ROS are generated as byproducts of oxygen metabolism [121]. Single BK channels recorded from *in situ* renal artery endothelium preparations revealed that NO increased channel open probability by two mechanisms: either by direct action on the channel proteins (probably by nitrosylation) or by indirect action via cGMP (mediated by phosphorylation) [132]. In contrast, ROS (H_2O_2 *etc.*) resulted in irreversible channel inactivation. The authors concluded that BK activation by NO creates a positive feedback by autocrine regulation of endothelial function, whereas ROS inhibits BK channels which impairs hyperpolarization of endothelial cells and consequently prevents vasodilation.

NO inhibits M-currents of rat sensory neurons from trigeminal ganglia [236]. M-type K^+ channels are subthreshold voltage-gated K^+ channels of the six TM type (designated Kv7, *KCNQ* gene family) which are inhibited by M1 muscarinic acetylcholine receptors and affect excitability of neurons in the central and peripheral nervous system or cardiac tissue. The authors identified a site of NO action within the cytosolic channel linker between transmembrane domains 2 and 3, which appears also to be a site of oxidative modification by ROS. However, NO and oxidative modifications exhibit opposing effects on M-currents. These channels therefore appear to contain a dynamic redox sensor that is responsible for dynamic M-current modulation by gasotransmitters and ROS and may play a role in trigeminal disorders such as headache and migraine. It appears interesting to probe this channel site for different types of gasotransmitters and ROS also at BK channels. Other types of K^+ channels and oxidative stress are discussed in [112,117,237,238].

In our experiments using GH3 pituitary tumor cells, sodium hydrogen sulfide (NaHS), a H_2S donor, increased channel open probability which was prevented by the reducing agent DTT, whereas the oxidizing agent thimerosal increased channel open probability in the presence of H_2S [129]. The effect was linked to the reducing action of H_2S on sulfhydryl groups of the channel protein. This finding is in concert with a report by Liu *et al.* 2009 [239] indicating that the H_2S donor NaHS prevents postischemic mitochondrial dysfunction by a BK channel dependent mechanism. The development of drugs interfering with H_2S signaling might be rewarding in the treatment of mitochondrial linked diseases or to BK channel dependent high blood pressure. Some reagents used to study the modulation of BK channels by gasotransmitters are listed in Table 1.

4.7. Oxidative Stress, Proliferation and Various Types of K^+ Channels

In pulmonary artery smooth muscle cells various types of K^+ channels, such as K_V, K_{ir}, BK, K_{2P}, participate in vascular remodeling (proliferation, apoptosis) [240]. In general, loss or inhibition of K^+ channel function contributes to pulmonary pathogenesis which leads to a decrease of proliferation and inhibition of apoptosis. Hypoxic inhibition of ROS production causing inhibition of K^+ channels leads to depolarization, opening of Ca^{2+} channels and augmentation of intracellular Ca^{2+} which finally results in vasoconstriction of small pulmonary arteries. K^+ channel expression is transcriptionally regulated and cells need K^+ channels to proceed through the cell cycle (in particular G1 progression) [240]. Hence, K^+ channels are appealing therapeutic targets in pulmonary arterial hypertension. Oxidized low-density lipoprotein lysophosphatidylcholine (LPC) increases BK channel open state probability by causing capacitative Ca^{2+} influx [241]. Ca^{2+} accumulation increases ROS production that causes reduction of NO generation which promotes proliferation of cultured human endothelial cells. In addition LPC-induced BK activation contributes to increased cGMP levels, if ROS production was prevented by transfection with antisense oligonucleotides against NAD(P)H oxidase. Cancer incidence also appears associated with a great variety of ion channels, in particular K^+ channelopathies [242–244]. In fact, almost all known types of K^+ channels have been implicated in oncogenic processes. The dysregulation of K^+ channel expression (mostly overexpression) and resulting dysfunction of K^+ channels correlates with dysregulation of proliferation, malignant growth and migration (metastatic spread) of tumorigenic cells. As a general phenomenon, the membrane potential of cancer cells is more depolarized [245] and the resting membrane potential oscillates during the cell cycle, being more depolarized during S and G2 phases [246,247].

Polyamines (mainly putrescine, spermidine and spermine) exhibit a wide array of functions from modulating ion channels, involvement in apoptosis, carcinogenicity, cell proliferation or development. Modulatory properties of polyamines concerning K^+ channels involved in cell proliferation has been summarized by Weiger and Hermann (2014) [248]. Here we briefly comment on some issues in the context of oxygen impact. Polyamine deficient yeast (*Saccharomyces cerevisiae*), for example, is very sensitive to oxygen. Polyamine depleted cells accumulate ROS, develop an apoptotic phenotype and die after incubation in polyamine-deficient medium [249]. Addition of spermine caused a marked decrease in ROS accumulation and some protection against cell death. The data indicate that part of the function of polyamines is protection of the cells from accumulation of ROS.

Polyamine biosynthesis is increased during cerebral ischemia through induction of ornithine decarboxylase (ODC), a key enzyme for their synthesis. Inhibition of ODC prevents ischemic brain injury [250]. Metabolization of polyamines by polyamine oxidases generates cytotoxic aldehydes and ROS. Polyamines may therefore constitute crucial players in oxidative stress. There is little information on polyamines, ion channels (BK) and oxidative stress which remains an interesting field for future investigations.

4.8. Oxidative Stress of BK Channels in Tumors

Tumors are remarkably tolerant to hypoxia whereas normal nerve cells have a high demand on oxygen and rapidly die on hypoxia. This of course raises the question concerning the mechanisms involved. Evidence suggests that the modulation of ion channels and intracellular signaling pathways may be a key in determining cellular tolerance to low oxidative stress [251]. Expression of BK channels is increased in cancerous compared to healthy cells and correlates with the malignancy of the tumors [252,253]. It was hypothesized that differences in the expression of BK channels in tumor cells compared to healthy cells could be the reason for the differences in their response to hypoxia. These findings may provide an approach for novel vistas in cancer therapy.

Hypoxia in a variety of healthy cells in general decreases open probability of BK channels in the plasma membrane (plBK), whereas mtBK channels are opened by hypoxia [254,255]. Studies of normoxic *vs.* hypoxic conditions of single mtBK and plBK channels from human glioma cells showed that plBK channels were insensitive to hypoxia whereas mtBK channels open probability was increased during hypoxia [254,256]. Activation of mtBK channels by specific agonists had no effect on cell viability. In contrast, activation of plBK channels by other specific agonists, impaired cell viability of tumor cells [255,257] and this effect was increased by hypoxia [255]. It was suggested that the mechanism leading to cell death/apoptosis in normoxia is based on Ca^{2+} toxicity, *i.e.*, it is mediated by an increase of cytosolic Ca^{2+} and activation of calpains [251,258]. The effect of hypoxia on mtBK channels appears to have different consequences on cell viability [254]. A cytoprotective effect has been attributed to (a) increased matrix K^+, (b) prevention of Ca^{2+} overload, and c) closing of the mPTP pore. Mitochondrial permeability transition pores (mPTP) close on hypoxia. mPTP in the inner mitochondrial membrane are considered to activate a pathway for the release of pro-apoptotic factors from mitochondria (for further information about this topic the reader is referred to [259]).

4.9. Clinical Relevance

Oxidative modulation of BK channels is important not only in order to understand and treat diseased conditions like myocardial infarction or stroke but, as it recently came also into focus, for transplantation surgery. For instance, in the case of lung transplantation the graft is subjected to ischemia followed by reperfusion during routine procedure. During ischemia a number of cellular processes eventually lead to membrane depolarization and ROS generation [260] which in turn favors inflammation and cell death [261]. Noda *et al.* 2014 [262] found that preconditioning lung grafts with inhaled hydrogen, a cytoprotective gaseous signaling molecule, reduced the proinflammatory changes and led to better post-transplant graft function. The molecular mechanism behind this effect is, among others, a stimulation of hemeoxygenase-1 (HO-1) expression by hydrogen [262]. As already discussed above HO-1 activity results in production of CO which then activates BK channels [263]. Active BK channels will cause repolarization of the membrane resting potential and this way may contribute to improved lung graft transplantation. These discoveries again underline the importance and protective role of BK channels. In conclusion, ROS modulation may exert conformational alterations at the channel proteins with two major implications: (1) impairment or dysfunction of channels which may lead to diseases or death, such as in vascular impairment or heart attack, or (2) modifications at channels may lead to physiological modulation of channel functioning, such as in sensing of oxygen tension.

4.10. Perspectives

In a recent publication the authors report that a point mutation of a phenylalanine at the F380 position in the S6 transmembrane helix of BK channels greatly obviates channel opening [264]. Based on further functional experiments and molecular dynamic simulations they proposed a model where in the process of channel conduction a hydrophobic ring structure forms that acts as an integration node which affects the interaction between the voltage sensor and the pore. It appears interesting to probe this site for ROS manipulation.

Sulcatone (6-Methyl-5-hepten-2-one (C8H14O)), a prominent volatile component of human body odor that is less abundant in non-human animals, appears to specifically attract some species of female mosquitos (*Aedes aegypti*) to feed on our blood [265]. A prokaryotic protein with features reminiscent to an ancestral single domain bacterial K^+ channel protein may have provided the evolutionary blueprint for pentameric ligand-gated ion channels which serve as odorant receptors (*Ore4*) for sulcatone [266]. Interestingly, sulcatone belongs to the family of oxidoreductases which are known to act on ion channels. It may be speculated that sulcatone acting on K^+ channels which have been found to alter cellular signaling and are involved in controlling proliferation [248] may be interesting targets for future investigations.

Various types of other K^+ channels, other than BK channels, such as voltage gated K^+ channels (K_V), ATP-gated K^+ channels (K_{ATP}) or (K_{ir}), appear to be involved in vasomodulation [267,268]. Only a few other K^+ channel studies in the context of oxidative stress, from which we may draw interesting information and extend our view, will be covered in this section. In a recent study, Park *et al.* (2015) [269] report that H_2O_2 relaxes rat mesenteric arteries which was reversed by application of the reducing agent dithiothreitol. The vasodilatory effect of H_2O_2 was reduced by the voltage gated K^+ channels (K_V) blocker 4-aminopyridine (4-AP) but was resistant to BK and inward rectifier K^+ channels (K_{ir}) channel blockers. Whole cell patch clamp studies further showed that K_V currents recorded from mesenteric smooth muscle cells were dose-dependently increased by H_2O_2 as well as by oxidized glutathione (GSSH) and prevented by glutathione reductase. Further studies showed that reduced glutathione (GSH) is incorporated into the K_V channel protein indicating S-glutathionylation of the channels by H_2O_2. Park *et al.* (2015) [269] now report that by an increased basal level of H_2O_2, *i.e.*, under conditions of persistent oxidative stress, K_V channels were not activated, but rather inhibited by the addition of H_2O_2. The findings suggest that the actual cellular redox status affects S-glutathionylation of the K_V channels and determines the response of these K_V channels to H_2O_2.

Oxidative stress is also a hallmark of vascular disorders such as diabetic retinopathy. In this case ATP-sensitive K channels (K_{ATP}) activity were found to be increased during exposure of retinal capillaries to H_2O_2 [270]. The effect on K_{ATP} was boosted by increasing the influx of Ca^{2+} into microvascular cells and oxidant-induced activation of Ca^{2+}-permeable nonspecific cation channels. Furthermore, it was found that inhibition of K_{ATP} channels by the specific K_{ATP} blocker glibenclamide significantly lessened H_2O_2 induced microvascular cell death. The findings were suggested to provide new targets for pharmacological treatments of retinal microvasculature during oxidative stress.

5. Conclusions

In conclusion, the overall status of the present studies indicates that redox modifications of cysteines/methionines cause conformational alterations at the BK channel protein which translate into modulation of channel pore openings (gating) in which interference with the Ca^{2+} activation mechanism also appears to play a major role. If the target alterations occur at single sites or in combination, whether other amino acid targets bear functional relevance and how these manipulations affect conformational changes of the 3D-protein structure needs further investigation. Primary effects, directly at the channels, and/or secondary effects on signaling pathways, may be both relevant, but have to be clearly separated. BK channel/ROS modulation will certainly be an important target for development of pharmacological agents which function as channel openers/blockers opposing negatively effective redox mechanisms.

Acknowledgments

This work was supported by travel grants from the Stiftungs- und Förderungsgesellschaft der Universität Salzburg, grants from the University of Salzburg to Guzel F. Sitdikova and by the Russian Scientific Fund No.14-15-00618 to Guzel F. Sitdikova.

Abbreviations

AA	Arachidonic acid
ACA	Acetaldehyde
Ach	Acetylcholine
AD	Alzheimer's disease
Ang	II angiotensin II
4-AP	4-aminopyridine
ATP	Adenosine triphosphate
AT_1R	Angiotensin-1 receptor
β-ME	Mercaptoethanol
BME	2-hydroxy-1-ethanethiol (β-mercaptoethanol)
BK	Big or maxi calcium-activated potassium channel
$BK^{-/-}$	Knock-out animal
CaMKII	Calcium/Calmodulin Kinase II
cAMP	Cyclic adenosinemonophosphate
cav-1	Caveolin-1
cGMP	Cyclic gCanosinemonophosphate
CBS	Cystathionine b-synthase
Ch-T	Chloramine-T
CO	Carbon monoxide
CORM	Carbon monoxide releasing molecule
CREB	Cyclic AMP response element-binding protein
CSE	Cystathionine c-lyase
Cys or C	Cysteine
diCl-DHAA	12,14-dichlorodehydroabietic acid
DTT	1,4-dithio-DL-threitol (short: dithiothretiol)
DTDP	2,2'-dithiodipyridine or 4,4'-dithiodipyridine
DTNB	5,5'-dithiobis(2-nitrobenzoic acid)

EDHF	Endothelium-derived hyperpolarizing factor
EET	Epoxyeicosatrienoic acid
ER	Estrogen receptor
EtOH	Ethanol
fEPSP	Field excitatory postsynaptic potential
G-protein	Guanosine triphosphate (GTP) binding protein
GSH	Glutathione-reduced form
GSSH	Glutathione-oxidized form
20-HETE	20-hydroxyeicosatetraenoic acid
H_2O_2	Hydrogen peroxide
H_2S	Hydrogen sulfide
HEK	Human Embryonic Kidney cells
HO	Heme oxygenase
HO-2	Hemoxygenase-2
HS	Hydrogen sulfide anion
hSlo	Human BK channel
IK	Intermediate conductance K^+ channel
IP	Ischemic preconditioning
K_{ATP}	Adenosine-triphosphate dependent K^+ channel
LPC	Lysophosphatidylcholine
LTP	Long-term potentiation
Met or M	Methionine
Met-O	Methionine sulfoxide
mM	Milli Molar
MPTP	1-methyl-4phenyl-1,2,3,6-tetrahydropyridine
MSRA	Methionine sulfoxide reductases
mtBK	Mitochondrial BK channel
MTSEA	Methanethiosulfonate ethylammonium
MVN	Medial vestibular nucleus
NAD	Nicotinamide adenine dinucleotide
NADH	Nicotinamide adenine dinucleotide
NaHS	Sodium hydrogen sulfide
nBK	Nuclear BK channel
NCS	*N*-chlorosuccinimid
NO	Nitric oxide
NOS	NO synthase
nM	Nano Molar
NMDA	*N*-methyl D-aspartate
O_2	Oxygen
$O_2^{-\bullet}$	Superoxide anion radical
•OH	Hydroxyl radical
$ONOO^-$	Peroxynitrite

ODC	Ornithine decarboxylase
P	Pore loop
PDE	Phosphodiesterase
PGI2	Prostacyclin 2 (Prostaglandin I_2)
PKA, PKC, PKG	Protein kinase A, C, G
plBK	Plasma membrane BK channel
PMA	Phorbol-12-myristate-13-acetate
PO_2	Oxygen tension
Po (Popen)	Open probability of channels
PP2A	Protein phosphatase
PS	Phosphatidylserine
pS	Piko Siemens
PTP	permeability transition pore
RCK	Regulatory domain of K^+ conductance
ROS	Reactive oxygen species
SK	Small conductance K^+ channel
sGC	Soluble guanylyl cyclase
Slob	Slo binding protein
STREX	Stress-axis regulated exon
T	Transmembrane
t-BHP	t-butyl hydroperoxide
TEA	tetraethylammonium
TRX	Thioredoxine
VBI	Vertebrobasilar insufficiency

References

1. Hille, B. *Ion Channels of Excitable Membranes*, 3rd ed.; Sinauer: Sunderland, MA, USA, 2001.
2. Hodgkin, A.L.; Huxley, A.F.; Katz, B. Measurements of current-voltage relations in the membrane of the giant axon of Loligo. *J. Physiol.* **1952**, *116*, 424–448.
3. Hamill, O.P.; Marty, A.; Neher, E.; Sakmann, B.; Sigworth, F.J. Improved patch-clamp techniques for high-resolution current recording from cells and cell-free membrane patches. *Pflüg. Arch.* **1981**, *391*, 85–100.
4. Doyle, D.A.; Cabral, J.M.; Pfuetzner, R.A.; Kuo, A.; Gulbis, J.M.; Cohen, S.L.; Chait, B.T.; MacKinnon, R. The structure of the potassium channel: Molecular basis of K^+ conduction and selectivity. *Science* **1998**, *280*, 69–77.
5. Brueggemann, L.I.; Gentile, S.; Byron, K.L. Social networking among voltage-activated potassium channels. *Prog. Mol. Biol. Transl. Sci.* **2013**, *117*, 269–302.
6. Levitan, I.B. Signaling protein complexes associated with neuronal ion channels. *Nat. Neurosci.* **2006**, *9*, 305–310.
7. Armstrong, C.M.; Hille, B. Voltage-gated ion channels and electrical excitability. *Neuron* **1998**, *20*, 371–380.

8. Yu, F.H.; Yarov-Yarovoy, V.; Gutman, G.A.; Catterall, W.A. Overview of molecular relationships in the voltage-gated ion channel superfamily. *Pharmacol. Rev.* **2005**, *57*, 387–395.
9. Bezanilla, F. Voltage-gated ion channels. IEEE Trans. *Nanobioscience* **2005**, *4*, 34–48.
10. Long, S.B.; Tao, X.; Campbell, E.B.; MacKinnon, R. Atomic structure of a voltage-dependent K^+ channel in a lipid membrane-like environment. *Nature* **2007**, *450*, 376–382.
11. Dai, S.; Hall, D.D.; Hell, J.W. Supramolecular assemblies and localized regulation of voltage-gated ion channels. *Physiol. Rev.* **2009**, *89*, 411–452.
12. Catterall, W.A. Structure and function of voltage-gated ion channels. *Annu. Rev. Biochem.* **1995**, *64*, 493–531.
13. Catterall, W.A. Structure and function of voltage-gated sodium channels at atomic resolution. *Exp. Physiol.* **2014**, *99*, 35–51.
14. Catterall, W.A. Voltage-gated calcium channels. *Cold Spring Harb. Perspect. Biol.* **2011**, *3*, doi:10.1101/cshperspect.a003947.
15. González, C.; Baez-Nieto, D.; Valencia, I.; Oyarzún, I.; Rojas, P.; Naranjo, D.; Latorre, R. K^+ channels: Function-structural overview. *Compr. Physiol.* **2012**, *2*, 2087–2149.
16. Stölting, G.; Fischer, M.; Fahlke, C. CLC channel function and dysfunction in health and disease. *Front. Physiol.* **2014**, doi:10.3389/fphys.2014.00378.
17. Gardos, G. The function of calcium in the potassium permeability of human erythrocytes. *Biochim. Biophys. Acta* **1958**, *30*, 653–654.
18. Wei, A.D.; Gutman, G.A.; Aldrich, R.; Chandy, K.G.; Grissmer, S.; Wulff, H. International union of pharmacology. LII. Nomenclature and molecular relationships of calcium-activated potassium channels. *Pharmacol. Rev.* **2005**, *57*, 463–472.
19. Ghatta, S.; Nimmagadda, D.; Xu, X.; O'Rourke, S.T. Large-conductance, calcium-activated potassium channels: Structural and functional implications. *Pharmacol. Ther.* **2006**, *110*, 103–116.
20. Shipston, M.J. S-acylation dependent post-translational cross-talk regulates large conductance calcium- and voltage- activated potassium (BK) channels. *Membr. Physiol. Membr. Biophys.* **2014**, doi:10.3389/fphys.2014.00281.
21. Salkoff, L.; Butler, A.; Ferreira, G.; Santi, C.; Wei, A. High-conductance potassium channels of the SLO family. *Nat. Rev. Neurosci.* **2006**, *7*, 921–931.
22. Cui, J.; Yang, H.; Lee, U.S. Molecular mechanisms of BK channel activation. *Cell. Mol. Life Sci.* **2009**, *66*, 852–875.
23. Berkefeld, H.; Fakler, B.; Schulte, U. Ca^{2+}-activated K^+ channels: From protein complexes to function. *Physiol. Rev.* **2010**, *90*, 1437–1459.
24. Cui, J. BK-type calcium-activated potassium channels: Coupling of metal ions and voltage sensing. *J. Physiol.* **2010**, *588*, 4651–4658.
25. Lee, U.S.; Cui, J. BK channel activation: Structural and functional insights. *Trends Neurosci.* **2010**, *33*, 415–423.
26. Grimm, P.R.; Sansom, S.C. BK channels and a new form of hypertension. *Kidney Int.* **2010**, *78*, 956–962.
27. Hill, M.A.; Yang, Y.; Ella, S.R.; Davis, M.J.; Braun, A.P. Large conductance, Ca^{2+}-activated K^+ channels (BKCa) and arteriolar myogenic signaling. *FEBS Lett.* **2010**, *584*, 2033–2042.

28. Wu, Y.; Yang, Y.; Ye, S.; Jiang, Y. Structure of the gating ring from the human large-conductance Ca^{2+}-gated K^{+} channel. *Nature* **2010**, *466*, 393–397.
29. Latorre, R.; Morera, F.J.; Zaelzer, C. Allosteric interactions and the modular nature of the voltage- and Ca^{2+}-activated (BK) channel. *J. Physiol.* **2010**, *588*, 3141–3148.
30. Hermann, A.; Sitdikova, G.F.; Weiger, T.M. BK Channels—Focus on Polyamines, Ethanol/Acetaldehyde and Hydrogen Sulfide (H2S). In *Patch Clamp Technique*; Shad Kaneez, F., Ed.; InTech: Rijeka, Croatia, 2012; pp. 109–142.
31. Rothberg, B.S. The BK channel: A vital link between cellular calcium and electrical signaling. *Protein Cell* **2012**, *3*, 883–892.
32. N'Gouemo, P. Targeting BK (big potassium) channels in epilepsy. *Expert Opin. Ther. Targets* **2011**, *15*, 1283–1295.
33. Hoshi, T.; Pantazis, A.; Olcese, R. Transduction of voltage and Ca^{2+} signals by Slo1 BK channels. *Physiology* **2013**, *28*, 172–189.
34. Toro, L.; Li, M.; Zhang, Z.; Singh, H.; Wu, Y.; Stefani, E. MaxiK channel and cell signalling. *Pflüg. Arch.* **2014**, *466*, 875–886.
35. Ge, L.; Hoa, N.T.; Wilson, Z.; Arismendi-Morillo, G.; Kong, X.-T.; Tajhya, R.B.; Beeton, C.; Jadus, M.R. Big Potassium (BK) ion channels in biology, disease and possible targets for cancer immunotherapy. *Int. Immunopharmacol.* **2014**, *22*, 427–443.
36. Kyle, B.D.; Braun, A.P. The regulation of BK channel activity by pre- and post-translational modifications. *Front. Physiol.* **2014**, doi:10.3389/fphys.2014.00316.
37. Yang, H.; Zhang, G.; Cui, J. BK channels: Multiple sensors, one activation gate. *Front. Physiol.* **2015**, doi:10.3389/fphys.2015.00029.
38. Fodor, A.A.; Aldrich, R.W. Convergent evolution of alternative splices at domain boundaries of the BK channel. *Annu. Rev. Physiol.* **2009**, *71*, 19–36.
39. Xie, J.; McCobb, D.P. Control of alternative splicing of potassium channels by stress hormones. *Science* **1998**, *280*, 443–446.
40. Erxleben, C.; Everhart, A.L.; Romeo, C.; Florance, H.; Bauer, M.B.; Alcorta, D.A.; Rossie, S.; Shipston, M.J.; Armstrong, D.L. Interacting effects of N-terminal variation and strex exon splicing on slo potassium channel regulation by calcium, phosphorylation, and oxidation. *J. Biol. Chem.* **2002**, *277*, 27045–27052.
41. Berthois, Y.; Katzenellenbogen, J.A.; Katzenellenbogen, B.S. Phenol red in tissue culture media is a weak estrogen: Implications concerning the study of estrogen-responsive cells in culture. *Proc. Natl. Acad. Sci. USA* **1986**, *83*, 2496–2500.
42. Holdiman, A.J.; Fergus, D.J.; England, S.K. 17β-Estradiol upregulates distinct maxi-K channel transcripts in mouse uterus. *Mol. Cell. Endocrinol.* **2002**, *192*, 1–6.
43. Norfleet, A.M.; Thomas, M.L.; Gametchu, B.; Watson, C.S. Estrogen receptor-alpha detected on the plasma membrane of aldehyde-fixed GH3/B6/F10 rat pituitary tumor cells by enzyme-linked immunocytochemistry. *Endocrinology* **1999**, *140*, 3805–3814.
44. Hermann, A.; Sitdikova, G.F.; Weiger, T.M. Modulated by Gasotransmitters: BK Channels. In *Gasotransmitters: Physiology and Pathophysiology*; Hermann, A., Sitdikova, G.F., Weiger, T.M., Eds.; Springer: Heidelberg, Germany, 2012; pp. 163–201.

45. Hermann, A.; Gorman, A.L. External and internal effects of tetraethylammonium on voltage-dependent and Ca-dependent K^+ currents components in molluscan pacemaker neurons. *Neurosci. Lett.* **1979**, *12*, 87–92.
46. Lang, D.G.; Ritchie, A.K. Tetraethylammonium blockade of apamin-sensitive and insensitive Ca^{2+}-activated K^+ channels in a pituitary cell line. *J. Physiol.* **1990**, *425*, 117–132.
47. Zhou, Y.; Lingle, C.J. Paxilline inhibits BK channels by an almost exclusively closed-channel block mechanism. *J. Gen. Physiol.* **2014**, *144*, 415–440.
48. Gribkoff, V.K.; Lum-Ragan, J.T.; Boissard, C.G.; Post-Munson, D.J.; Meanwell, N.A.; Starrett, J.E.; Kozlowski, E.S.; Romine, J.L.; Trojnacki, J.T.; Mckay, M.C.; *et al*. Effects of channel modulators on cloned large-conductance calcium-activated potassium channels. *Mol. Pharmacol.* **1996**, *50*, 206–217.
49. Thompson, J.; Begenisich, T. Mechanistic details of BK channel inhibition by the intermediate conductance, Ca^{2+}-activated K channel. *Channels* **2009**, *3*, 194–204.
50. Gola, M.; Crest, M. Colocalization of active KCa channels and Ca^{2+} channels within Ca^{2+} domains in helix neurons. *Neuron* **1993**, *10*, 689–699.
51. Issa, N.P.; Hudspeth, A.J. Clustering of Ca^{2+} channels and Ca^{2+}-activated K^+ channels at fluorescently labeled presynaptic active zones of hair cells. *Proc. Natl. Acad. Sci. USA* **1994**, *91*, 7578–7582.
52. Levitan, I.B. Modulation of ion channels by protein phosphorylation and dephosphorylation. *Annu. Rev. Physiol.* **1994**, *56*, 193–212.
53. Levitan, I.B. Modulation of ion channels by protein phosphorylation. How the brain works. *Adv. Second Messenger Phosphoprotein Res.* **1999**, *33*, 3–22.
54. Tian, L.; Duncan, R.R.; Hammond, M.S.L.; Coghill, L.S.; Wen, H.; Rusinova, R.; Clark, A.G.; Levitan, I.B.; Shipston, M.J. Alternative splicing switches potassium channel sensitivity to protein phosphorylation. *J. Biol. Chem.* **2001**, *276*, 7717–7720.
55. Zhou, X.B.; Arntz, C.; Kamm, S.; Motejlek, K.; Sausbier, U.; Wang, G.X.; Ruth, P.; Korth, M. A molecular switch for specific stimulation of the BKCa channel by cGMP and cAMP kinase. *J. Biol. Chem.* **2001**, *276*, 43239–43245.
56. Zhou, X.-B.; Wulfsen, I.; Utku, E.; Sausbier, U.; Sausbier, M.; Wieland, T.; Ruth, P.; Korth, M. Dual role of protein kinase C on BK channel regulation. *Proc. Natl. Acad. Sci. USA* **2010**, *107*, 8005–8010.
57. Weiger, T.M.; Hermann, A.; Levitan, I.B. Modulation of calcium-activated potassium channels. *J. Comp. Physiol. Neuroethol. Sens. Neural. Behav. Physiol.* **2002**, *188*, 79–87.
58. Alioua, A.; Kumar, Y.; Eghbali, M.; Stefani, E. MaxiK channel partners: Physiological impact. *J. Physiol.* **2006**, *570*, 65–72.
59. Hou, S.; Heinemann, S.H.; Hoshi, T. Modulation of BKCa channel gating by endogenous signaling molecules. *Physiology* **2009**, *24*, 26–35.
60. Sitdikova, G.F.; Fuchs, R.; Kainz, V.; Weiger, T.M.; Hermann, A. Phosphorylation of BK channels modulates the sensitivity to hydrogen sulfide (H2S). *Front. Physiol.* **2014**, doi:10.3389/fphys. 2014.00431.
61. Tang, X.D.; Garcia, M.L.; Heinemann, S.H.; Hoshi, T. Reactive oxygen species impair Slo1 BK channel function by altering cysteine-mediated calcium sensing. *Nat. Struct. Mol. Biol.* **2004**, *11*, 171–178.

62. Tang, G.; Wu, L.; Wang, R. Interaction of hydrogen sulfide with ion channels. *Clin. Exp. Pharmacol. Physiol.* **2010**, *37*, 753–763.
63. Wilkinson, W.J.; Kemp, P.J. Carbon monoxide: An emerging regulator of ion channels. *J. Physiol.* **2011**, *589*, 3055–3062.
64. Peers, C.; Bauer, C.C.; Boyle, J.P.; Scragg, J.L.; Dallas, M.L. Modulation of ion channels by hydrogen sulfide. *Antioxid. Redox Signal.* **2012**, *17*, 95–105.
65. Weiger, T.M.; Holmqvist, M.H.; Levitan, I.B.; Clark, F.T.; Sprague, S.; Huang, W.J.; Ge, P.; Wang, C.; Lawson, D.; Jurman, M.E.; *et al.* A novel nervous system beta subunit that downregulates human large conductance calcium-dependent potassium channels. *J. Neurosci.* **2000**, *20*, 3563–3570.
66. Zhang, J.; Yan, J. Regulation of BK channels by auxiliary γ subunits. *Membr. Physiol. Membr. Biophys.* **2014**, doi:10.3389/fphys.2014.00401.
67. Torres, Y.P.; Granados, S.T.; Latorre, R. Pharmacological consequences of the coexpression of BK channel α and auxiliary β subunits. *Membr. Physiol. Membr. Biophys.* **2014**, doi:10.3389/fphys.2014.00383.
68. McCormack, T.; McCormack, K. Shaker K^+ channel beta subunits belong to an NAD(P)H-dependent oxidoreductase superfamily. *Cell* **1994**, *79*, 1133–1135.
69. Gulbis, J.M.; Mann, S.; MacKinnon, R. Structure of a voltage-dependent K^+ channel beta subunit. *Cell* **1999**, *97*, 943–952.
70. Wallner, M.; Meera, P.; Toro, L. Determinant for beta-subunit regulation in high-conductance voltage-activated and Ca^{2+}-sensitive K^+ channels: An additional transmembrane region at the N terminus. *Proc. Natl. Acad. Sci. USA* **1996**, *93*, 14922–14927.
71. Hanner, M.; Schmalhofer, W.A.; Munujos, P.; Knaus, H.G.; Kaczorowski, G.J.; Garcia, M.L. The beta subunit of the high-conductance calcium-activated potassium channel contributes to the high-affinity receptor for charybdotoxin. *Proc. Natl. Acad. Sci. USA* **1997**, *94*, 2853–2858.
72. Meera, P.; Wallner, M.; Jiang, Z.; Toro, L. A calcium switch for the functional coupling between alpha (hslo) and beta subunits (KV,Ca beta) of maxi K channels. *FEBS Lett.* **1996**, *382*, 84–88.
73. Orio, P.; Rojas, P.; Ferreira, G.; Latorre, R. New disguises for an old channel: MaxiK channel beta-subunits. *News Physiol. Sci.* **2002**, *17*, 156–161.
74. Wallner, M.; Meera, P.; Toro, L. Molecular basis of fast inactivation in voltage and Ca^{2+}-activated K^+ channels: A transmembrane beta-subunit homolog. *Proc. Natl. Acad. Sci. USA* **1999**, *96*, 4137–4142.
75. Tseng-Crank, J.; Godinot, N.; Johansen, T.E.; Ahring, P.K.; Strøbaek, D.; Mertz, R.; Foster, C.D.; Olesen, S.P.; Reinhart, P.H. Cloning, expression, and distribution of a Ca^{2+}-activated K^+ channel beta-subunit from human brain. *Proc. Natl. Acad. Sci. USA* **1996**, *93*, 9200–9205.
76. Bentrop, D.; Beyermann, M.; Wissmann, R.; Fakler, B. NMR structure of the "ball-and-chain" domain of KCNMB2, the beta 2-subunit of large conductance Ca^{2+}- and voltage-activated potassium channels. *J. Biol. Chem.* **2001**, *276*, 42116–42121.
77. Xia, X.-M.; Ding, J.P.; Lingle, C.J. Inactivation of BK channels by the NH2 terminus of the beta2 auxiliary subunit: An essential role of a terminal peptide segment of three hydrophobic residues. *J. Gen. Physiol.* **2003**, *121*, 125–148.
78. Ha, T.S.; Heo, M.-S.; Park, C.-S. Functional effects of auxiliary beta4-subunit on rat large-conductance Ca^{2+}-activated K^+ channel. *Biophys. J.* **2004**, *86*, 2871–2882.

79. Brenner, R.; Chen, Q.H.; Vilaythong, A.; Toney, G.M.; Noebels, J.L.; Aldrich, R.W. BK channel beta4 subunit reduces dentate gyrus excitability and protects against temporal lobe seizures. *Nat. Neurosci.* **2005**, *8*, 1752–1759.
80. Brenner, R.; Jegla, T.J.; Wickenden, A.; Liu, Y.; Aldrich, R.W. Cloning and functional characterization of novel large conductance calcium-activated potassium channel β subunits, hKCNMB3 and hKCNMB4. *J. Biol. Chem.* **2000**, *275*, 6453–6461.
81. Meera, P.; Wallner, M.; Toro, L. A neuronal beta subunit (KCNMB4) makes the large conductance, voltage- and Ca^{2+}-activated K^+ channel resistant to charybdotoxin and iberiotoxin. *Proc. Natl. Acad. Sci. USA* **2000**, *97*, 5562–5567.
82. Behrens, R.; Nolting, A.; Reimann, F.; Schwarz, M.; Waldschütz, R.; Pongs, O. hKCNMB3 and hKCNMB4, cloning and characterization of two members of the large-conductance calcium-activated potassium channel beta subunit family. *FEBS Lett.* **2000**, *474*, 99–106.
83. Yan, J.; Aldrich, R.W. LRRC26 auxiliary protein allows BK channel activation at resting voltage without calcium. *Nature* **2010**, *466*, 513–516.
84. Yan, J.; Aldrich, R.W. BK potassium channel modulation by leucine-rich repeat-containing proteins. *Proc. Natl. Acad. Sci. USA* **2012**, *109*, 7917–7922.
85. Schopperle, W.M.; Holmqvist, M.H.; Zhou, Y.; Wang, J.; Wang, Z.; Griffith, L.C.; Keselman, I.; Kusinitz, F.; Dagan, D.; Levitan, I.B. Slob, a novel protein that interacts with the Slowpoke calcium-dependent potassium channel. *Neuron* **1998**, *20*, 565–573.
86. Zhou, Y.; Schopperle, W.M.; Murrey, H.; Jaramillo, A.; Dagan, D.; Griffith, L.C.; Levitan, I.B. A dynamically regulated 14–3–3, Slob, and Slowpoke potassium channel complex in *Drosophila* presynaptic nerve terminals. *Neuron* **1999**, *22*, 809–818.
87. Zeng, H.; Weiger, T.M.; Fei, H.; Levitan, I.B. Mechanisms of two modulatory actions of the channel-binding protein Slob on the *Drosophila* Slowpoke calcium-dependent potassium channel. *J. Gen. Physiol.* **2006**, *128*, 583–591.
88. Zeng, H.; Weiger, T.M.; Fei, H.; Jaramillo, A.M.; Levitan, I.B. The amino terminus of Slob, Slowpoke channel binding protein, critically influences its modulation of the channel. *J. Gen. Physiol.* **2005**, *125*, 631–640.
89. Jaramillo, A.M.; Zheng, X.; Zhou, Y.; Amado, D.A.; Sheldon, A.; Sehgal, A.; Levitan, I.B. Pattern of distribution and cycling of SLOB, Slowpoke channel binding protein, in *Drosophila*. *BMC Neurosci.* **2004**, doi:10.1186/1471-2202-5-3.
90. Kawakubo, T.; Naruse, K.; Matsubara, T.; Hotta, N.; Sokabe, M. Characterization of a newly found stretch-activated KCa, ATP channel in cultured chick ventricular myocytes. *Am. J. Physiol.* **1999**, *276*, H1827–H1838.
91. Gasull, X.; Ferrer, E.; Llobet, A.; Castellano, A.; Nicolás, J.M.; Palés, J.; Gual, A. Cell membrane stretch modulates the high-conductance Ca^{2+}-activated K^+ channel in bovine trabecular meshwork cells. *Invest. Ophthalmol. Vis. Sci.* **2003**, *44*, 706–714.
92. Tang, Q.Y.; Qi, Z.; Naruse, K.; Sokabe, M. Characterization of a functionally expressed stretch-activated BKca channel cloned from chick ventricular myocytes. *J. Membr. Biol.* **2003**, *196*, 185–200.
93. Wang, W.; Huang, H.; Hou, D.; Liu, P.; Wei, H.; Fu, X.; Niu, W. Mechanosensitivity of STREX-lacking BKCa channels in the colonic smooth muscle of the mouse. *Am. J. Physiol. Gastrointest. Liver Physiol.* **2010**, *299*, G1231–G1240.

94. Pattillo, J.M.; Yazejian, B.; DiGregorio, D.A.; Vergara, J.L.; Grinnell, A.D.; Meriney, S.D. Contribution of presynaptic calcium-activated potassium currents to transmitter release regulation in cultured Xenopus nerve-muscle synapses. *Neuroscience* **2001**, *102*, 229–240.
95. Meredith, A.L.; Wiler, S.W.; Miller, B.H.; Takahashi, J.S.; Fodor, A.A.; Ruby, N.F.; Aldrich, R.W. BK calcium-activated potassium channels regulate circadian behavioral rhythms and pacemaker output. *Nat. Neurosci.* **2006**, *9*, 1041–1049.
96. Chen, L.; Jeffries, O.; Rowe, I.C.M.; Liang, Z.; Knaus, H.-G.; Ruth, P.; Shipston, M.J. Membrane trafficking of large conductance calcium-activated potassium channels is regulated by alternative splicing of a transplantable, acidic trafficking motif in the RCK1-RCK2 linker. *J. Biol. Chem.* **2010**, *285*, 23265–23275.
97. Singh, H.; Stefani, E.; Toro, L. Intracellular BK(Ca) (iBK(Ca)) channels. *J. Physiol.* **2012**, *590*, 5937–5947.
98. Li, B.; Jie, W.; Huang, L.; Wei, P.; Li, S.; Luo, Z.; Friedman, A.K.; Meredith, A.L.; Han, M.-H.; Zhu, X.-H.; *et al*. Nuclear BK channels regulate gene expression via the control of nuclear calcium signaling. *Nat. Neurosci.* **2014**, *17*, 1055–1063.
99. Nardi, A.; Olesen, S.-P. BK channel modulators: A comprehensive overview. *Curr. Med. Chem.* **2008**, *15*, 1126–1146.
100. Rolim, A.L.R.; Lindsey, S.C.; Kunii, I.S.; Fujikawa, A.M.; Soares, F.A.; Chiamolera, M.I.; Maciel, R.M.B.; da Silva, M.R.D. Ion channelopathies in endocrinology: Recent genetic findings and pathophysiological insights. *Arq. Bras. Endocrinol. Metabol.* **2010**, *54*, 673–681.
101. Catterall, W.A. Ion channel voltage sensors: Structure, function, and pathophysiology. *Neuron* **2010**, *67*, 915–928.
102. Kullmann, D.M.; Waxman, S.G. Neurological channelopathies: New insights into disease mechanisms and ion channel function: Neurological channelopathies. *J. Physiol.* **2010**, *588*, 1823–1827.
103. Zhang, L.; Li, X.; Zhou, R.; Xing, G. Possible role of potassium channel, big K in etiology of schizophrenia. *Med. Hypotheses* **2006**, *67*, 41–43.
104. Laumonnier, F.; Roger, S.; Guérin, P.; Molinari, F.; M'rad, R.; Cahard, D.; Belhadj, A.; Halayem, M.; Persico, A.M.; Elia, M.; *et al*. Association of a functional deficit of the BKCa channel, a synaptic regulator of neuronal excitability, with autism and mental retardation. *Am. J. Psychiatry* **2006**, *163*, 1622–1629.
105. Lorenz, S.; Heils, A.; Kasper, J.M.; Sander, T. Allelic association of a truncation mutation of the KCNMB3 gene with idiopathic generalized epilepsy. *Am. J. Med. Genet. Part B Neuropsychiatr. Genet.* **2007**, *144*, 10–13.
106. Du, W.; Bautista, J.F.; Yang, H.; Diez-Sampedro, A.; You, S.-A.; Wang, L.; Kotagal, P.; Lüders, H.O.; Shi, J.; Cui, J.; *et al*. Calcium-sensitive potassium channelopathy in human epilepsy and paroxysmal movement disorder. *Nat. Genet.* **2005**, *37*, 733–738.
107. Kourie, J.I. Interaction of reactive oxygen species with ion transport mechanisms. *Am. J. Physiol.* **1998**, *275*, C1–C24.
108. Hoshi, T.; Heinemann, S.H. Regulation of cell function by methionine oxidation and reduction. *J. Physiol.* **2001**, *531*, 1–11.

109. Tang, X.D.; Daggett, H.; Hanner, M.; Garcia, M.L.; McManus, O.B.; Brot, N.; Weissbach, H.; Heinemann, S.H.; Hoshi, T. Oxidative regulation of large conductance calcium-activated potassium channels. *J. Gen. Physiol.* **2001**, *117*, 253–274.
110. Annunziato, L.; Pannaccione, A.; Cataldi, M.; Secondo, A.; Castaldo, P.; di Renzo, G.; Taglialatela, M. Modulation of ion channels by reactive oxygen and nitrogen species: A pathophysiological role in brain aging? *Neurobiol. Aging* **2002**, *23*, 819–834.
111. Chinopoulos, C.; Adam-Vizi, V. Calcium, mitochondria and oxidative stress in neuronal pathology. *FEBS J.* **2006**, *273*, 433–450.
112. Sesti, F.; Liu, S.; Cai, S.-Q. Oxidation of potassium channels by ROS: A general mechanism of aging and neurodegeneration? *Trends Cell Biol.* **2010**, *20*, 45–51.
113. Sahoo, N.; Hoshi, T.; Heinemann, S.H. Oxidative modulation of voltage-gated potassium channels. *Antioxid. Redox Signal.* **2014**, *21*, 933–952.
114. Ray, P. D.; Huang, B.-W.; Tsuji, Y. Reactive oxygen species (ROS) homeostasis and redox regulation in cellular signaling. *Cell. Signal.* **2012**, *24*, 981–990.
115. Circu, M.L.; Aw, T.Y. Reactive oxygen species, cellular redox systems, and apoptosis. *Free Radic. Biol. Med.* **2010**, *48*, 749–762.
116. Bertram, C.; Hass, R. Cellular responses to reactive oxygen species-induced DNA damage and aging. *Biol. Chem.* **2008**, *389*, 211–220.
117. Peers, C.; Boyle, J.P. Oxidative modulation of K^+ channels in the central nervous system in neurodegenerative diseases and aging. *Antioxid. Redox Signal.* **2015**, *22*, 505–521.
118. Arnér, E.S.J.; Holmgren, A. Physiological functions of thioredoxin and thioredoxin reductase. *Eur. J. Biochem.* **2000**, *267*, 6102–6109.
119. Levine, R.L.; Mosoni, L.; Berlett, B.S.; Stadtman, E.R. Methionine residues as endogenous antioxidants in proteins. *Proc. Natl. Acad. Sci. USA* **1996**, *93*, 15036–15040.
120. Barlow, R.S.; White, R.E. Hydrogen peroxide relaxes porcine coronary arteries by stimulating BKCa channel activity. *Am. J. Physiol.* **1998**, *275*, H1283–H1289.
121. Gutterman, D.D.; Miura, H.; Liu, Y. Redox modulation of vascular tone: Focus of potassium channel mechanisms of dilation. *Arterioscler. Thromb. Vasc. Biol.* **2005**, *25*, 671–678.
122. Stowe, D.F.; Aldakkak, M.; Camara, A.K.S.; Riess, M.L.; Heinen, A.; Varadarajan, S.G.; Jiang, M.-T. Cardiac mitochondrial preconditioning by Big Ca^{2+}-sensitive K^+ channel opening requires superoxide radical generation. *Am. J. Physiol. Heart Circ. Physiol.* **2006**, *290*, H434–H440.
123. Thuringer, D.; Findlay, I. Contrasting effects of intracellular redox couples on the regulation of maxi-K channels in isolated myocytes from rabbit pulmonary artery. *J. Physiol.* **1997**, *500*, 583–592.
124. Lang, R.J.; Harvey, J.R.; McPhee, G.J.; Klemm, M.F. Nitric oxide and thiol reagent modulation of Ca^{2+}-activated K^+ (BKCa) channels in myocytes of the guinea-pig taenia caeci. *J. Physiol.* **2000**, *525*, 363–376.
125. Wang, Z.-W.; Nara, M.; Wang, Y.-X.; Kotlikoff, M.I. Redox regulation of large conductance Ca^{2+}-activated K^+ channels in smooth muscle cells. *J. Gen. Physiol.* **1997**, *110*, 35–44.
126. Riesco-Fagundo, A.M.; Pérez-García, M.T.; González, C.; López-López, J.R. O_2 modulates large-conductance Ca^{2+}-dependent K^+ channels of rat chemoreceptor cells by a membrane-restricted and CO-sensitive mechanism. *Circ. Res.* **2001**, *89*, 430–436.

127. Gong, L.-W.; Gao, T.M.; Huang, H.; Tong, Z. Redox modulation of large conductance calcium-activated potassium channels in CA1 pyramidal neurons from adult rat hippocampus. *Neurosci. Lett.* **2000**, *286*, 191–194.

128. Zhang, G.; Horrigan, F.T. Cysteine modification alters voltage- and Ca^{2+}-dependent gating of large conductance (BK) potassium channels. *J. Gen. Physiol.* **2005**, *125*, 213–236.

129. Sitdikova, G.F.; Weiger, T.M.; Hermann, A. Hydrogen sulfide increases calcium-activated potassium (BK) channel activity of rat pituitary tumor cells. *Pflüg. Arch.* **2010**, *459*, 389–397.

130. Haugland, R.P.; Spence, M.T.Z.; Johnson, I.D. *Handbook of Fluorescent Probes and Research Chemicals*; Molecular Probes: Eugene, OR, USA, 1996.

131. Brzezinska, A.K.; Gebremedhin, D.; Chilian, W.M.; Kalyanaraman, B.; Elliott, S.J. Peroxynitrite reversibly inhibits Ca^{2+}-activated K^{+} channels in rat cerebral artery smooth muscle cells. Am. *J. Physiol. Heart Circ. Physiol.* **2000**, *278*, H1883–H1890.

132. Brakemeier, S.; Eichler, I.; Knorr, A.; Fassheber, T.; Köhler, R.; Hoyer, J. Modulation of Ca^{2+}-activated K^{+} channel in renal artery endothelium *in situ* by nitric oxide and reactive oxygen species. *Kidney Int.* **2003**, *64*, 199–207.

133. Jiang, C.; Haddad, G.G. Oxygen deprivation inhibits a K^{+} channel independently of cytosolic factors in rat central neurons. *J. Physiol.* **1994**, *481*, 15–26.

134. Lewis, A.; Peers, C.; Ashford, M.L.J.; Kemp, P.J. Hypoxia inhibits human recombinant large conductance, Ca^{2+}-activated K^{+} (maxi-K) channels by a mechanism which is membrane delimited and Ca^{2+} sensitive. *J. Physiol.* **2002**, *540*, 771–780.

135. DiChiara, T.J.; Reinhart, P.H. Redox Modulation of hslo Ca^{2+}-Activated K^{+} Channels. *J. Neurosci.* **1997**, *17*, 4942–4955.

136. Soh, H.; Jung, W.; Uhm, D.Y.; Chung, S. Modulation of large conductance calcium-activated potassium channels from rat hippocampal neurons by glutathione. *Neurosci. Lett.* **2001**, *298*, 115–118.

137. Bednarczyk, P.; Wieckowski, M.R.; Broszkiewicz, M.; Skowronek, K.; Siemen, D.; Szewczyk, A. Putative structural and functional coupling of the mitochondrial BKCa channel to the respiratory chain. *PLoS One* **2013**, *8*, e68125.

138. Moncada, S.; Palmer, R.M.; Higgs, E.A. Nitric oxide: Physiology, pathophysiology, and pharmacology. *Pharmacol. Rev.* **1991**, *43*, 109–142.

139. Yoo, D.; Jupiter, R.C.; Pankey, E.A.; Reddy, V.G.; Edward, J.A.; Swan, K.W.; Peak, T.C.; Mostany, R.; Kadowitz, P.J. Analysis of cardiovascular responses to the H2S donors Na2S and NaHS in the rat. *Am. J. Physiol.* **2015**, doi:10.1152/ajpheart.00171.2015.

140. Wang, R.; Wu, L. The chemical modification of KCa channels by carbon monoxide in vascular smooth muscle cells. *J. Biol. Chem.* **1997**, *272*, 8222–8226.

141. Basuroy, S.; Leffler, C.W.; Parfenova, H. CORM-A1 prevents blood-brain barrier dysfunction caused by ionotropic glutamate receptor-mediated endothelial oxidative stress and apoptosis. *Am. J. Physiol. Cell Physiol.* **2013**, *304*, C1105–C1115.

142. Ryan, M.J.; Jernigan, N.L.; Drummond, H.A.; McLemore, G.R., Jr; Rimoldi, J.M.; Poreddy, S.R.; Gadepalli, R.S.V.; Stec, D.E. Renal vascular responses to CORM-A1 in the mouse. *Pharmacol. Res.* **2006**, *54*, 24–29.

143. Soni, H.; Pandya, G.; Patel, P.; Acharya, A.; Jain, M.; Mehta, A.A. Beneficial effects of carbon monoxide-releasing molecule-2 (CORM-2) on acute doxorubicin cardiotoxicity in mice: Role of oxidative stress and apoptosis. *Toxicol. Appl. Pharmacol.* **2011**, *253*, 70–80.
144. Smith, H.; Mann, B.E.; Motterlini, R.; Poole, R.K. The carbon monoxide-releasing molecule, CORM-3 (RU(CO)$_3$ CL(glycinate)), targets respiration and oxidases in *Campylobacter jejuni*, generating hydrogen peroxide. *IUBMB Life* **2011**, *63*, 363–371.
145. Stadtman, E.R.; Berlett, B.S. Reactive oxygen-mediated protein oxidation in aging and disease. *Drug Metab. Rev.* **1998**, *30*, 225–243.
146. Manning, G. Genomic Overview of Protein Kinases. Available online: http://www.wormbook.org (accessed on 20 July 2015).
147. Levitan, I.B. Phosphorylation of ion channels. *J. Membr. Biol.* **1985**, *87*, 177–190.
148. Park, K.-S.; Yang, J.-W.; Seikel, E.; Trimmer, J.S. Potassium channel phosphorylation in excitable cells: Providing dynamic functional variability to a diverse family of ion channels. *Physiology* **2008**, *23*, 49–57.
149. Ismailov, I.I.; Benos, D.J. Effects of phosphorylation on ion channel function. *Kidney Int.* **1995**, *48*, 1167–1179.
150. Hille, B. Modulation of ion-channel function by G-protein-coupled receptors. *Trends Neurosci.* **1994**, *17*, 531–536.
151. Shipston, M.J. Ion channel regulation by protein S-acylation. *J. Gen. Physiol.* **2014**, *143*, 659–678.
152. Schubert, R.; Nelson, M.T. Protein kinases: Tuners of the BKCa channel in smooth muscle. *Trends Pharmacol. Sci.* **2001**, *22*, 505–512.
153. Newton, P.M.; Messing, R.O. Intracellular signaling pathways that regulate behavioral responses to ethanol. *Pharmacol. Ther.* **2006**, *109*, 227–237.
154. Lorca, R.A.; Prabagaran, M.; England, S.K. Functional insights into modulation of BKCa channel activity to alter myometrial contractility. *Front. Physiol.* **2014**, doi:10.3389/fphys.2014.00289.
155. Ashcroft, F.M.; Rorsman, P. K(ATP) channels and islet hormone secretion: New insights and controversies. *Nat. Rev. Endocrinol.* **2013**, *9*, 660–669.
156. Friedman, J. Why Is the Nervous System Vulnerable to Oxidative Stress? In *Oxidative Stress and Free Radical Damage in Neurology, Oxidative Stress in Applied Basic Research and Clinical Practice*; Gadoth, N., Göbel, H.H., Eds.; Humana Press: New York, NY, USA, 2011; pp. 19–27.
157. Gao, T.-M.; Fung, M.-L. Decreased large conductance Ca^{2+}-activated K^+ channel activity in dissociated CA1 hippocampal neurons in rats exposed to perinatal and postnatal hypoxia. *Neurosci. Lett.* **2002**, *332*, 163–166.
158. Ji, L.L.; Fu, R.; Mitchell, E.W. Glutathione and antioxidant enzymes in skeletal muscle: Effects of fiber type and exercise intensity. *J. Appl. Physiol.* **1992**, *73*, 1854–1859.
159. Meister, A. Glutathione metabolism. *Methods Enzymol.* **1995**, *251*, 3–7.
160. Langeveld, C.H.; Schepens, E.; Jongenelen, C.A.; Stoof, J.C.; Hjelle, O.P.; Ottersen, O.P.; Drukarch, B. Presence of glutathione immunoreactivity in cultured neurones and astrocytes. *Neuroreport* **1996**, *7*, 1833–1836.
161. Dringen, R.; Kussmaul, L.; Gutterer, J.M.; Hirrlinger, J.; Hamprecht, B. The glutathione system of peroxide detoxification is less efficient in neurons than in astroglial cells. *J. Neurochem.* **1999**, *72*, 2523–2530.

162. Fernandez-Fernandez, S.; Almeida, A.; Bolaños, J.P. Antioxidant and bioenergetic coupling between neurons and astrocytes. *Biochem. J.* **2012**, *443*, 3–11.

163. Martin, H.L.; Teismann, P. Glutathione—A review on its role and significance in Parkinson's disease. *Fed. Am. Soc. Exp. Biol.* **2009**, *23*, 3263–3272.

164. Smeyne, M.; Smeyne, R.J. Glutathione metabolism and Parkinson's disease. *Free Radic. Biol. Med.* **2013**, *62*, 13–25.

165. Liu, H.; Moczydlowski, E.; Haddad, G.G. O_2 deprivation inhibits Ca^{2+}-activated K^+ channels via cytosolic factors in mice neocortical neurons. *J. Clin. Invest.* **1999**, *104*, 577–588.

166. Wyatt, C.N.; Wright, C.; Bee, D.; Peers, C. O_2-sensitive K^+ currents in carotid body chemoreceptor cells from normoxic and chronically hypoxic rats and their roles in hypoxic chemotransduction. *Proc. Natl. Acad. Sci. USA* **1995**, *92*, 295–299.

167. Clark, A.G.; Hall, S.K.; Shipston, M.J. ATP inhibition of a mouse brain large-conductance K^+ (mslo) channel variant by a mechanism independent of protein phosphorylation. *J. Physiol.* **1999**, *516*, 45–53.

168. López-López, J.R.; González, C. Time course of K^+ current inhibition by low oxygen in chemoreceptor cells of adult rabbit carotid body. Effects of carbon monoxide. *FEBS Lett.* **1992**, *299*, 251–254.

169. Gleichmann, M.; Mattson, M.P. Neuronal calcium homeostasis and dysregulation. *Antioxid. Redox Signal.* **2011**, *14*, 1261–1273.

170. Storm, J.F. Action potential repolarization and a fast after-hyperpolarization in rat hippocampal pyramidal cells. *J. Physiol.* **1987**, *385*, 733–759.

171. Hu, H.; Shao, L.R.; Chavoshy, S.; Gu, N.; Trieb, M.; Behrens, R.; Laake, P.; Pongs, O.; Knaus, H.G.; Ottersen, O.P.; *et al.* Presynaptic Ca^{2+}-activated K^+ channels in glutamatergic hippocampal terminals and their role in spike repolarization and regulation of transmitter release. *J. Neurosci.* **2001**, *21*, 9585–9597.

172. Mancini, M.; Soldovieri, M.V.; Gessner, G.; Wissuwa, B.; Barrese, V.; Boscia, F.; Secondo, A.; Miceli, F.; Franco, C.; Ambrosino, P.; *et al.* Critical role of large-conductance calcium- and voltage-activated potassium channels in leptin-induced neuroprotection of N-methyl-D-aspartate-exposed cortical neurons. *Pharmacol. Res.* **2014**, *87*, 80–86.

173. Rundén-Pran, E.; Haug, F.M.; Storm, J.F.; Ottersen, O.P. BK channel activity determines the extent of cell degeneration after oxygen and glucose deprivation: A study in organotypical hippocampal slice cultures. *Neuroscience* **2002**, *112*, 277–288.

174. Liao, Y.; Kristiansen, A.-M.; Oksvold, C.P.; Tuvnes, F.A.; Gu, N.; Rundén-Pran, E.; Ruth, P.; Sausbier, M.; Storm, J.F. Neuronal Ca^{2+}-activated K^+ channels limit brain infarction and promote survival. *PLoS ONE* **2010**, *5*, e15601.

175. Kulawiak, B.; Szewczyk, A. Glutamate-induced cell death in HT22 mouse hippocampal cells is attenuated by paxilline, a BK channel inhibitor. *Mitochondrion* **2012**, *12*, 169–172.

176. Ye, H.; Jalini, S.; Mylvaganam, S.; Carlen, P. Activation of large-conductance Ca^{2+}-activated K^+ channels depresses basal synaptic transmission in the hippocampal CA1 area in APP (swe/ind) TgCRND8 mice. *Neurobiol. Aging* **2010**, *31*, 591–604.

177. Gáspár, T.; Domoki, F.; Lenti, L.; Katakam, P.V.G.; Snipes, J.A.; Bari, F.; Busija, D.W. Immediate neuronal preconditioning by NS1619. *Brain Res.* **2009**, *1285*, 196–207.

178. Gáspár, T.; Katakam, P.; Snipes, J.A.; Kis, B.; Domoki, F.; Bari, F.; Busija, D.W. Delayed neuronal preconditioning by NS1619 is independent of calcium activated potassium channels. *J. Neurochem.* **2008**, *105*, 1115–1128.
179. Xie, H.; Zhang, Y.-Q.; Pan, X.-L.; Wu, S.-H.; Chen, X.; Wang, J.; Liu, H.; Qian, X.-Z.; Liu, Z.-G.; Liu, L.-J. Decreased calcium-activated potassium channels by hypoxia causes abnormal firing in the spontaneous firing medial vestibular nuclei neurons. *Eur. Arch. Otorhinolaryngol.* **2014**, doi:10.1007/s00405-014-3158-4.
180. Tang, X.D.; Xu, R.; Reynolds, M.F.; Garcia, M.L.; Heinemann, S.H.; Hoshi, T. Haem can bind to and inhibit mammalian calcium-dependent Slo1 BK channels. *Nature* **2003**, *425*, 531–535.
181. Kulawiak, B.; Kudin, A.P.; Szewczyk, A.; Kunz, W.S. BK channel openers inhibit ROS production of isolated rat brain mitochondria. *Exp. Neurol.* **2008**, *212*, 543–547.
182. Augustynek, B.; Kudin, A.P.; Bednarczyk, P.; Szewczyk, A.; Kunz, W.S. Hemin inhibits the large conductance potassium channel in brain mitochondria: A putative novel mechanism of neurodegeneration. *Exp. Neurol.* **2014**, *257*, 70–75.
183. Hayabuchi, Y.; Nakaya, Y.; Matsuoka, S.; Kuroda, Y. Hydrogen peroxide-induced vascular relaxation in porcine coronary arteries is mediated by Ca^{2+}-activated K^+ channels. *Heart Vessels* **1998**, *13*, 9–17.
184. Zeng, X.-H.; Xia, X.-M.; Lingle, C.J. Redox-sensitive extracellular gates formed by auxiliary beta subunits of calcium-activated potassium channels. *Nat. Struct. Biol.* **2003**, *10*, 448–454.
185. Gribkoff, V.K.; Starrett, J.E.; Dworetzky, S.I.; Hewawasam, P.; Boissard, C.G.; Cook, D.A.; Frantz, S.W.; Heman, K.; Hibbard, J.R.; Huston, K.; *et al.* Targeting acute ischemic stroke with a calcium-sensitive opener of maxi-K potassium channels. *Nat. Med.* **2001**, *7*, 471–477.
186. Shintani, Y.; Node, K.; Asanuma, H.; Sanada, S.; Takashima, S.; Asano, Y.; Liao, Y.; Fujita, M.; Hirata, A.; Shinozaki, Y.; *et al.* Opening of Ca^{2+}-activated K^+ channels is involved in ischemic preconditioning in canine hearts. *J. Mol. Cell. Cardiol.* **2004**, *37*, 1213–1218.
187. Bentzen, B.H.; Olesen, S.-P.; Rønn, L.C.B.; Grunnet, M. BK channel activators and their therapeutic perspectives. *Front. Physiol.* **2014**, doi:10.3389/fphys.2014.00389.
188. Wojtovich, A.P.; Nadtochiy, S.M.; Urciuoli, W.R.; Smith, C.O.; Grunnet, M.; Nehrke, K.; Brookes, P.S. A non-cardiomyocyte autonomous mechanism of cardioprotection involving the SLO1 BK channel. *Peer J.* **2013**, doi:10.7717/peerj.48.
189. Wagner, S.; Rokita, A.G.; Anderson, M.E.; Maier, L.S. Redox regulation of sodium and calcium handling. *Antioxid. Redox Signal.* **2013**, *18*, 1063–1077.
190. Jovanović, S.; Crawford, R.M.; Ranki, H.J.; Jovanović, A. Large conductance Ca^{2+}-activated K^+ channels sense acute changes in oxygen tension in alveolar epithelial cells. *Am. J. Respir. Cell Mol. Biol.* **2003**, *28*, 363–372.
191. Williams, S.E.J.; Wootton, P.; Mason, H.S.; Bould, J.; Iles, D.E.; Riccardi, D.; Peers, C.; Kemp, P.J. Hemoxygenase-2 is an oxygen sensor for a calcium-sensitive potassium channel. *Science* **2004**, *306*, 2093–2097.
192. Sheu, S.-J.; Wu, S.-N. Mechanism of inhibitory actions of oxidizing agents on calcium-activated potassium current in cultured pigment epithelial cells of the human retina. *Invest. Ophthalmol. Vis. Sci.* **2003**, *44*, 1237–1244.

193. Mathias, S.; Peña, L.A.; Kolesnick, R.N. Signal transduction of stress via ceramide. *Biochem. J.* **1998**, *335*, 465–480.
194. Li, X.; Becker, K.A.; Zhang, Y. Ceramide in redox signaling and cardiovascular diseases. *Cell. Physiol. Biochem.* **2010**, *26*, 41–48.
195. Li, P.L.; Zhang, D.X.; Zou, A.P.; Campbell, W.B. Effect of ceramide on KCa channel activity and vascular tone in coronary arteries. *Hypertension* **1999**, *33*, 1441–1446.
196. Hanner, M.; Vianna-Jorge, R.; Kamassah, A.; Schmalhofer, W.A.; Knaus, H.G.; Kaczorowski, G.J.; Garcia, M.L. The beta subunit of the high conductance calcium-activated potassium channel. Identification of residues involved in charybdotoxin binding. *J. Biol. Chem.* **1998**, *273*, 16289–16296.
197. Knaus, H.G.; Folander, K.; Garcia-Calvo, M.; Garcia, M.L.; Kaczorowski, G.J.; Smith, M.; Swanson, R. Primary sequence and immunological characterization of beta-subunit of high conductance Ca^{2+}-activated K^+ channel from smooth muscle. *J. Biol. Chem.* **1994**, *269*, 17274–17278.
198. Santarelli, L.C.; Chen, J.; Heinemann, S.H.; Hoshi, T. The beta1 subunit enhances oxidative regulation of large-conductance calcium-activated K^+ channels. *J. Gen. Physiol.* **2004**, *124*, 357–370.
199. Santarelli, L.C.; Wassef, R.; Heinemann, S.H.; Hoshi, T. Three methionine residues located within the regulator of conductance for K^+ (RCK) domains confer oxidative sensitivity to large-conductance Ca^{2+}-activated K^+ channels. *J. Physiol.* **2006**, *571*, 329–348.
200. Lu, T.; Wang, X.-L.; He, T.; Zhou, W.; Kaduce, T.L.; Katusic, Z.S.; Spector, A.A.; Lee, H.-C. Impaired arachidonic acid-mediated activation of large-conductance Ca^{2+}-activated K^+ channels in coronary arterial smooth muscle cells in Zucker Diabetic Fatty rats. *Diabetes* **2005**, *54*, 2155–2163.
201. Lu, T.; He, T.; Katusic, Z.S.; Lee, H.-C. Molecular mechanisms mediating inhibition of human large conductance Ca^{2+}-activated K^+ channels by high glucose. *Circ. Res.* **2006**, *99*, 607–616.
202. Lu, T.; Zhang, D.-M.; Wang, X.-L.; He, T.; Wang, R.-X.; Chai, Q.; Katusic, Z.S.; Lee, H.-C. Regulation of coronary arterial BK channels by caveolae-mediated angiotensin II signaling in diabetes mellitus. *Circ. Res.* **2010**, *106*, 1164–1173.
203. Burgoyne, J.R.; Madhani, M.; Cuello, F.; Charles, R.L.; Brennan, J.P.; Schröder, E.; Browning, D.D.; Eaton, P. Cysteine redox sensor in PKGIa enables oxidant-induced activation. *Science* **2007**, *317*, 1393–1397.
204. Zhang, D.X.; Borbouse, L.; Gebremedhin, D.; Mendoza, S.A.; Zinkevich, N.S.; Li, R.; Gutterman, D.D. H2O2-induced dilation in human coronary arterioles: Role of protein kinase G dimerization and large-conductance Ca^{2+}-activated K^+ channel activation. *Circ. Res.* **2012**, *110*, 471–480.
205. Rodgers-Garlick, C.I.; Hogg, D.W.; Buck, L.T. Oxygen-sensitive reduction in Ca^{2+}-activated K^+ channel open probability in turtle cerebrocortex. *Neuroscience* **2013**, *237*, 243–254.
206. O'Reilly, J.P.; Cummins, T.R.; Haddad, G.G. Oxygen deprivation inhibits Na^+ current in rat hippocampal neurones via protein kinase C. *J. Physiol.* **1997**, *503*, 479–488.
207. Gebremedhin, D.; Bonnet, P.; Greene, A.S.; England, S.K.; Rusch, N.J.; Lombard, J.H.; Harder, D.R. Hypoxia increases the activity of Ca^{2+}-sensitive K^+ channels in cat cerebral arterial muscle cell membranes. *Pflüg. Arch.* **1994**, *428*, 621–630.
208. Gebremedhin, D.; Yamaura, K.; Harder, D.R. Role of 20-HETE in the hypoxia-induced activation of Ca^{2+}-activated K^+ channel currents in rat cerebral arterial muscle cells. *Am. J. Physiol. Heart Circ. Physiol.* **2008**, *294*, H107–H120.

209. Hu, L.-F.; Lu, M.; Hon Wong, P.T.; Bian, J.-S. Hydrogen sulfide: Neurophysiology and neuropathology. *Antioxid. Redox Signal.* **2011**, *15*, 405–419.
210. Hu, X.-Q.; Xiao, D.; Zhu, R.; Huang, X.; Yang, S.; Wilson, S.M.; Zhang, L. Chronic hypoxia suppresses pregnancy-induced upregulation of large-conductance Ca^{2+}-activated K^+ channel activity in uterine arteries. *Hypertension* **2012**, *60*, 214–222.
211. Zhu, R.; Xiao, D.; Zhang, L. Potassium channels and uterine vascular adaptation to pregnancy and chronic hypoxia. *Curr. Vasc. Pharmacol.* **2013**, *11*, 737–747.
212. Xiao, D.; Zhu, R.; Zhang, L. Gestational hypoxia up-regulates protein kinase C and inhibits calcium-activated potassium channels in ovine uterine arteries. *Int. J. Med. Sci.* **2014**, *11*, 886–892.
213. Wong, C.-M.; Tsang, S.-Y.; Yao, X.; Chan, F.L.; Huang, Y. Differential effects of estrogen and progesterone on potassium channels expressed in Xenopus oocytes. *Steroids* **2008**, *73*, 272–279.
214. Zhu, R.; Huang, X.; Hu, X.-Q.; Xiao, D.; Zhang, L. Gestational hypoxia increases reactive oxygen species and inhibits steroid hormone-mediated upregulation of Ca^{2+}-activated K^+ channel function in uterine arteries. *Hypertension* **2014**, *64*, 415–422.
215. Wareing, M. Oxygen sensitivity, potassium channels, and regulation of placental vascular tone. *Microcirculation* **2014**, *21*, 58–66.
216. Szewczyk, A.; Skalska, J.; Głab, M.; Kulawiak, B.; Malińska, D.; Koszela-Piotrowska, I.; Kunz, W.S. Mitochondrial potassium channels: From pharmacology to function. *Biochim. Biophys. Acta* **2006**, *1757*, 715–720.
217. Szewczyk, A.; Jarmuszkiewicz, W.; Kunz, W.S. Mitochondrial potassium channels. *IUBMB Life* **2009**, *61*, 134–143.
218. Kalogeris, T.; Bao, Y.; Korthuis, R.J. Mitochondrial reactive oxygen species: A double edged sword in ischemia/reperfusion *vs.* preconditioning. *Redox Biol.* **2014**, *2*, 702–714.
219. Balderas, E.; Zhang, J.; Stefani, E.; Toro, L. Mitochondrial BKCa channel. *Front. Physiol.* **2015**, doi:10.3389/fphys.2015.00104.
220. Siemen, D.; Loupatatzis, C.; Borecky, J.; Gulbins, E.; Lang, F. Ca^{2+}-activated K channel of the BK-type in the inner mitochondrial membrane of a human glioma cell line. *Biochem. Biophys. Res. Commun.* **1999**, *257*, 549–554.
221. Xu, W.; Liu, Y.; Wang, S.; McDonald, T.; van Eyk, J.E.; Sidor, A.; O'Rourke, B. Cytoprotective role of Ca^{2+}- activated K^+ channels in the cardiac inner mitochondrial membrane. *Science* **2002**, *298*, 1029–1033.
222. Singh, H.; Lu, R.; Bopassa, J.C.; Meredith, A.L.; Stefani, E.; Toro, L. MitoBK(Ca) is encoded by the Kcnma1 gene, and a splicing sequence defines its mitochondrial location. *Proc. Natl. Acad. Sci. USA* **2013**, *110*, 10836–10841.
223. Heinen, A.; Camara, A.K.S.; Aldakkak, M.; Rhodes, S.S.; Riess, M.L.; Stowe, D.F. Mitochondrial Ca^{2+}-induced K^+ influx increases respiration and enhances ROS production while maintaining membrane potential. *Am. J. Physiol. Cell Physiol.* **2007**, *292*, C148–C156.
224. Heinen, A.; Aldakkak, M.; Stowe, D.F.; Rhodes, S.S.; Riess, M.L.; Varadarajan, S.G.; Camara, A.K.S. Reverse electron flow-induced ROS production is attenuated by activation of mitochondrial Ca^{2+}-sensitive K^+ channels. *Am. J. Physiol. Heart Circ. Physiol.* **2007**, *293*, H1400–H1407.

225. Bentzen, B.H.; Osadchii, O.; Jespersen, T.; Hansen, R.S.; Olesen, S.-P.; Grunnet, M. Activation of big conductance Ca^{2+}-activated K^+ channels (BK) protects the heart against ischemia-reperfusion injury. *Pflüg. Arch.* **2009**, *457*, 979–988.
226. Sakamoto, K.; Ohya, S.; Muraki, K.; Imaizumi, Y. A novel opener of large-conductance Ca^{2+}-activated K^+ (BK) channel reduces ischemic injury in rat cardiac myocytes by activating mitochondrial K(Ca) channel. *J. Pharmacol. Sci.* **2008**, *108*, 135–139.
227. Ohya, S.; Kuwata, Y.; Sakamoto, K.; Muraki, K.; Imaizumi, Y. Cardioprotective effects of estradiol include the activation of large-conductance Ca^{2+}-activated K^+ channels in cardiac mitochondria. *Am. J. Physiol. Heart Circ. Physiol.* **2005**, *289*, H1635–H1642.
228. Tano, J.-Y.; Gollasch, M. Calcium-activated potassium channels in ischemia reperfusion: A brief update. *Front. Physiol.* **2014**, doi:10.3389/fphys.2014.00381.
229. Hagen, B.M.; Sanders, K.M. Deglycosylation of the beta1-subunit of the BK channel changes its biophysical properties. *Am. J. Physiol. Cell Physiol.* **2006**, *291*, C750–C756.
230. Borchert, G.H.; Yang, C.; Kolár, F. Mitochondrial BKCa channels contribute to protection of cardiomyocytes isolated from chronically hypoxic rats. *Am. J. Physiol. Heart Circ. Physiol.* **2011**, *300*, H507–H513.
231. Borchert, G.H.; Hlaváčková, M.; Kolář, F. Pharmacological activation of mitochondrial BK(Ca) channels protects isolated cardiomyocytes against simulated reperfusion-induced injury. *Exp. Biol. Med. Maywood* **2013**, *238*, 233–241.
232. Neckár, J.; Borchert, G.H.; Hlousková, P.; Mícová, P.; Nováková, O.; Novák, F.; Hroch, M.; Papousek, F.; Ost'ádal, B.; Kolár, F. Brief daily episode of normoxia inhibits cardioprotection conferred by chronic continuous hypoxia. Role of oxidative stress and BKCa channels. *Curr. Pharm. Des.* **2013**, *19*, 6880–6889.
233. Noursadeghi, E.; Jafari, A.; Saghiri, R.; Sauve, R.; Eliassi, A. Impairment of brain mitochondrial charybdotoxin- and ATP-insensitive BK channel activities in diabetes. *Neuromol. Med.* **2014**, *16*, 862–871.
234. Leffler, C.W.; Parfenova, H.; Jaggar, J.H. Carbon monoxide as an endogenous vascular modulator. *Am. J. Physiol. Heart Circ. Physiol.* **2011**, *301*, H1–H11.
235. Li, H.; Förstermann, U. Prevention of atherosclerosis by interference with the vascular nitric oxide system. *Curr. Pharm. Des.* **2009**, *15*, 3133–3145.
236. Ooi, L.; Gigout, S.; Pettinger, L.; Gamper, N. Triple cysteine module within M-type K^+ channels mediates reciprocal channel modulation by nitric oxide and reactive oxygen species. *J. Neurosci.* **2013**, *33*, 6041–6046.
237. Liu, Y.; Gutterman, D.D. Oxidative stress and potassium channel function. *Clin. Exp. Pharmacol. Physiol.* **2002**, *29*, 305–311.
238. Drews, G.; Düfer, M. Role of K(ATP) channels in β-cell resistance to oxidative stress. *Diabetes Obes. MeTable* **2012**, *14*, 120–128.
239. Liu, Y.; Kalogeris, T.; Wang, M.; Zuidema, M.Y.; Wang, Q.; Dai, H.; Davis, M.J.; Hill, M.A.; Korthuis, R.J. Hydrogen sulfide preconditioning or neutrophil depletion attenuates ischemia-reperfusion-induced mitochondrial dysfunction in rat small intestine. *Am. J. Physiol. Gastrointest. Liver Physiol.* **2012**, *302*, G44–G54.

240. Moudgil, R.; Michelakis, E.D.; Archer, S.L. The role of K^+ channels in determining pulmonary vascular tone, oxygen sensing, cell proliferation, and apoptosis: Implications in hypoxic pulmonary vasoconstriction and pulmonary arterial hypertension. *Microcirculation* **2006**, *13*, 615–632.
241. Wolfram Kuhlmann, C.R.; Wiebke Lüdders, D.; Schaefer, C.A.; Kerstin Most, A.; Backenköhler, U.; Neumann, T.; Tillmanns, H.; Erdogan, A. Lysophosphatidylcholine-induced modulation of Ca^{2+}-activated K^+ channels contributes to ROS-dependent proliferation of cultured human endothelial cells. *J. Mol. Cell. Cardiol.* **2004**, *36*, 675–682.
242. Pardo, L.A. Voltage-gated potassium channels in cell proliferation. *Physiology* **2004**, *19*, 285–292.
243. Huang, X.; Jan, L.Y. Targeting potassium channels in cancer. *J. Cell Biol.* **2014**, *206*, 151–162.
244. Pardo, L.A.; Stühmer, W. The roles of K^+ channels in cancer. *Nat. Rev. Cancer* **2014**, *14*, 39–48.
245. Yang, M.; Brackenbury, W.J. Membrane potential and cancer progression. *Front. Physiol.* **2013**, doi:10.3389/fphys.2013.00185.
246. Barghouth, P.G.; Thiruvalluvan, M.; Oviedo, N.J. Bioelectrical regulation of cell cycle and the planarian model system. *Biochim. Biophys. Acta* **2015**, doi:10.1016/j.bbamem.2015.02.024.
247. Blackiston, D.J.; McLaughlin, K.A.; Levin, M. Bioelectric controls of cell proliferation: Ion channels, membrane voltage and the cell cycle. *Cell Cycle* **2009**, *8*, 3519–3528.
248. Weiger, T.M.; Hermann, A. Cell proliferation, potassium channels, polyamines and their interactions: A mini review. *Amino Acids* **2014**, *46*, 681–688.
249. Chattopadhyay, M.K.; Tabor, C.W.; Tabor, H. Polyamine deficiency leads to accumulation of reactive oxygen species in a spe2Delta mutant of Saccharomyces cerevisiae. *Yeast Chichester. Engl.* **2006**, *23*, 751–761.
250. Takano, K.; Ogura, M.; Nakamura, Y.; Yoneda, Y. Neuronal and glial responses to polyamines in the ischemic brain. *Curr. Neurovasc. Res.* **2005**, *2*, 213–223.
251. Pamenter, M.E.; Haddad, G.G. Do BK channels mediate glioma hypoxia-tolerance? *Channels* **2014**, *8*, 176–177.
252. Liu, X.; Chang, Y.; Reinhart, P.H.; Sontheimer, H.; Chang, Y. Cloning and characterization of glioma BK, a novel BK channel isoform highly expressed in human glioma cells. *J. Neurosci.* **2002**, *22*, 1840–1849.
253. Sontheimer, H. An unexpected role for ion channels in brain tumor metastasis. *Exp. Biol. Med.* **2008**, *233*, 779–791.
254. Cheng, Y.; Gu, X.Q.; Bednarczyk, P.; Wiedemann, F.R.; Haddad, G.G.; Siemen, D. Hypoxia increases activity of the BK-channel in the inner mitochondrial membrane and reduces activity of the permeability transition pore. *Cell. Physiol. Biochem.* **2008**, *22*, 127–136.
255. Gu, X.Q.; Pamenter, M.E.; Siemen, D.; Sun, X.; Haddad, G.G. Mitochondrial but not plasmalemmal BK channels are hypoxia-sensitive in human glioma. *Glia* **2014**, *62*, 504–513.
256. Gu, X.Q.; Siemen, D.; Parvez, S.; Cheng, Y.; Xue, J.; Zhou, D.; Sun, X.; Jonas, E.A.; Haddad, G.G. Hypoxia increases BK channel activity in the inner mitochondrial membrane. *Biochem. Biophys. Res. Commun.* **2007**, *358*, 311–316.
257. Han, X.; Xi, L.; Wang, H.; Huang, X.; Ma, X.; Han, Z.; Wu, P.; Ma, X.; Lu, Y.; Wang, G.; *et al.* The potassium ion channel opener NS1619 inhibits proliferation and induces apoptosis in A2780 ovarian cancer cells. *Biochem. Biophys. Res. Commun.* **2008**, *375*, 205–209.

258. Debska-Vielhaber, G.; Godlewski, M.M.; Kicinska, A.; Skalska, J.; Kulawiak, B.; Piwonska, M.; Zablocki, K.; Kunz, W.S.; Szewczyk, A.; Motyl, T. Large-conductance K^+ channel openers induce death of human glioma cells. *J. Physiol. Pharmacol.* **2009**, *60*, 27–36.

259. Halestrap, A.P. What is the mitochondrial permeability transition pore? *J. Mol. Cell. Cardiol.* **2009**, *46*, 821–831.

260. Chatterjee, S.; Nieman, G.F.; Christie, J.D.; Fisher, A.B. Shear stress-related mechanosignaling with lung ischemia: Lessons from basic research can inform lung transplantation. *Am. J. Physiol. Lung Cell. Mol. Physiol.* **2014**, *307*, L668–L680.

261. Mohamed, M.S.A. Calcium-activated potassium channels in ischemia-reperfusion: Learning for the clinical application. *Pulm. Med.* **2015**, doi:10.3389/fmed.2015.00021.

262. Noda, K.; Shigemura, N.; Tanaka, Y.; Bhama, J.; D'Cunha, J.; Kobayashi, H.; Luketich, J.D.; Bermudez, C.A. Hydrogen preconditioning during *ex vivo* lung perfusion improves the quality of lung grafts in rats. *Transplantation* **2014**, *98*, 499–506.

263. Dong, D.-L.; Zhang, Y.; Lin, D.-H.; Chen, J.; Patschan, S.; Goligorsky, M.S.; Nasjletti, A.; Yang, B.-F.; Wang, W.-H. Carbon monoxide stimulates the Ca^{2+}-activated big conductance K channels in cultured human endothelial cells. *Hypertension* **2007**, *50*, 643–651.

264. Carrasquel-Ursulaez, W.; Contreras, G.F.; Sepúlveda, R.V.; Aguayo, D.; González-Nilo, F.; González, C.; Latorre, R. Hydrophobic interaction between contiguous residues in the S6 transmembrane segment acts as a stimuli integration node in the BK channel. *J. Gen. Physiol.* **2015**, *145*, 61–74.

265. McBride, C.S.; Baier, F.; Omondi, A.B.; Spitzer, S.A.; Lutomiah, J.; Sang, R.; Ignell, R.; Vosshall, L.B. Evolution of mosquito preference for humans linked to an odorant receptor. *Nature* **2014**, *515*, 222–227.

266. Adler, E.M. Of ELIC and evolution. *J. Gen. Physiol.* **2015**, *145*, 1–2.

267. Korovkina, V.P.; England, S.K. Molecular diversity of vascular potassium channel isoforms. *Clin. Exp. Pharmacol. Physiol.* **2002**, *29*, 317–323.

268. Ko, E.A.; Han, J.; Jung, I.D.; Park, W.S. Physiological roles of K^+ channels in vascular smooth muscle cells. *J. Smooth Muscle Res.* **2008**, *44*, 65–81.

269. Park, S.W.; Noh, H.J.; Sung, D.J.; Kim, J.G.; Kim, J.M.; Ryu, S.-Y.; Kang, K.; Kim, B.; Bae, Y.M.; Cho, H. Hydrogen peroxide induces vasorelaxation by enhancing 4-aminopyridine-sensitive Kv currents through S-glutathionylation. *Pflüg. Arch.* **2015**, *467*, 285–297.

270. Fukumoto, M.; Nakaizumi, A.; Zhang, T.; Lentz, S.I.; Shibata, M.; Puro, D.G. Vulnerability of the retinal microvasculature to oxidative stress: Ion channel-dependent mechanisms. *Am. J. Physiol. Cell Physiol.* **2012**, *302*, C1413–C1420.

2

Advanced Glycation End Products and Oxidative Stress in Type 2 Diabetes Mellitus

Kerstin Nowotny, Tobias Jung, Annika Höhn, Daniela Weber and Tilman Grune *

Department of Molecular Toxicology, German Institute of Human Nutrition Potsdam-Rehbruecke, Arthur-Scheunert-Allee 114-116, 14558 Nuthetal, Germany; E-Mails: Kerstin.Nowotny@dife.de (K.N.); Tobias.Jung@dife.de (T.J.); Annika.Hoehn@dife.de (A.H.); Daniela.Weber@dife.de (D.W.)

* Author to whom correspondence should be addressed; E-Mail: scientific.director@dife.de;

Academic Editors: Michael Breitenbach and Peter Eckl

Abstract: Type 2 diabetes mellitus (T2DM) is a very complex and multifactorial metabolic disease characterized by insulin resistance and β cell failure leading to elevated blood glucose levels. Hyperglycemia is suggested to be the main cause of diabetic complications, which not only decrease life quality and expectancy, but are also becoming a problem regarding the financial burden for health care systems. Therefore, and to counteract the continually increasing prevalence of diabetes, understanding the pathogenesis, the main risk factors, and the underlying molecular mechanisms may establish a basis for prevention and therapy. In this regard, research was performed revealing further evidence that oxidative stress has an important role in hyperglycemia-induced tissue injury as well as in early events relevant for the development of T2DM. The formation of advanced glycation end products (AGEs), a group of modified proteins and/or lipids with damaging potential, is one contributing factor. On the one hand it has been reported that AGEs increase reactive oxygen species formation and impair antioxidant systems, on the other hand the formation of some AGEs is induced *per se* under oxidative conditions. Thus, AGEs contribute at least partly to chronic stress conditions in diabetes. As AGEs are not only formed endogenously, but also derive from exogenous sources, *i.e.*, food, they have been assumed as risk factors for T2DM. However, the role of AGEs in the pathogenesis of T2DM and diabetic complications—if they are causal or simply an effect—is only partly understood. This review will highlight the involvement of AGEs in the development and progression of T2DM and their role in diabetic complications.

Keywords: advanced glycation end products; oxidative stress; type 2 diabetes mellitus; insulin resistance; β cell dysfunction; diabetic complications

1. Introduction

With around 350 million cases in 2014 [1], type 2 diabetes mellitus (T2DM) is one of the most frequent diseases throughout the world. This number is predicted to increase dramatically in the coming years, resulting in serious health and economic challenges. In general, diabetes mellitus is a group of metabolic diseases in which the pancreas is not able to produce insulin, insulin production is not sufficient or cells cannot effectively use this hormone [2]. In T2DM, the body is *per se* able to produce insulin. However, several processes induce abnormalities in a manner that either hormone production is insufficient or cells are unable to mediate the effects of insulin. As insulin is required for an efficient cellular uptake of glucose to convert it into energy, the ineffectiveness of insulin causes elevated blood glucose levels (hyperglycemia). Hyperglycemia is estimated to be one major factor contributing to diabetic complications including diseases which affect the cardiovascular and nervous system, eyes or kidneys. Every year, 4.9 million people die from diabetes [1] some 50% of them by cardiovascular complications [3]. To prevent diabetic complications, an early diagnosis is very important. However, the high prevalence of T2DM, its morbidity as well as mortality rates are at least partly due to the fact, that most people only recognize this disease in a state when symptoms already occur. This indicates that the diagnostic and therefore therapeutic possibilities are limited at the moment. For therapeutic approaches it is essential to gain further knowledge about the main risk factors and molecular mechanisms in the pathogenesis of T2DM and related complications.

The formation of reactive oxygen species (ROS) is an inevitable byproduct of metabolism. The main source of ROS in mammalian cells is the "dripping" of electrons from the mitochondrial respiratory chain, and their subsequent transfer to molecular oxygen, resulting in the formation of the superoxide anion ($O_2^{\bullet -}$). Together with hydrogen peroxide (H_2O_2) and nitric oxide ($^{\bullet}NO$), superoxide is considered to be one of the main "primary" ROS, forming the bulk variety of other ROS found in cells in further reactions. Due to the highly reducing cellular environment, powerful antioxidative systems are needed, that are capable of scavenging ROS or transforming them into less reactive products. Another task of the cell's antioxidative machinery is the "repair" of already (oxidatively) damaged structures or their degradation. Thus, "three lines of defense" can be defined roughly (as shown in Figure 1): low molecular ROS-scavengers, antioxidative enzymes, and repairing or degrading ones. If both, the antioxidative machinery of a cell is overwhelmed by the present amount of ROS and the cellular redox-signaling is disturbed, the cell is found in a stage defined as "oxidative stress" [4]. Oxidative stress, induced by an abundance of ROS or failure in the antioxidative machinery, is the cause of many pathologies.

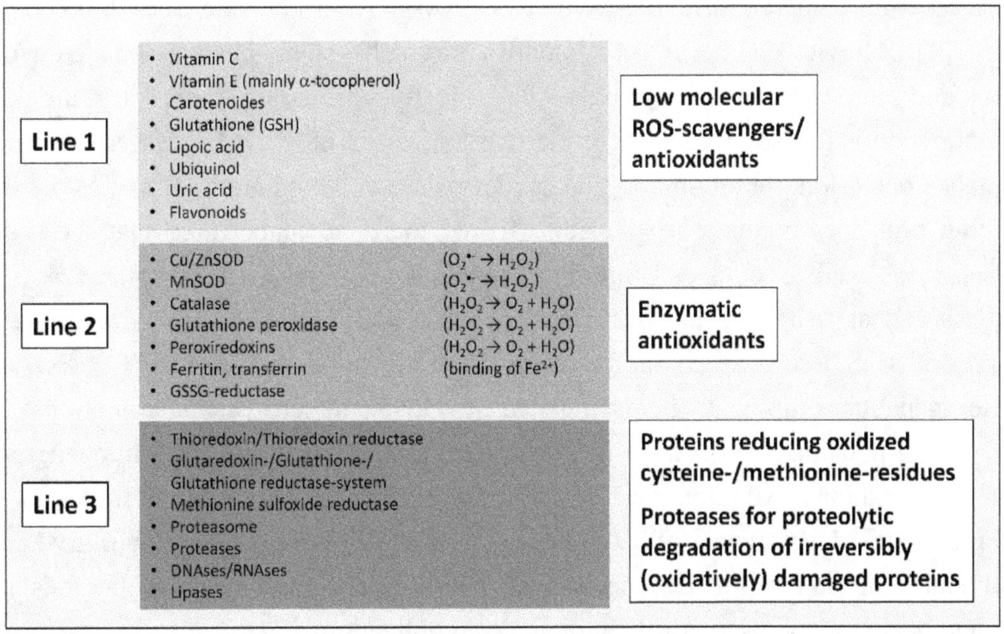

Figure 1. "Three lines of antioxidative defense" in mammalian cells, modified according to [5]. The first line contains low molecular antioxidants that can scavenge ROS/reactive particles in a purely competitive way, preventing damage from cellular structures like proteins, nucleobases and lipids. The reaction products are mostly significantly less reactive and may in several cases be restored by cellular systems (like vitamin C and tocopherol in a glutathione (GSH)-consuming manner). The most important and abundant low molecular intracellular scavenger is GSH, thus determining the cellular redox-state, defined as the ratio of GSH to its oxidized form, glutathione disulfide (GSSG). Under normal physiological conditions, this ratio is about 1:1000 (GSSG:GSH) or even higher, providing a strongly reducing cellular environment. The second line of defense contains antioxidative enzymes that are able to convert ROS into less reactive particles. This includes the superoxide dismutases (Cu, ZnSOD and MnSOD) as well as catalase. Further important enzymes are glutathione peroxidases, catalyzing the reaction of peroxides (R-OOH) to hydroxyls (R-OH) via GSH-consumption. Besides catalase, glutathione peroxidases are the most important H_2O_2-detoxifying enzymes. In this group enzymes are also found, which bind redox-active metals—iron is the most important transition metal in mammalian cells—in an inert form. Otherwise, metals like iron (Fe^{2+}) or copper (Cu^{+}) are able to transfer an electron to H_2O_2 (Fenton reaction), releasing both OH^- and the highly reactive hydroxyl radical ($^{\bullet}OH$). $^{\bullet}OH$ is able to oxidize virtually every organic molecule. The oxidized forms of those metals (Fe^{3+}/Cu^{2+}) are quickly reduced in the cytosolic environment, fuelling the vicious circle. In the last line of defense is a summary of enzymes that are able to restore oxidatively modified amino acids (only methionine and cysteine), thus preventing proteolytic degradation of the whole damaged protein. If repair is not possible, several proteases are available, that can recognize and remove dysfunctional proteins in a proteolytic manner, preventing their intracellular accumulation. The most important one is the proteasomal system, responsible for the degradation of more than 90% of all (oxidatively) damaged proteins, as well as the cathepsins of the lysosomal system. Other catabolic enzymes might also play some role in these defense lines.

There is increasing evidence that oxidative stress also plays a key role in pathological processes observed in T2DM (reviewed in [6–8]). Oxidative stress has long been associated with diabetic complications and more recent studies indicate that oxidative stress is also causal in the development of β cell dysfunction and insulin resistance, the two hallmarks of T2DM. Beta cell dysfunction and insulin resistance occur long before blood glucose levels reach the amount defined as prediabetes [9]. Moreover, both processes mediate progression of prediabetes to diabetes so that the prevention of insulin resistance as well as of β cell failure is essential to prevent T2DM. Diabetes is related to oxidative stress at least partly to the overproduction of ROS. Under diabetic conditions, there are several sources of ROS described, among them advanced glycation end products (AGEs). AGEs are a group of heterogeneous compounds increasingly formed under hyperglycemic conditions. Due to this and their damaging potential, AGEs have been assumed to be involved in the pathogenesis of T2DM and diabetic complications. Additionally, it was recently proposed that another source of AGEs, the diet, contributes to the development of T2DM (reviewed in [10]). This review will give an overview on the involvement of AGEs in T2DM, in particular in the development of insulin resistance, β cell dysfunction and death as well as their role in diabetic complications.

2. AGEs in Diabetes: An Overview

The first link between glycated proteins and diabetes was made in 1968 with the discovery of an altered form of hemoglobin (meanwhile known as $HbA1_c$) in red blood cells of patients with diabetes [11]. It became clear, that glycation occurs predominantly on the N-terminal valine of the β chain and that this Amadori product is formed non-enzymatically in a reaction which was before only known to take place in food [12,13]. In this so-called Maillard reaction, the carbonyl group of a reducing sugar reacts with the amino group of a protein, lipid or nucleic acid generating Schiff bases which rearrange to Amadori products (Figure 2). However, Amadori products are relatively unstable so that further consecutive and parallel reactions occur, eventually leading to the formation of irreversible AGEs. The Maillard reaction is the most common pathway known to form AGEs. Not only during all stages of the Maillard reaction, but also as intermediates or byproducts of glucose autoxidation, lipid peroxidation or the polyol pathway, high reactive carbonyl compounds, including glyoxal, methylglyoxal or 3-deoxyglucosone are formed [14–16]. Increased concentrations of glyoxal, methylglyoxal as well as 3-deoxyglucosone have been found in plasma of patients with T2DM [17]. Glyoxal, for example, causes the formation of Nε-(carboxymethyl) lysine (CML) [18] which is at present the best characterized AGE. Further AGEs formed by glyoxal are glyoxal-derived lysyl dimer (GOLD) [19], Nω-(carboxymethyl) arginine (CMA) [20] or S-carboxymethylcysteine [21]. Methylglyoxal causes the generation of, for example, Nε-(carboxyethyl) lysine (CEL) [22], methylglyoxal-derived lysyl dimer (MOLD) [23], argpyrimidine [24] or methylglyoxal-derived hydroimidazolone MG-H1 [25] whereas 3-deoxyglucosone leads to the formation of pyrraline [26], pentosidine [27], imidazolone or also CML [28].

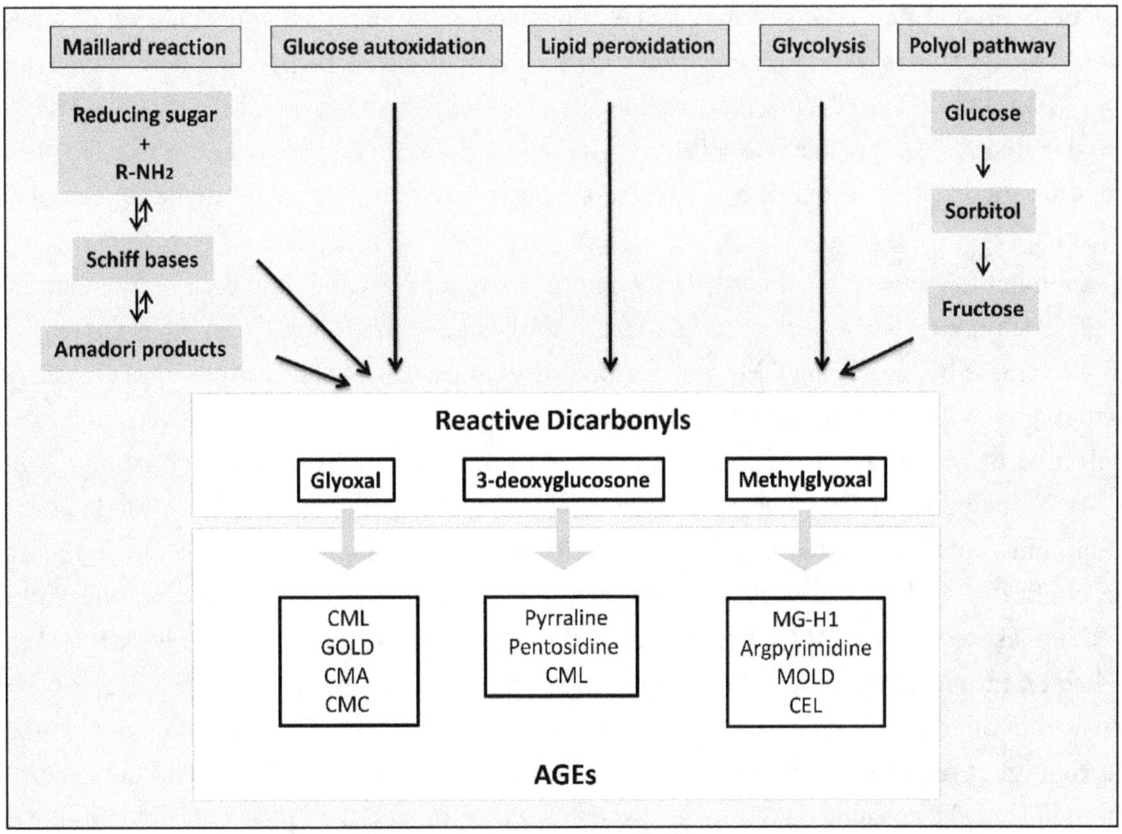

Figure 2. Formation of reactive dicarbonyls and AGEs, modified according to [29]. Reactive dicarbonyls including methylglyoxal, glyoxal and 3-deoxyglucosone are formed through several pathways: the Maillard reaction, the polyol pathway, glycolysis, lipid peroxidation or glucose autoxidation. Dicarbonyl compounds react further to form irreversible products, the so-called AGEs.

Even if oxidation is not always necessary, many AGEs are generated by a combination of oxidation and glycation so that the formation of so-called glycoxidation products is triggered by oxidative stress [30]. Two important AGEs produced by glycoxidation are pentosidine and CML. The complexity and diversity of AGE formation makes clear why substances belonging to the group of AGEs are so heterogeneous regarding their chemical and physical properties. Some AGEs are fluorescent; a few induce protein cross-linking. There are compounds which show both properties, other AGEs are either fluorescent or cross-linkers. AGE formation occurs intra- as well as extracellularly as part of physiological metabolism. To detect AGE formation in fluids and tissues, AGE-specific fluorescence can be measured. The majority of so far identified AGEs are characterized by fluorescence in the area around an excitation wavelength of 370 nm and an emission of 440 nm [24,31,32]. Additionally, pentosidine emits light at 385 nm when excited at 335 nm [33]. Further methods for AGE detection used in *in vitro* and *in vivo* studies are immunohistochemical staining or enzyme-linked immunosorbent assay (ELISA) using antibodies against different AGEs, for example, CML or pentosidine. However, application of these methods is often limited due to lack of reliable antibodies. More sensitive methods for AGE detection include high performance liquid chromatography (HPLC), gas chromatography or liquid chromatography with different detectors (reviewed in [34]).

In people with diabetes, AGE formation is accelerated due to increased concentration of circulating glucose, AGE precursors and oxidative stress. In serum and tissues of patients with type 1 diabetes mellitus (T1DM) as well as T2DM, increased levels of AGEs, among them CML [35–37], MG-derived hydroimidazolone [38], pentosidine [39] or glucosepane [40,41] have been found. Furthermore, accumulation of AGEs in diabetic tissue was shown to correlate with diabetic complications (reviewed in [42]).

At this point it should be noted, that another source of AGEs, the diet, may contribute to pathological features related to diabetes. Different studies have investigated the AGE content in food items, for example by measuring CML concentrations with ELISA [43] or further ultra pressure liquid chromatography with mass spectrometry detection (UPLC-MS) [44]. That the detection of AGEs with only one method is not sufficient and can result in under- or overestimation of AGEs in food items [34], is demonstrated by comparing these studies. Goldberg *et al.* reported that food items rich in fat and protein contain particularly high concentrations of CML in contrast to carbohydrate-rich food in which only low CML concentrations were measured. In contrast, CML detection with UPLC-MS observed highest levels of CML in bread crust and evaporated full-fat milk while lowest CML levels were detected in uncooked minced beef, raw full-fat milk and pasteurized skimmed milk. No CML was detected in olive oil. Nevertheless, people are continuously exposed to AGEs through their diet, and this source might even be greater than the amount formed endogenously [45]. Around one-tenth [46] or even more [47] of the consumed AGEs are absorbed in the gut and, therefore, contribute to the body's AGE pool. Studies reported that the dietary uptake of AGEs correlates with serum AGE levels [46,48] and that restriction of food-derived AGEs can lower AGE concentrations in serum [49,50]. There is a growing body of evidence that the amount of food-derived AGEs which contributes to the AGE pool is sufficient to promote the development of T2DM. Studies in humans and mice revealed that an AGE-rich diet elevates AGE concentrations such as MG-H1 also in non-diabetic subjects and increases biomarkers of inflammation and oxidative stress associated with insulin resistance [51,52]. Furthermore, it was shown that a high-AGE diet causes decreased insulin secretion and increased β cell death in rats [53]. Taken together, AGEs are proposed to play a role in the development and progression of T2DM as well as in diabetic complications.

The damaging potential of AGEs results from direct alterations on protein structures and functions due to AGEs *per se* or the cross-linking effect of some AGEs. AGEs are often found in the extracellular matrix (ECM) and thus modified matrix proteins impair matrix-matrix as well as matrix-cell interactions. This may cause cell death, cell differentiation or reduced cell adhesion and migration. Intracellular proteins are also targets of modifications and AGE formation was shown to impair their functions. Besides direct changes in protein structures and functions, AGE-mediated damage occurs via binding of AGEs to the receptor of advanced glycation end products (RAGE). RAGE belongs to the immunoglobulin superfamily and additionally interacts with a wide range of ligands including S100 calgranulins [54], high mobility group box 1 [55], or Mac-1 [56] which is why RAGE is also classified as a multi-ligand receptor. In recent years the interaction of AGEs with RAGE was studied *in vitro* demonstrating the activation of Janus kinase, rho-GTPases, extracellular-signal-regulated kinase 1/2 and p38 mitogen-activated protein kinase due to AGE-RAGE interaction [57–60]. It should be pointed out that ligand binding to RAGE activates NAPDH oxidases and thus increases intracellular ROS formation [61,62]. Increased ROS in turn leads to AGE formation, which triggers all described damaging mechanisms mediated by AGEs, but also activates the transcription factor nuclear factor kappa B (NFκB) [63]. Activation of NFκB increases the expression of proinflammatory cytokines such as interleukin 6 [64] and monocyte chemoattractant peptide 1 (MCP-1) [65] as well as RAGE itself [66] thus intensifying the inflammatory response. There are two possibilities to protect tissues against AGE-mediated damage: AGEs are eliminated or cells activate compensatory mechanisms. It has been shown that cells process other AGE receptors which are able to bind extracellular AGEs and mediate their cellular uptake (reviewed in [67]). In cells, AGEs can be degraded by the endosomal-lysosomal system. Cathepsin D and L were identified as two enzymes involved in the detoxification of AGEs [68,69]. After digestion, the AGE degradation products are released and circulate in the bloodstream until renal elimination. One possibility to prevent RAGE-mediated damage, is enabled by the upregulation of advanced glycation end products receptor 1 (AGER1) after AGE exposure. Although AGER1 was first only associated with the AGE turnover, studies observed that its upregulation suppresses RAGE-mediated pathways: AGER1 inhibits the activity of NADPH oxidase and weakens oxidative stress generation as well as ROS-mediated signaling [70–73]. Moreover, it has been described that AGER1 is linked to sirtuin1 (SIRT1) [50,52], a NAD^+-dependent deacetylase. For example, through deacetylation of NFκB SIRT1 suppresses NFκB-mediated proinflammatory processes. However, long term exposure to AGEs depletes AGER1 and SIRT1 expression [52] so that the regulation of RAGE signaling fails and oxidative stress and inflammation increase.

3. Advanced Glycation End Products and Insulin Resistance

Insulin resistance describes the condition when cells are no longer able to appropriately respond to the hormone insulin, which mediates the uptake of glucose. Although not all persons with insulin resistance develop T2DM, it is one relevant factor which increases the risk to develop diabetes [74]. In turn, genetic as well as environmental factors, especially obesity and lack of physical activity, increase the risk that cells become insulin resistant. There is more and more evidence that AGEs are another risk factor for the development of insulin resistance. For example, by multiple regression analysis Tan *et al.* reported that AGE levels are independently correlated with insulin resistance in healthy subjects [75]. Insulin resistance was estimated by the homeostatic model assessment index (HOMA-IR) which is based on the calculation of fasting insulin and glucose concentration [76]. Another study by Tahara *et al.* in which more than 300 non-diabetic persons were examined confirmed that serum AGE levels were independently correlated with the HOMA-IR [77].

The underlying molecular mechanisms leading to AGE-induced insulin resistance are still poorly understood, but some *in vitro* and *in vivo* studies were performed to gain knowledge in this research area. Figure 3 summarizes the principle mechanisms of AGE contribution to insulin resistance which have been observed in the described studies. As serum proteins are frequent targets of modifications by sugars and reactive carbonyl compounds such as methylglyoxal and glyoxal, *in vitro* glycated albumin is often used for cell culture experiments. Glycated albumin was shown to induce the expression of tumor necrosis factor alpha (TNFα) which suppresses insulin signaling [78]. Furthermore, protein kinase C alpha (PKCα) was reported as a target of glycated human serum albumin leading to increased serine/threonine phosphorylation of insulin receptor substrate (IRS) 1 and 2 but reduced IRS tyrosine phosphorylation. This resulted in impaired insulin signaling (phosphatidylinositol 3-kinase/protein kinase B pathway) and inhibition of insulin-mediated glucose metabolism [79]. As the induction of PKCα via protein tyrosine kinase Src and the inhibition of insulin receptor substrate 1 (IRS1) was mediated by RAGE after the incubation of skeletal muscle cells with glycated albumin, it is assumed that RAGE plays a role in this process [80]. Previous studies already showed that both, TNFα (reviewed in [81]) and proteins of the PKC family [82,83], are involved in insulin resistance.

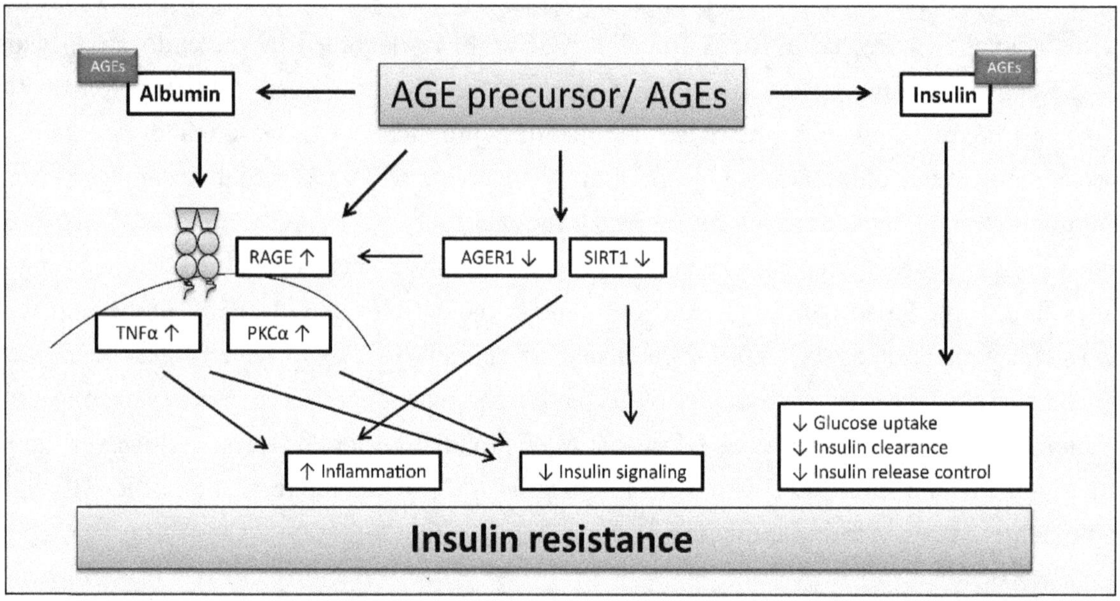

Figure 3. Mechanisms of AGEs leading to insulin resistance in insulin-sensitive tissues according to [52,78–80,84–86]. AGEs are involved in mechanisms contributing to insulin resistance due to direct modification of insulin which alters insulin action resulting in impaired glucose uptake, inhibited insulin clearance or further increased insulin secretion. Furthermore, AGEs may contribute to insulin resistance via increased expression of RAGE and reduced expression of AGER1 and SIRT1. AGEs affect insulin signaling and trigger inflammation via stimulation of PKCα and upregulation of TNFα. SIRT1 depletion causes changes in insulin signaling and induces inflammation.

Another possible factor contributing to insulin resistance is the direct glycation of insulin. Although insulin has a very short half-life and is not a typical target for modification, glycation sites on insulin have been found *in vivo* [85,87] as well as when cells and islets were cultured under hyperglycemic conditions [87,88]. *In vitro*, glycation of human insulin was shown to occur in the phenylalanine position in the amino terminus of the insulin β chain [89]. Moreover, it was demonstrated that the reaction of glyoxal with insulin results in the formation of N-terminal pyrazinones, products belonging to AGEs [90]. Studies using *in vitro* prepared monoglycated insulin revealed that glycation affects insulin function. When *in vitro* monoglycated insulin was infused to mice the glucose lowering potential was around 20% lower compared to unmodified insulin [84]. Cell culture experiments with isolated muscle and glycated insulin additionally showed that glucose uptake, oxidation and glycogen production is reduced. Monoglycated insulin had similar effects in humans. After infusion to healthy humans, around 70% more glycated insulin was needed to induce an equal amount of glucose uptake compared to unmodified insulin [85]. The role of glycated insulin in T2DM was supported by the measurement of the modified insulin in the plasma of T2DM subjects which was found to comprise around 9% of the total insulin level [85]. Furthermore, it was shown that methylglyoxal causes modifications on the arginine residue of the insulin β chain decreasing glucose uptake in insulin-sensitive cells including 3T3-L1 adipocytes and L8 skeletal muscle cells [86]. Usually, extracellular insulin concentrations regulate insulin secretion from β cells, however, methylglyoxal-modified insulin can no longer inhibit insulin secretion. Moreover, Jia *et al.* reported that methylglyoxal-modified insulin triggers hyperinsulinemia through decreased insulin clearance by liver cells [86].

The potential role of AGEs in insulin resistance is supported by studies examining the influence of antiglycating agents on the manifestation of insulin resistance. The agents pyridoxamine or TM2002 inhibit the formation of AGEs and were shown to improve insulin resistance in rodents which already suffered from diabetes [91] or in which insulin resistance was induced by the administration of methylglyoxal [92]. Antioxidants reduced insulin resistance in the same manner making it difficult to distinguish between oxidative stress and AGE formation but making clear that there is an important link between both factors.

There are studies which indicate that predominantly prooxidative AGEs derived from the diet increase the risk for insulin resistance. Cai *et al.* reported that a diet enriched with methylglyoxal-modified albumin (MG-diet) and fed to non-obese C57BL6 mice led to weight gain, adiposity and the development of insulin resistance in F3/MG^{+} mice [52]. The MG-diet increased serum AGE levels and AGE-lipids. Furthermore, the chronic AGE intake caused elevated oxidative stress and inflammation and resulted in an insulin-resistant state (increased level of fasting plasma insulin and leptin and reduced adiponectin level). The phenotypical change was mediated by reduced expression of AGER1 and SIRT1 and upregulation of RAGE in skeletal muscle, liver and white adipose tissue as well as an increase in NFκB acetylated p65 in adipocytes. This resulted in changes in insulin signaling and increased inflammation leading to elevated levels of oxidative stress. In this context, another study showed that restriction of AGEs in a cohort of patients with T2DM reduced insulin resistance and increased expression of SIRT1 and AGER1 [50]. Due to its deacetylase activity, SIRT1 seems to have a significant role in insulin actions. On the one hand, SIRT1 is involved in the induction of insulin secretion from β cells, on the other hand SIRT1 induces insulin signaling by inhibiting negative regulators or further regulating the activation of IRS1 and 2 and Akt. Moreover, as SIRT1 also regulates inflammatory processes, adiponectin secretion or ROS formation, it indirectly contributes to the development of insulin resistance [93].

4. The Role of AGEs in β Cell Dysfunction and β Cell Death

There is a growing body of evidence that AGEs not only contribute to insulin resistance, but also damage β cells directly, leading to impaired functions or even cell death. The cytotoxic potential of AGEs on pancreatic β cells was investigated in a number of studies. Lim *et al.* reported that AGE treatment caused apoptosis in β cells [94]. Additionally, they showed that AGEs stimulate ROS production and induce the expression of RAGE. Inhibition of RAGE as well as antioxidant treatment prevented these changes so that AGEs might induce apoptosis via RAGE-induced ROS formation. However, they also demonstrated that proliferation increases after AGE treatment. In contrast, decreased proliferation but also ROS-induced cell death in HIT-T15 cells due to treatment with ribose-modified serum was observed by Viviani *et al.* [95]. Moreover, Zhu *et al.* indicated that apoptosis in β cells characterized by caspase activation, cytochrome c release and reduced expression of anti-apoptotic bcl2 might be due to RAGE [96]. In the study of Lin *et al.*, INS-1 cells were treated with AGEs resulting in cell apoptosis [97]. They concluded that the AGE-induced ROS generation occurs primarily through the mitochondrial electron transport chain but also through stress-related signaling pathways (Jun N-terminal kinase and p38) which activates ROS production via the NAPDH oxidase.

In addition to β cell death, most of the studies showed that AGEs affect insulin secretion [95–97]. Further evidence for the AGE-mediated decline in insulin secretion was given by Zhao *et al.* [98]. They showed that AGEs block the activity of cytochrome c oxidase and production of adenosine triphosphate (ATP) in islets isolated from mice. Impaired insulin secretion increases plasma glucose levels which were accompanied by increased formation of •NO and elevated expression of inducible nitric oxide synthase (iNOS) suggesting that AGEs cause the induction of iNOS so that increasing concentrations of •NO inhibit cytochrome c oxidase activity and ATP production (Figure 4). ATP is necessary for insulin secretion as ATP causes the shutdown of ATP-sensitive potassium channels leading to membrane depolarization and the influx of Ca^{2+}. Increased intracellular Ca^{2+}-concentrations trigger the exocytosis of insulin granules [99]. Low ATP levels inhibit this process. More recently, Hachiya *et al.* tested the influence of bovine serum albumin (BSA) modified with glucose and glyceraldehyde on insulin secretion of isolated rat pancreatic islets [100]. Both, glucose-BSA and glyceraldehyde-BSA impaired insulin secretion induced by high concentrations of glucose. However, the authors were not able to detect any changes in the expression of oxidative response genes including iNOS. They concluded the impaired insulin secretion to be due to defects in the tricarboxylic acid cycle (TCA). Moreover, gene expression of the NADH (reduced nicotinamide adenine dinucleotide) shuttle enzymes malate dehydrogenase 1/2 was reduced after AGE treatment. However, as the expression of glycerol phosphate shuttle was unchanged, and inhibition of both have been linked to decreased insulin secretion [101], further investigations must clarify whether AGEs affect NADH shuttle function and thus ATP production. The different results reveal that more research to understand the causal mechanisms leading to impaired insulin secretion needs to be performed.

Another factor contributing to reduced insulin secretion is the decline in insulin gene transcription. In this respect, AGEs are also proposed to play a role (Figure 4). Shu *et al.* reported the impaired insulin secretion of β cells as a result of the downregulation of insulin transcription [102]. They identified that the transcription factor FoxO1 (Forkhead box protein O1) accumulates in the nucleus which in turn decreases the expression of the transcription factor PDX-1 (pancreatic and duodenal homeobox-1) by reducing protein stability. This is in agreement with the study of Puddu *et al.* who observed that the reduced insulin content in the pancreatic islet cell line HIT-T15 after AGE treatment is linked to reduced expression of PDX-1 and an increase of FoxO1 in the nucleus [103]. They suggested that AGEs decrease phosphorylation of FoxO1, thus, inducing the translocation into the nucleus. In addition, they were able to show that AGEs induce FoxO1 acetylation and that PDX-1 translocates into the cytoplasm. Acetylation of the transcription factor protects against proteasomal degradation, and nucleocytoplasmic translocation of PDX-1 decreases its availability for insulin transcription. For therapeutic approaches, it is important to note that even if hyperglycemia is restored, AGEs can still contribute to β cell failure and thus T2DM. Therefore, destroying AGE structures, preventing endogenous AGE formation as well as dietary AGE uptake may all be considered as part of a therapy to completely "remove" sources of β cell dysfunction.

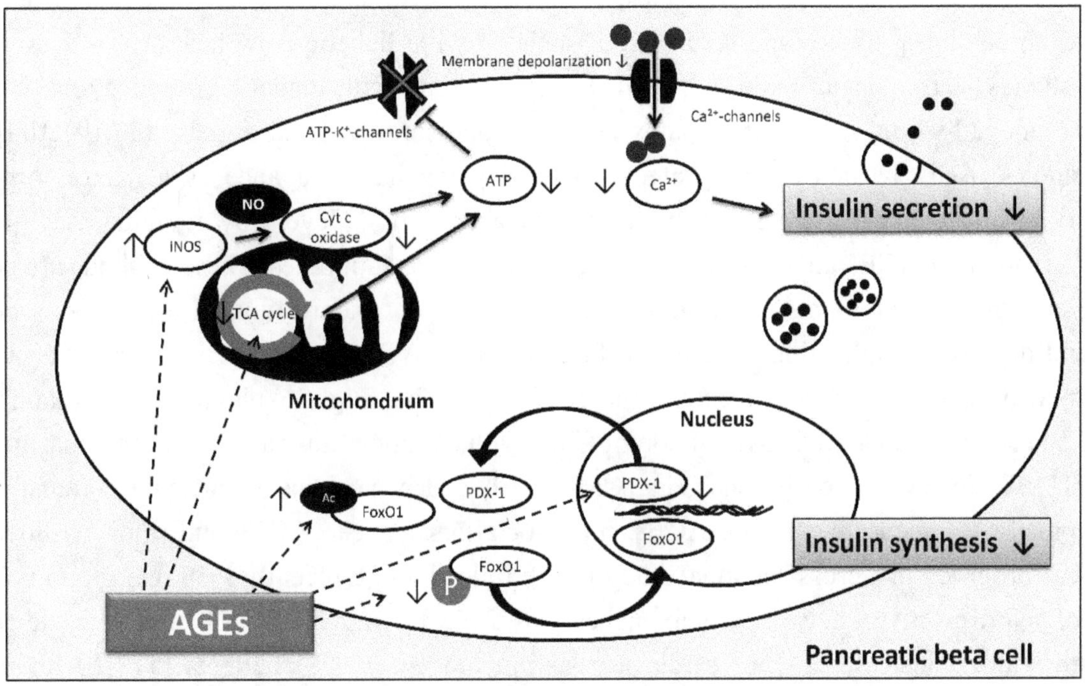

Figure 4. AGE-induced pathways involved in β cell dysfunction according to [98,100,102,103]. Decreased insulin synthesis and reduced insulin secretion are both involved in β cell failure contributing to hyperglycemia. AGEs reduce phosphorylation (P) and induce acetylation (Ac) of FoxO1, thus, FoxO1 translocates into the nucleus and is protected against proteasomal degradation, respectively. In addition, AGEs induce PDX-1 translocation into the cytoplasm and decrease PDX-1 protein expression, finally affecting insulin gene transcription and insulin synthesis. Regarding insulin secretion, AGEs cause inhibition by activation of iNOS and consequent blocking of cytochrome c oxidase activity and ATP depletion. Moreover, AGEs decrease insulin secretion through alterations in the TCA cycle which limits ATP production. ATP depletion inhibits closure of ATP-dependent potassium channels which leads to reduced membrane depolarization and decrease of intracellular calcium concentration inhibiting insulin secretion. (Arrows illustrate direct interactions; dashed arrows illustrate possible targets of AGEs).

5. The Role of AGEs in Diabetic Complications

Diabetics have an increased risk for the development of infections and several diseases including cardiovascular, kidney, eye, nerves and skin diseases. Referring to the International Diabetes Federation, diabetes is the main cause of blindness, amputation of the lower limb, kidney failure and cardiovascular diseases in many countries [2]. Most relevant for the development of diabetic complications is the exposure to hyperglycemia. One of the most important injuries which arise from hyperglycemia is damage to the vascular system. If injury occurs on large or small blood vessel most diabetic diseases can be accordingly grouped into macro- or microvascular complications, respectively. Several mechanisms leading from hyperglycemia to diabetic diseases have been described, among them the formation of AGEs.

AGE formation and accumulation increasingly occurs under diabetic conditions, and even if glycemic control is restored, AGEs can remain in tissues of diabetic subjects for a long time. Diabetic complications occur in both types of diabetes, T1DM and T2DM, and in general, there is evidence that AGEs accumulating in tissues and serum are associated with diabetic complications [42]. Regarding T2DM, studies were able to show that AGE-levels correlate with diabetic complications including diabetic retinopathy [104–106], nephropathy [107,108] and cardiovascular disease [35,107,109]. Most of the studies used glycated serum proteins to investigate their relevance as a marker or even a predictor for diabetic diseases. Frequently, they measured AGE-levels by immunoassay (ELISA, DELFIA) using anti-AGE, anti-CML or anti-hydroimidazolone antibodies. These methods are not very specific and the measurements of plasma AGEs may lead to underestimated associations between AGEs and diabetic complications due to the fact that most AGEs are formed or further accumulate intracellular or in tissues without reaching the circulation. This might explain why, for example, the study of Busch *et al.* reported no association of CML with cardiovascular and renal outcomes in T2DM patients [110]. However, even by analyzing CML, CEL and pentosidine with liquid chromatography, a more precise technique to measure AGEs, no association between AGEs and prior cardiovascular events was observed in a cross-sectional study [111]. Indeed, more recently the same group published results of a prospective study in which higher plasma levels of CML, CEL and pentosidine were associated with an increased risk of cardiovascular outcome in T2DM patients [112]. Although the causal relationship needs to be further clarified there is increasing evidence observed in these clinical studies that AGEs serve as potential biomarkers for diabetic complications.

5.1. Modified ECM Proteins and the Relevance for Diabetic Complications

One main target of increased concentrations of reducing sugars and dicarbonyl compounds found in diabetes is collagen. There are a few studies which investigated the effect of modified collagen on cell functions. According to the subsequent studies, a simplified scheme describing which cell functions are affected by methylglyoxal-modified collagen, contributing to diabetes-related changes related to diabetic complications, is illustrated in Figure 5. Firstly, it was reported that modification of basement membrane collagen type IV by methylglyoxal reduces attachment of vascular endothelial cells and angiogenesis [113]. AGEs such as MG-H1 were found on integrin-binding sites of collagen and were assumed to cause decreased cell attachment. As demonstrated by Chong *et al.*, methylglyoxal modifications of arginine residues in the integrin-binding sequence also reduce collagen binding so that collagen degradation by phagocytosis is impaired, thus, promoting fibrosis [114]. Regarding the collagen turnover, it was shown that collagen modified with glyoxal and methylglyoxal is *per se* less degradable by proteases due to the formation of cross-links [115]. Yuen *et al.* observed that methylglyoxal-modified collagen type I inhibits cell adhesion of cardiac fibroblasts while stimulating myofibroblast differentiation and their migration [116]. Further investigations of Talior-Volodarsky *et al.* reported that myofibroblast differentiation is stimulated by upregulation of $\alpha 11$ integrin expression which was found to be induced by transforming growth factor (TGF) $\beta 2$/Smad3 signaling [117,118].

Myofibroblasts mediate fibrosis, a pathological response to tissue damage which has been associated with diabetic complication and heart failure due to fibrotic lesions. That modified collagen alters cell-matrix interaction was also shown by Pozzi *et al.* [119]. On the one hand, cell adhesion and migration but not proliferation of mesangial cells plated on collagen type IV modified with methylglyoxal is inhibited. On the other hand, long term glucose-modified collagen reduces migration and proliferation and increases collagen synthesis. The authors proposed that these findings may explain pathological mechanisms seen in diabetic nephropathy: early proliferation of mesangial cells followed by impaired proliferation, mesangial matrix expansion and mesangial cell dysfunction. In the study of Sassi-Gaha *et al.* [120], after modification of type I collagen lattices with methylglyoxal and 3-deoxyglucosone, dermal fibroblast were seeded on the modified collagen. For methylglyoxal an increased expression of TGFβ1, collagen and β1 integrin, and decreased expression of the transcription factor Smad7 was shown. The opposite effects were found for fibroblasts grown on 3-deoxyglucosone-modified collagen. To understand the pathological mechanisms mediated by AGEs it is quite important to investigate the effect of different AGE precursors as well as individual AGEs.

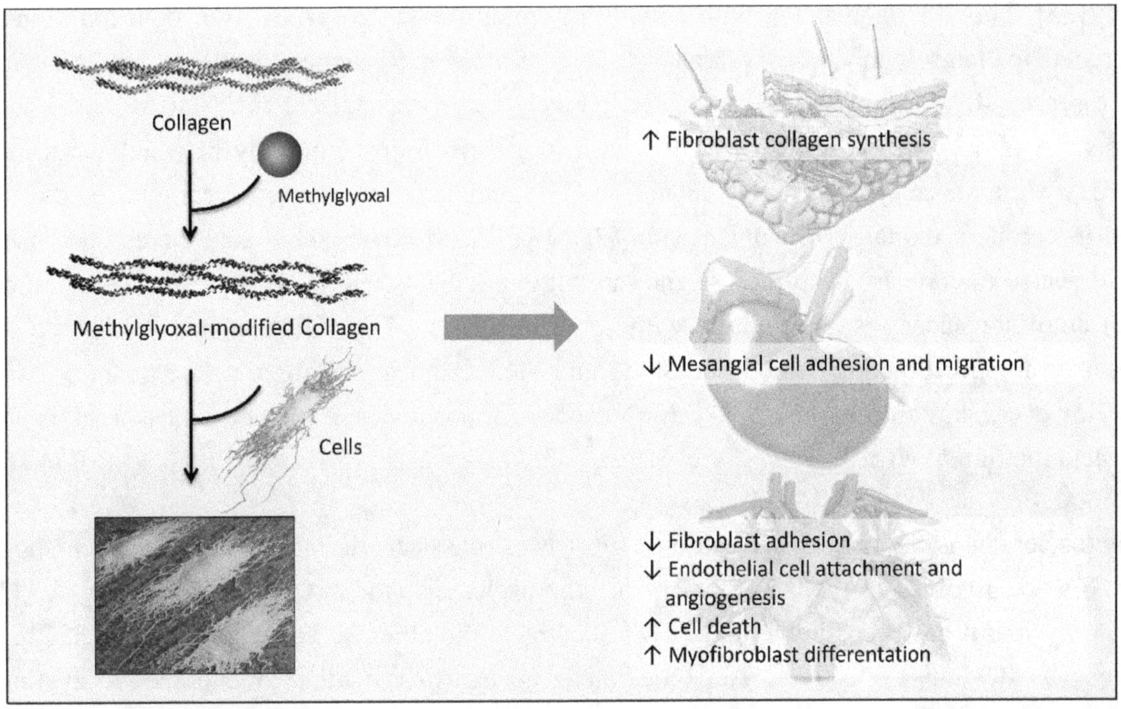

Figure 5. Influence of methylglyoxal-modified collagen on cellular functions relevant to diabetic complications, according to [113,116,119,120]. Cells were cultured on methylglyoxal-modified collagen and the effect of AGE-collagen on cellular functions was studied. Dermal fibroblasts grown on modified collagen increase collagen synthesis. Methylglyoxal-modified collagen reduces cell adhesion and migration of mesangial cells, cardiac fibroblast are less adherent. Myofibroblast differentiation is stimulated in cardiac fibroblasts cultured on modified collagen and modifications of basement membrane collagen causes detachment, cell death and reduced angiogenesis of vascular endothelial cells.

In addition to collagen, AGE formation on further ECM protein including laminin or fibronectin has been studied showing impaired self-assembly and interactions with other matrix components as well as matrix-cell interactions [121–125]. It is assumed that glycation of these ECM proteins, therefore, may contribute to diabetic complications. Just to give an example, bovine retinal pericytes cultured on fibronectin modified with glyoxal and methylglyoxal were shown to induce apoptosis which may contribute to cell loss in diabetic retinopathy [124]. Regarding the development of diabetic neuropathy, glycated laminin and fibronectin reduced neurite outgrowth providing a mechanism for the failure of axonal regeneration and collateral sprouting [125].

5.2. Ischemic Heart Disease and Atherosclerosis

As already mentioned, patients with diabetes are at high risk of developing cardiovascular disease [126]. One common type of cardiovascular diseases is ischemic heart disease in which the blood supply to the heart is reduced often due to plaque formation in the arterial wall. The association between AGEs and ischemic heart disease of patients with T2DM was investigated in some clinical studies. Kilhovd *et al.* reported increased serum AGEs in diabetic persons with ischemic heart disease compared to patients without [35]. This is in agreement with a study in which increased serum AGE concentrations were measured in patients with coronary artery disease compared to diabetic patients without coronary artery disease [109]. Furthermore, serum AGE levels may predict ischemic heart disease mortality in persons with T2DM. Evidence was given in a long-term follow-up study in which serum AGEs, measured by immunoassay using a polyclonal anti-AGE antibody, were predictors for total, cardiovascular as well as coronary mortality in woman with T2DM [127]. Moreover, skin autofluorescence has been strongly related to coronary heart disease and mortality in patients with T2DM [128]. The measurement of skin autofluorescence is a technique with some limitations for AGE analysis: not all AGEs are fluorescent, no specific compound is measured and there is no information on the quantity of AGEs at all. However, skin autofluorescence can be used as a non-invasive method and is associated with AGE accumulation *in vivo* [129].

Atherosclerosis is the underlying cause of most ischemic heart diseases and it has been suggested that AGEs are involved in the development of atherosclerosis (for more detailed review see [130]). On the one hand it has been shown that AGEs accumulate in atherosclerotic lesions [131,132] and on the other hand studies identified AGE-mediated mechanisms related to endothelial dysfunction, inflammation and lipid modifications. In particular, it was shown that AGEs contribute to endothelial dysfunction through their pro-apoptotic effect on endothelial cells [133,134] and endothelial progenitor cells [135]. Moreover, AGEs stimulate the expression of genes such as MCP-1, intercellular adhesion molecule 1, vascular cell adhesion molecule 1 and plasminogen activator inhibitor 1 [136–141]. This mediates recruitment and adhesion of inflammatory cells to the vessel wall or further inhibition of fibrinolysis. The expression of MCP-1 and adhesion molecules can be induced by endothelin 1, its expression in turn is also stimulated by AGEs [142]. In porcine coronary fibroblasts, it was observed that AGEs induce the mRNA expression of interleukin 6, vascular cell adhesion molecule 1 and MCP-1 followed by increased interleukin secretion and leukocyte adhesion [143]. The atherogenic properties of AGEs could be further attributed to the ·NO quenching effect of AGEs [144]. Furthermore, AGEs impair the synthesis of ·NO by decreasing the expression as well as the activity of endothelial ·NO synthase [145,146]. ·NO mediates a series of intracellular effects that lead to endothelial regeneration, vasodilation and inhibition of platelet adhesion and leukocyte chemotaxis and therefore defects in ·NO production and activity are supposed to be major mechanisms of endothelial dysfunction and atherosclerosis [147].

To further clarify the role of AGEs in diabetic complications, *in vitro* and *in vivo* studies showing AGE-mediated pathomechanisms which contribute to diabetic retino- and nephropathy will be highlighted below.

5.3. Diabetic Retinopathy

AGE-albumin-induced pericyte death has been assumed to be caused by AGE-RAGE interaction [148]. Moreover, studies reported that AGEs induce ROS formation and reduce protein kinase B/Akt [124] or further platelet-derived growth factor signaling [149] contributing to reduced pericyte survival. Pericytes are cells which cover capillaries in the retina and their loss is an early event in diabetic retinopathy followed by the formation of acellular capillaries (capillaries devoid of cells), microaneurysms and vascular basement membrane thickening. The interaction of pericyctes with endothelial cells is necessary for the maintenance of the blood-retina barrier [150] so that pericycte loss is also associated with the breakdown of the barrier.

The blood-retina barrier regulates the flux of nutrients, fluids and other blood components into the retina and its breakdown may cause the development of macular edema which is a main cause for vision loss in diabetes. Another mechanism leading to the breakdown of the blood-retinal barrier is the induction of retinal leukostasis, a state of chronic inflammation, which may contribute to endothelial cell death and increased vascular permeability. The involvement of AGEs in this process has been shown by Moore *et al.* [151]. AGEs induce NFκB DNA binding, intercellular adhesion molecule 1 expression and further leukocyte adhesion to retinal endothelial cells. *In vivo*, AGE-albumin injection into mice resulted in a breakdown of the blood-retina barrier [151]. Moreover, the breakdown of the blood-retina barrier has been shown to be dependent on VEGF. AGE-upregulation of vascular endothelial growth factor (VEGF) expression was observed after AGE-albumin injection into rodents, as well as after AGE-albumin incubation of cultured retinal cells [148,152–154], leading to vascular permeability and angiogenesis.

5.4. Diabetic Nephropathy

Similar to pericytes in the retina, AGEs induce apoptosis and VEGF expression in mesangial cells [155], specialized pericytes which are located in blood vessels of the kidney. Mesangial cells in the glomerulus regulate glomerular filtration and provide structural support [156]. Their loss or further VEGF upregulation contributes to increased vascular permeability associated with hyperfiltration and proteinuria in kidney diseases, suggesting that AGEs play a role here. Furthermore, it has been shown, that AGEs upregulate MCP-1 expression in mesangial cells [155]. MCP-1 is a chemokine which regulates macrophage/monocyte migration and infiltration, therefore, MCP-1 upregulation by AGEs may trigger inflammation in renal tissue. Moreover, AGE-induced expression of TGFβ was demonstrated *in vitro* [157,158] and as TGFβ1 upregulation was associated with enhanced expression of ECM proteins this might be an explanation for glomerular hypertrophy *in vivo* [159]. In this context, studies reported that TGFβ expression is induced via AGE-RAGE interaction [160,161]. Elevated expression of TGFβ is also known to induce epithelial-to-mesenchymal transition and AGEs were shown to stimulate transdifferentiation in a rat kidney epithelial cell line [161]. Another protein which plays a role in AGE-induced epithelial-to-mesenchymal transition is prosclerotic connective tissue growth factor (CTGF), a target gene of TGFβ signaling which is elevated in plasma of persons with diabetic nephropathy [162,163]. Because AGEs induce epithelial-to-mesenchymal transition and CTGF is a downstream target of TGFβ it is generally believed that increased CTGF expression is mediated through TGFβ/Smad signaling. More recently, AGE stimulation of tubular CTGF expression was also found to be independent of TGFβ/Smad signaling. This pathway is RAGE dependent and leads to the activation of Smad3 through the induction of extracellular-signal-regulated kinase 1/2 and p38 [164].

6. Conclusions

Extensive research has been performed in the last years to clarify the impact of AGEs as well as to identify causal pathological mechanisms in the development and progression of T2DM (insulin resistance, β cell death and dysfunction) and in diabetic diseases. Increasing concentrations of different AGEs in serum and tissues of diabetic persons have been detected and the association of AGEs with diabetic complications has been shown in several studies suggesting AGEs as biomarkers and even predictors for diabetic complications. Moreover, studies reported a wide range of effects and reactions induced by AGEs which were proposed to play a role in insulin resistance, β cell failure and diabetic complications. In general, it was found that AGE formation occurs on extra- and intracellular proteins leading to protein cross-linking, structural and functional changes, e.g., loss of enzyme activity. Modifications on ECM proteins have been shown to affect not only matrix-matrix but especially matrix-cell interactions. Many studies proposed that the damaging effect of AGEs is predominantly induced by RAGE signaling thus increasing ROS formation and inflammatory processes. However, there are also some critical aspects which must be considered when obtaining general conclusions from the data. Many studies used AGEs prepared *in vitro* by incubating a protein, often albumin, with a reducing sugar or dicarbonyl compound. According to Henle, only a small amount of AGEs is formed in glucose-modified BSA in contrast to Amadori products such as fructoselysine which accounts for about 90% of the detectable lysine derivatization [45]. Therefore, it is questionable whether the induced biological effects are stimulated by AGEs, Amadori products or even other modification products. Improvement of analytical methods may contribute to more precise results regarding the involvement of AGEs in general but also individual AGE structures. Because it is estimated that the wide range of AGEs is at present only partially identified, this further impedes the research about the biological effects of AGEs. Moreover, the use of different AGE precursors, different concentration and varying target proteins hinders not only comparability of studies but also complicates the understanding of AGE-induced mechanisms. Most frequently, the products are not characterized, the AGE content was measured by unspecific methods, and the individual AGEs were not identified. Without knowing the specifity of the products it is difficult to assess the findings observed in these studies and to determine the role of AGEs in several aspects of diabetes.

Another point which should be noted is that the relevance of AGE-RAGE interactions is still controversially discussed [165–167]. Only aggregates of highly modified AGE-proteins which are rare in tissue and body fluids are able to activate RAGE. However, these highly modified proteins can be formed exogenously in thermally processed food, thus the relevance for AGE-RAGE interaction needs to be further clarified. Nevertheless, the reduction of hyperglycemia-induced AGE formation as well as the restriction of AGE-rich food items both present targets for therapeutic approaches against the pathological events in diabetes and associated diseases. Finally, the use of AGEs as biomarkers/predictors for diabetic complications may help to reduce health problems in persons with diabetes.

Abbreviations

AGEs	Advanced glycation end products
AGER1	Advanced glycation end product receptor 1
ATP	Adenosine triphosphate
BSA	Bovine serum albumin
CEL	Nε-(carboxyethyl) lysine
CMA	Nω-(carboxymethyl) arginine
CML	Nε-(carboxymethyl) lysine
CTGF	Connective tissue growth factor
ELISA	Enzyme-linked immunosorbent assay
ECM	Extracellular matrix
FoxO1	Forkhead box protein O1
GOLD	Glyoxal-derived lysyl dimer
GSH	Glutathione
GSSG	Glutathione disulfide
H_2O_2	Hydrogen peroxide
HPLC	High performance liquid chromatography
iNOS	Inducible nitric oxide synthase
IRS	Insulin receptor substrate
MCP-1	Monocyte chemoattractant peptide 1
MOLD	Methylglyoxal-derived lysyl dimer
NAD^+	Oxidized nicotinamide adenine dinucleotide
NADH	Reduced nicotinamide adenine dinucleotide
NADPH	Reduced nicotinamide adenine dinucleotide phosphate
NFκB	Nuclear factor kappa B
$^{\bullet}NO$	Nitric oxide
$O_2^{\bullet-}$	Superoxide anion
PDX-1	Pancreatic and duodenal homeobox-1
PKC α	Protein kinase c alpha
RAGE	Receptor of advanced glycation end products
ROS	Reactive oxygen species
SIRT1	Sirtuin1
T1DM	Type 1 diabetes mellitus
T2DM	Type 2 diabetes mellitus
TCA	Tricarboxylic acid cycle
TNFα	Tumor necrosis factor alpha
UPLC-MS	Ultra pressure liquid chromatography with mass spectrometry detection
VEGF	Vascular endothelial growth factor

References

1. International Diabetes Federation. *IDF Diabetes Atlas Update Poster*, 6th ed.; International Diabetes Federation: Brussels, Belgium, 2014.
2. International Diabetes Federation. *IDF Diabetes Atlas*, 6th ed.; International Diabetes Federation: Brussels, Belgium, 2013.
3. Morrish, N.; Wang, S.-L.; Stevens, L.; Fuller, J.; Keen, H. Mortality and causes of death in the who multinational study of vascular disease in diabetes. *Diabetologia* **2001**, *44*, S14–S21.
4. Sies, H. Oxidative stress: A concept in redox biology and medicine. *Redox Biol.* **2015**, *4*, 180–183.
5. Jung, T.; Catalgol, B.; Grune, T. The proteasomal system. *Mol. Asp. Med.* **2009**, *30*, 191–296.
6. Giacco, F.; Brownlee, M. Oxidative stress and diabetic complications. *Circ. Res.* **2010**, *107*, 1058–1070.
7. Abdul-Ghani, M.A.; DeFronzo, R.A. Oxidative stress in type 2 diabetes mellitus. In *Oxidative stress in aging*, Miwa, S., Beckman, K., Muller, F., Eds. Humana Press: Totowa, NJ, USA, 2008; pp. 191–211.
8. Henriksen, E.J.; Diamond-Stanic, M.K.; Marchionne, E.M. Oxidative stress and the etiology of insulin resistance and type 2 diabetes. *Free Radic. Biol. Med.* **2011**, *51*, 993–999.
9. Leahy, J.L. Pathogenesis of type 2 diabetes mellitus. *Arch. Med. Res.* **2005**, *36*, 197–209.
10. Vlassara, H.; Uribarri, J. Advanced glycation end products (AGE) and diabetes: Cause, effect, or both? *Curr. Diab. Rep.* **2014**, doi:10.1007/s11892-013-0453-1.
11. Rahbar, S. An abnormal hemoglobin in red cells of diabetics. *Clin. Chim. Acta* **1968**, *22*, 296–298.
12. Bunn, H.F.; Haney, D.N.; Gabbay, K.H.; Gallop, P.M. Further identification of the nature and linkage of the carbohydrate in hemoglobin A_{1c}. *Biochem. Biophys. Res. Commun.* **1975**, *67*, 103–109.
13. Hodge, J.E. Dehydrated foods—Chemistry of browning reactions in model systems. *J. Agric. Food Chem.* **1953**, *1*, 928–943.
14. Fu, M.X.; Requena, J.R.; Jenkins, A.J.; Lyons, T.J.; Baynes, J.W.; Thorpe, S.R. The advanced glycation end product, nepsilon-(carboxymethyl)lysine, is a product of both lipid peroxidation and glycoxidation reactions. *J. Biol. Chem.* **1996**, *271*, 9982–9986.
15. Thornalley, P.J.; Langborg, A.; Minhas, H.S. Formation of glyoxal, methylglyoxal and 3-deoxyglucosone in the glycation of proteins by glucose. *Biochem. J.* **1999**, *344*, 109–116.
16. Wells-Knecht, K.J.; Zyzak, D.V.; Litchfield, J.E.; Thorpe, S.R.; Baynes, J.W. Mechanism of autoxidative glycosylation: Identification of glyoxal and arabinose as intermediates in the autoxidative modification of proteins by glucose. *Biochemistry* **1995**, *34*, 3702–3709.
17. Scheijen, J.L.; Schalkwijk, C.G. Quantification of glyoxal, methylglyoxal and 3-deoxyglucosone in blood and plasma by ultra performance liquid chromatography tandem mass spectrometry: Evaluation of blood specimen. *Clin. Chem. Lab. Med.* **2014**, *52*, 85–91.
18. Al-Abed, Y.; Bucala, R. Nε-carboxymethyllysine formation by direct addition of glyoxal to lysine during the maillard reaction. *Bioorg. Med. Chem. Lett.* **1995**, *5*, 2161–2162.
19. Wells-Knecht, K.J.; Brinkmann, E.; Baynes, J.W. Characterization of an imidazolium salt formed from glyoxal and N.alpha.-hippuryllysine: A model for maillard reaction crosslinks in proteins. *J. Org. Chem.* **1995**, *60*, 6246–6247.

20. Glomb, M.A.; Lang, G. Isolation and characterization of glyoxal-arginine modifications. *J. Agric. Food Chem.* **2001**, *49*, 1493–1501.
21. Thorpe, S.R.; Baynes, J.W. Maillard reaction products in tissue proteins: New products and new perspectives. *Amino Acids* **2003**, *25*, 275–281.
22. Ahmed, M.U.; Brinkmann Frye, E.; Degenhardt, T.P.; Thorpe, S.R.; Baynes, J.W. N-epsilon-(carboxyethyl)lysine, a product of the chemical modification of proteins by methylglyoxal, increases with AGE in human lens proteins. *Biochem. J.* **1997**, *324*, 565–570.
23. Nagaraj, R.H.; Shipanova, I.N.; Faust, F.M. Protein cross-linking by the maillard reaction: Isolation, characterization, and *in vivo* detection of a lysine-lysine cross-link derived from methylglyoxal. *J. Biol. Chem.* **1996**, *271*, 19338–19345.
24. Shipanova, I.N.; Glomb, M.A.; Nagaraj, R.H. Protein modification by methylglyoxal: Chemical nature and synthetic mechanism of a major fluorescent adduct. *Arch. Biochem. Biophys.* **1997**, *344*, 29–36.
25. Henle, T.; Walter, A.; Haeßner, R.; Klostermeyer, H. Detection and identification of a protein-bound imidazolone resulting from the reaction of arginine residues and methylglyoxal. *Z. Lebensm. Unters. Forsch.* **1994**, *199*, 55–58.
26. Portero-Otin, M.; Nagaraj, R.H.; Monnier, V.M. Chromatographic evidence for pyrraline formation during protein glycation *in vitro* and *in vivo*. *Biochim. Biophys. Acta* **1995**, *1247*, 74–80.
27. Dyer, D.; Blackledge, J.; Thorpe, S.; Baynes, J. Formation of pentosidine during nonenzymatic browning of proteins by glucose. Identification of glucose and other carbohydrates as possible precursors of pentosidine *in vivo*. *J. Biol. Chem.* **1991**, *266*, 11654–11660.
28. Jono, T.; Nagai, R.; Lin, X.; Ahmed, N.; Thornalley, P.J.; Takeya, M.; Horiuchi, S. Nε-(carboxymethyl) lysine and 3-DG-imidazolone are major AGE structures in protein modification by 3-deoxyglucosone. *J. Biochem.* **2004**, *136*, 351–358.
29. Singh, R.; Barden, A.; Mori, T.; Beilin, L. Advanced glycation end-products: A review. *Diabetologia* **2001**, *44*, 129–146.
30. Baynes, J.W. Role of oxidative stress in development of complications in diabetes. *Diabetes* **1991**, *40*, 405–412.
31. Nakamura, K.; Nakazawa, Y.; Ienaga, K. Acid-stable fluorescent advanced glycation end products: Vesperlysines A, B, and C are formed as crosslinked products in the maillard reaction between lysine or proteins with glucose. *Biochem. Biophys. Res. Commun.* **1997**, *232*, 227–230.
32. Pongor, S.; Ulrich, P.C.; Bencsath, F.A.; Cerami, A. Aging of proteins: Isolation and identification of a fluorescent chromophore from the reaction of polypeptides with glucose. *Proc. Natl. Acad. Sci. USA* **1984**, *81*, 2684–2688.
33. Sell, D.R.; Monnier, V.M. Structure elucidation of a senescence cross-link from human extracellular matrix. Implication of pentoses in the aging process. *J. Biol. Chem.* **1989**, *264*, 21597–21602.
34. Poulsen, M.W.; Hedegaard, R.V.; Andersen, J.M.; de Courten, B.; Bügel, S.; Nielsen, J.; Skibsted, L.H.; Dragsted, L.O. Advanced glycation endproducts in food and their effects on health. *Food Chem. Toxicol.* **2013**, *60*, 10–37.
35. Kilhovd, B.K.; Berg, T.J.; Birkeland, K.I.; Thorsby, P.; Hanssen, K.F. Serum levels of advanced glycation end products are increased in patients with type 2 diabetes and coronary heart disease. *Diabetes Care* **1999**, *22*, 1543–1548.

36. Schleicher, E.D.; Wagner, E.; Nerlich, A.G. Increased accumulation of the glycoxidation product N(epsilon)-(carboxymethyl) lysine in human tissues in diabetes and aging. *J. Clin. Investig.* **1997**, *99*, 457–468.
37. Schalkwijk, C.G.; Baidoshvili, A.; Stehouwer, C.D.A.; van Hinsbergh, V.W.M.; Niessen, H.W.M. Increased accumulation of the glycoxidation product Nε-(carboxymethyl)lysine in hearts of diabetic patients: Generation and characterisation of a monoclonal anti-CML antibody. *Biochim. Biophys. Acta* **2004**, *1636*, 82–89.
38. Kilhovd, B.K.; Giardino, I.; Torjesen, P.A.; Birkeland, K.I.; Berg, T.J.; Thornalley, P.J.; Brownlee, M.; Hanssen, K.F. Increased serum levels of the specific AGE-compound methylglyoxal-derived hydroimidazolone in patients with type 2 diabetes. *Metabolism* **2003**, *52*, 163–167.
39. Sell, D.R.; Carlson, E.C.; Monnier, V.M. Differential effects of type 2 (non-insulin-dependent) diabetes mellitus on pentosidine formation in skin and glomerular basement membrane. *Diabetologia* **1993**, *36*, 936–941.
40. Biemel, K.M.; Friedl, D.A.; Lederer, M.O. Identification and quantification of major maillard cross-links in human serum albumin and lens protein. Evidence for glucosepane as the dominant compound. *J. Biol. Chem.* **2002**, *277*, 24907–24915.
41. Sell, D.R.; Biemel, K.M.; Reihl, O.; Lederer, M.O.; Strauch, C.M.; Monnier, V.M. Glucosepane is a major protein cross-link of the senescent human extracellular matrix. Relationship with diabetes. *J. Biol. Chem.* **2005**, *280*, 12310–12315.
42. Monnier, V.M.; Sell, D.R.; Genuth, S. Glycation products as markers and predictors of the progression of diabetic complications. *Ann. NY Acad. Sci.* **2005**, *1043*, 567–581.
43. Goldberg, T.; Cai, W.; Peppa, M.; Dardaine, V.; Baliga, B.S.; Uribarri, J.; Vlassara, H. Advanced glycoxidation end products in commonly consumed foods. *J. Am. Diet. Assoc.* **2004**, *104*, 1287–1291.
44. Assar, S.; Moloney, C.; Lima, M.; Magee, R.; Ames, J. Determination of Nε-(carboxymethyl)lysine in food systems by ultra performance liquid chromatography-mass spectrometry. *Amino Acids* **2009**, *36*, 317–326.
45. Henle, T. AGEs in foods: Do they play a role in uremia? *Kidney Int. Suppl.* **2003**, S145–S147.
46. Koschinsky, T.; He, C.J.; Mitsuhashi, T.; Bucala, R.; Liu, C.; Buenting, C.; Heitmann, K.; Vlassara, H. Orally absorbed reactive glycation products (glycotoxins): An environmental risk factor in diabetic nephropathy. *Proc. Natl. Acad. Sci. USA* **1997**, *94*, 6474–6479.
47. Delgado-Andrade, C.; Tessier, F.J.; Niquet-Leridon, C.; Seiquer, I.; Pilar Navarro, M. Study of the urinary and faecal excretion of nepsilon-carboxymethyllysine in young human volunteers. *Amino Acids* **2012**, *43*, 595–602.
48. Vlassara, H.; Cai, W.; Crandall, J.; Goldberg, T.; Oberstein, R.; Dardaine, V.; Peppa, M.; Rayfield, E.J. Inflammatory mediators are induced by dietary glycotoxins, a major risk factor for diabetic angiopathy. *Proc. Natl. Acad. Sci. USA* **2002**, *99*, 15596–15601.
49. Uribarri, J.; Peppa, M.; Cai, W.; Goldberg, T.; Lu, M.; He, C.; Vlassara, H. Restriction of dietary glycotoxins reduces excessive advanced glycation end products in renal failure patients. *J. Am. Soc. Nephrol.* **2003**, *14*, 728–731.

50. Uribarri, J.; Cai, W.; Ramdas, M.; Goodman, S.; Pyzik, R.; Chen, X.; Zhu, L.; Striker, G.E.; Vlassara, H. Restriction of advanced glycation end products improves insulin resistance in human type 2 diabetes: Potential role of AGER1 and SIRT1. *Diabetes Care* **2011**, *34*, 1610–1616.
51. Uribarri, J.; Cai, W.; Peppa, M.; Goodman, S.; Ferrucci, L.; Striker, G.; Vlassara, H. Circulating glycotoxins and dietary advanced glycation endproducts: Two links to inflammatory response, oxidative stress, and aging. *J. Gerontol. Ser. A Biol. Sci. Med. Sci.* **2007**, *62*, 427–433.
52. Cai, W.; Ramdas, M.; Zhu, L.; Chen, X.; Striker, G.E.; Vlassara, H. Oral advanced glycation endproducts (AGEs) promote insulin resistance and diabetes by depleting the antioxidant defenses AGE receptor-1 and Sirtuin 1. *Proc. Natl. Acad. Sci. USA* **2012**, *109*, 15888–15893.
53. Coughlan, M.T.; Yap, F.Y.; Tong, D.C.; Andrikopoulos, S.; Gasser, A.; Thallas-Bonke, V.; Webster, D.E.; Miyazaki, J.; Kay, T.W.; Slattery, R.M.; *et al.* Advanced glycation end products are direct modulators of beta-cell function. *Diabetes* **2011**, *60*, 2523–2532.
54. Hofmann, M.A.; Drury, S.; Fu, C.; Qu, W.; Taguchi, A.; Lu, Y.; Avila, C.; Kambham, N.; Bierhaus, A.; Nawroth, P.; *et al.* RAGE mediates a novel proinflammatory axis: A central cell surface receptor for S100/calgranulin polypeptides. *Cell* **1999**, *97*, 889–901.
55. Hori, O.; Brett, J.; Slattery, T.; Cao, R.; Zhang, J.; Chen, J.X.; Nagashima, M.; Lundh, E.R.; Vijay, S.; Nitecki, D.; *et al.* The receptor for advanced glycation end products (RAGE) is a cellular binding site for amphoterin. Mediation of neurite outgrowth and co-expression of RAGE and amphoterin in the developing nervous system. *J. Biol. Chem.* **1995**, *270*, 25752–25761.
56. Chavakis, T.; Bierhaus, A.; Al-Fakhri, N.; Schneider, D.; Witte, S.; Linn, T.; Nagashima, M.; Morser, J.; Arnold, B.; Preissner, K.T.; *et al.* The pattern recognition receptor (RAGE) is a counterreceptor for leukocyte integrins: A novel pathway for inflammatory cell recruitment. *J. Exp. Med.* **2003**, *198*, 1507–1515.
57. Grimm, S.; Ott, C.; Horlacher, M.; Weber, D.; Hohn, A.; Grune, T. Advanced-glycation-end-product-induced formation of immunoproteasomes: Involvement of RAGE and Jak2/Stat1. *Biochem. J.* **2012**, *448*, 127–139.
58. Hirose, A.; Tanikawa, T.; Mori, H.; Okada, Y.; Tanaka, Y. Advanced glycation end products increase endothelial permeability through the RAGE/Rho signaling pathway. *FEBS Lett.* **2010**, *584*, 61–66.
59. Tanikawa, T.; Okada, Y.; Tanikawa, R.; Tanaka, Y. Advanced glycation end products induce calcification of vascular smooth muscle cells through RAGE/p38 mapk. *J. Vasc. Res.* **2009**, *46*, 572–580.
60. Li, J.H.; Wang, W.; Huang, X.R.; Oldfield, M.; Schmidt, A.M.; Cooper, M.E.; Lan, H.Y. Advanced glycation end products induce tubular epithelial-myofibroblast transition through the RAGE-ERK1/2 MAP kinase signaling pathway. *Am. J. Pathol.* **2004**, *164*, 1389–1397.
61. Guimaraes, E.L.; Empsen, C.; Geerts, A.; van Grunsven, L.A. Advanced glycation end products induce production of reactive oxygen species via the activation of NADPH oxidase in murine hepatic stellate cells. *J. Hepatol.* **2010**, *52*, 389–397.
62. Zhang, M.; Kho, A.L.; Anilkumar, N.; Chibber, R.; Pagano, P.J.; Shah, A.M.; Cave, A.C. Glycated proteins stimulate reactive oxygen species production in cardiac myocytes: Involvement of Nox2 (gp91phox)-containing NADPH oxidase. *Circulation* **2006**, *113*, 1235–1243.

63. Bierhaus, A.; Schiekofer, S.; Schwaninger, M.; Andrassy, M.; Humpert, P.M.; Chen, J.; Hong, M.; Luther, T.; Henle, T.; Kloting, I.; *et al.* Diabetes-associated sustained activation of the transcription factor nuclear factor-κB. *Diabetes* **2001**, *50*, 2792–2808.
64. Libermann, T.A.; Baltimore, D. Activation of interleukin-6 gene expression through the NF-κB transcription factor. *Mol. Cell. Biol.* **1990**, *10*, 2327–2334.
65. Ueda, A.; Ishigatsubo, Y.; Okubo, T.; Yoshimura, T. Transcriptional regulation of the human monocyte chemoattractant protein-1 gene: Cooperation of two NF-κB sites and NF-κB/Rel subunit specificity. *J. Biol. Chem.* **1997**, *272*, 31092–31099.
66. Li, J.; Schmidt, A.M. Characterization and functional analysis of the promoter of RAGE, the receptor for advanced glycation end products. *J. Biol. Chem.* **1997**, *272*, 16498–16506.
67. Ott, C.; Jacobs, K.; Haucke, E.; Navarrete Santos, A.; Grune, T.; Simm, A. Role of advanced glycation end products in cellular signaling. *Redox Biol.* **2014**, *2*, 411–429.
68. Grimm, S.; Ernst, L.; Grotzinger, N.; Hohn, A.; Breusing, N.; Reinheckel, T.; Grune, T. Cathepsin D is one of the major enzymes involved in intracellular degradation of AGE-modified proteins. *Free Radic. Res.* **2010**, *44*, 1013–1026.
69. Grimm, S.; Horlacher, M.; Catalgol, B.; Hoehn, A.; Reinheckel, T.; Grune, T. Cathepsins D and L reduce the toxicity of advanced glycation end products. *Free Radic. Biol. Med.* **2012**, *52*, 1011–1023.
70. Cai, W.; Torreggiani, M.; Zhu, L.; Chen, X.; He, J.C.; Striker, G.E.; Vlassara, H. AGER1 regulates endothelial cell NADPH oxidase-dependent oxidant stress via PKC-delta: Implications for vascular disease. *Am. J. Physiol. Cell Physiol.* **2010**, *298*, C624–C634.
71. Lu, C.; He, J.C.; Cai, W.; Liu, H.; Zhu, L.; Vlassara, H. Advanced glycation endproduct (AGE) receptor 1 is a negative regulator of the inflammatory response to AGE in mesangial cells. *Proc. Natl. Acad. Sci. USA* **2004**, *101*, 11767–11772.
72. Vlassara, H.; Cai, W.; Goodman, S.; Pyzik, R.; Yong, A.; Chen, X.; Zhu, L.; Neade, T.; Beeri, M.; Silverman, J.M.; *et al.* Protection against loss of innate defenses in adulthood by low advanced glycation end products (AGE) intake: Role of the antiinflammatory AGE receptor-1. *J. Clin. Endocrinol. Metab.* **2009**, *94*, 4483–4491.
73. Torreggiani, M.; Liu, H.; Wu, J.; Zheng, F.; Cai, W.; Striker, G.; Vlassara, H. Advanced glycation end product receptor-1 transgenic mice are resistant to inflammation, oxidative stress, and post-injury intimal hyperplasia. *Am. J. Pathol.* **2009**, *175*, 1722–1732.
74. Zimmet, P.; Alberti, K.G.M.M.; Shaw, J. Global and societal implications of the diabetes epidemic. *Nature* **2001**, *414*, 782–787.
75. Tan, K.C.; Shiu, S.W.; Wong, Y.; Tam, X. Serum advanced glycation end products (AGEs) are associated with insulin resistance. *Diabetes Metab. Res. Rev.* **2011**, *27*, 488–492.
76. Matthews, D.; Hosker, J.; Rudenski, A.; Naylor, B.; Treacher, D.; Turner, R. Homeostasis model assessment: Insulin resistance and β-cell function from fasting plasma glucose and insulin concentrations in man. *Diabetologia* **1985**, *28*, 412–419.
77. Tahara, N.; Yamagishi, S.; Matsui, T.; Takeuchi, M.; Nitta, Y.; Kodama, N.; Mizoguchi, M.; Imaizumi, T. Serum levels of advanced glycation end products (AGEs) are independent correlates of insulin resistance in nondiabetic subjects. *Cardiovasc. Ther.* **2012**, *30*, 42–48.

78. Naitoh, T.; Kitahara, M.; Tsuruzoe, N. Tumor necrosis factor-alpha is induced through phorbol ester—And glycated human albumin-dependent pathway in Thp-1 cells. *Cell Signal.* **2001**, *13*, 331–334.
79. Miele, C.; Riboulet, A.; Maitan, M.A.; Oriente, F.; Romano, C.; Formisano, P.; Giudicelli, J.; Beguinot, F.; van Obberghen, E. Human glycated albumin affects glucose metabolism in l6 skeletal muscle cells by impairing insulin-induced insulin receptor substrate (IRS) signaling through a protein kinase C alpha-mediated mechanism. *J. Biol. Chem.* **2003**, *278*, 47376–47387.
80. Cassese, A.; Esposito, I.; Fiory, F.; Barbagallo, A.P.; Paturzo, F.; Mirra, P.; Ulianich, L.; Giacco, F.; Iadicicco, C.; Lombardi, A.; *et al.* In skeletal muscle advanced glycation end products (AGEs) inhibit insulin action and induce the formation of multimolecular complexes including the receptor for AGEs. *J. Biol. Chem.* **2008**, *283*, 36088–36099.
81. Borst, S. The role of TNF-α in insulin resistance. *Endocrine* **2004**, *23*, 177–182.
82. Nakajima, K.; Yamauchi, K.; Shigematsu, S.; Ikeo, S.; Komatsu, M.; Aizawa, T.; Hashizume, K. Selective attenuation of metabolic branch of insulin receptor down-signaling by high glucose in a hepatoma cell line, Hepg2 cells. *J. Biol. Chem.* **2000**, *275*, 20880–20886.
83. Ravichandran, L.V.; Esposito, D.L.; Chen, J.; Quon, M.J. Protein kinase c-ζ phosphorylates insulin receptor substrate-1 and impairs its ability to activate phosphatidylinositol 3-kinase in response to insulin. *J. Biol. Chem.* **2001**, *276*, 3543–3549.
84. Boyd, A.C.; Abdel-Wahab, Y.H.; McKillop, A.M.; McNulty, H.; Barnett, C.R.; O'Harte, F.P.; Flatt, P.R. Impaired ability of glycated insulin to regulate plasma glucose and stimulate glucose transport and metabolism in mouse abdominal muscle. *Biochim. Biophys. Acta* **2000**, *1523*, 128–134.
85. Hunter, S.J.; Boyd, A.C.; O'Harte, F.P.; McKillop, A.M.; Wiggam, M.I.; Mooney, M.H.; McCluskey, J.T.; Lindsay, J.R.; Ennis, C.N.; Gamble, R.; *et al.* Demonstration of glycated insulin in human diabetic plasma and decreased biological activity assessed by euglycemic-hyperinsulinemic clamp technique in humans. *Diabetes* **2003**, *52*, 492–498.
86. Jia, X.; Olson, D.J.; Ross, A.R.; Wu, L. Structural and functional changes in human insulin induced by methylglyoxal. *FASEB J.* **2006**, *20*, 1555–1557.
87. Abdel-Wahab, Y.H.A.; O'Harte, F.P.; Ratcliff, H.; McClenaghan, N.H.; Barnett, C.R.; Flatt, P.R. Glycation of insulin in the islets of langerhans of normal and diabetic animals. *Diabetes* **1996**, *45*, 1489–1496.
88. Abdel-Wahab, Y.H.A.; O'Harte, F.P.M.; Barnett, C.R.; Flatt, P.R. Characterization of insulin glycation in insulin-secreting cells maintained in tissue culture. *J. Endocrinol.* **1997**, *152*, 59–67.
89. O'Harte, F.P.; Hojrup, P.; Barnett, C.R.; Flatt, P.R. Identification of the site of glycation of human insulin. *Peptides* **1996**, *17*, 1323–1330.
90. Krause, R.; Kühn, J.; Penndorf, I.; Knoll, K.; Henle, T. N-terminal pyrazinones: A new class of peptide-bound advanced glycation end-products. *Amino Acids* **2004**, *27*, 9–18.
91. Unoki-Kubota, H.; Yamagishi, S.; Takeuchi, M.; Bujo, H.; Saito, Y. Pyridoxamine, an inhibitor of advanced glycation end product (AGE) formation ameliorates insulin resistance in obese, type 2 diabetic mice. *Protein Pept. Lett.* **2010**, *17*, 1177–1181.
92. Guo, Q.; Mori, T.; Jiang, Y.; Hu, C.; Osaki, Y.; Yoneki, Y.; Sun, Y.; Hosoya, T.; Kawamata, A.; Ogawa, S.; *et al.* Methylglyoxal contributes to the development of insulin resistance and salt sensitivity in sprague-dawley rats. *J. Hypertens.* **2009**, *27*, 1664–1671.

93. Liang, F.; Kume, S.; Koya, D. Sirt1 and insulin resistance. *Nat. Rev. Endocrinol.* **2009**, *5*, 367–373.
94. Lim, M.; Park, L.; Shin, G.; Hong, H.; Kang, I.; Park, Y. Induction of apoptosis of beta cells of the pancreas by advanced glycation end-products, important mediators of chronic complications of diabetes mellitus. *Ann. NY Acad. Sci.* **2008**, *1150*, 311–315.
95. Luciano Viviani, G.; Puddu, A.; Sacchi, G.; Garuti, A.; Storace, D.; Durante, A.; Monacelli, F.; Odetti, P. Glycated fetal calf serum affects the viability of an insulin-secreting cell line *in vitro*. *Metabolism* **2008**, *57*, 163–169.
96. Zhu, Y.; Shu, T.; Lin, Y.; Wang, H.; Yang, J.; Shi, Y.; Han, X. Inhibition of the receptor for advanced glycation endproducts (RAGE) protects pancreatic beta-cells. *Biochem. Biophys. Res. Commun.* **2011**, *404*, 159–165.
97. Lin, N.; Zhang, H.; Su, Q. Advanced glycation end-products induce injury to pancreatic beta cells through oxidative stress. *Diabetes Metab.* **2012**, *38*, 250–257.
98. Zhao, Z.; Zhao, C.; Zhang, X.H.; Zheng, F.; Cai, W.; Vlassara, H.; Ma, Z.A. Advanced glycation end products inhibit glucose-stimulated insulin secretion through nitric oxide-dependent inhibition of cytochrome C oxidase and adenosine triphosphate synthesis. *Endocrinology* **2009**, *150*, 2569–2576.
99. Jitrapakdee, S.; Wutthisathapornchai, A.; Wallace, J.C.; MacDonald, M.J. Regulation of insulin secretion: Role of mitochondrial signalling. *Diabetologia* **2010**, *53*, 1019–1032.
100. Hachiya, H.; Miura, Y.; Inoue, K.; Park, K.H.; Takeuchi, M.; Kubota, K. Advanced glycation end products impair glucose-induced insulin secretion from rat pancreatic beta-cells. *J. Hepatobiliary Pancreat. Sci.* **2014**, *21*, 134–141.
101. Eto, K.; Tsubamoto, Y.; Terauchi, Y.; Sugiyama, T.; Kishimoto, T.; Takahashi, N.; Yamauchi, N.; Kubota, N.; Murayama, S.; Aizawa, T.; *et al.* Role of NADH shuttle system in glucose-induced activation of mitochondrial metabolism and insulin secretion. *Science* **1999**, *283*, 981–985.
102. Shu, T.; Zhu, Y.; Wang, H.; Lin, Y.; Ma, Z.; Han, X. AGEs decrease insulin synthesis in pancreatic beta-cell by repressing Pdx-1 protein expression at the post-translational level. *PLOS ONE* **2011**, *6*, e18782.
103. Puddu, A.; Storace, D.; Odetti, P.; Viviani, G.L. Advanced glycation end-products affect transcription factors regulating insulin gene expression. *Biochem. Biophys. Res. Commun.* **2010**, *395*, 122–125.
104. Wautier, M.; Massin, P.; Guillausseau, P.; Huijberts, M.; Levy, B.; Boulanger, E.; Laloi-Michelin, M.; Wautier, J. N(carboxymethyl) lysine as a biomarker for microvascular complications in type 2 diabetic patients. *Diabetes Metab.* **2003**, *29*, 44–52.
105. Boehm, B.; Schilling, S.; Rosinger, S.; Lang, G.; Lang, G.; Kientsch-Engel, R.; Stahl, P. Elevated serum levels of Nε-carboxymethyl-lysine, an advanced glycation end product, are associated with proliferative diabetic retinopathy and macular oedema. *Diabetologia* **2004**, *47*, 1376–1379.
106. Fosmark, D.S.; Torjesen, P.A.; Kilhovd, B.K.; Berg, T.J.; Sandvik, L.; Hanssen, K.F.; Agardh, C.-D.; Agardh, E. Increased serum levels of the specific advanced glycation end product methylglyoxal-derived hydroimidazolone are associated with retinopathy in patients with type 2 diabetes mellitus. *Metabolism* **2006** *55*, 232–236.

107. Aso, Y.; Inukai, T.; Tayama, K.; Takemura, Y. Serum concentrations of advanced glycation endproducts are associated with the development of atherosclerosis as well as diabetic microangiopathy in patients with type 2 diabetes. *Acta Diabetol.* **2000**, *37*, 87–92.
108. Ono, Y.; Aoki, S.; Ohnishi, K.; Yasuda, T.; Kawano, K.; Tsukada, Y. Increased serum levels of advanced glycation end-products and diabetic complications. *Diabetes Res. Clin. Pract.* **1998**, *41*, 131–137.
109. Kiuchi, K.; Nejima, J.; Takano, T.; Ohta, M.; Hashimoto, H. Increased serum concentrations of advanced glycation end products: A marker of coronary artery disease activity in type 2 diabetic patients. *Heart* **2001**, *85*, 87–91.
110. Busch, M.; Franke, S.; Wolf, G.; Brandstädt, A.; Ott, U.; Gerth, J.; Hunsicker, L.G.; Stein, G. The advanced glycation end product Nε-carboxymethyllysine is not a predictor of cardiovascular events and renal outcomes in patients with type 2 diabetic kidney disease and hypertension. *Am. J. Kidney Dis.* **2006**, *48*, 571–579.
111. Hanssen, N.M.; Engelen, L.; Ferreira, I.; Scheijen, J.L.; Huijberts, M.S.; van Greevenbroek, M.M.; van der Kallen, C.J.; Dekker, J.M.; Nijpels, G.; Stehouwer, C.D. Plasma levels of advanced glycation endproducts Nε-(carboxymethyl) lysine, Nε-(carboxyethyl) lysine, and pentosidine are not independently associated with cardiovascular disease in individuals with or without type 2 diabetes: The hoorn and codam studies. *J. Clin. Endocrinol. Metab.* **2013**, *98*, E1369–E1373.
112. Hanssen, N.M.; Beulens, J.W.; van Dieren, S.; Scheijen, J.L.; Spijkerman, A.M.; van der Schouw, Y.T.; Stehouwer, C.D.; Schalkwijk, C.G. Plasma advanced glycation endproducts are associated with incident cardiovascular events in individuals with type 2 diabetes: A case-cohort study with a median follow-up of 10 years (EPIC-NL). *Diabetes* **2014**, *64*, 257–265.
113. Dobler, D.; Ahmed, N.; Song, L.; Eboigbodin, K.E.; Thornalley, P.J. Increased dicarbonyl metabolism in endothelial cells in hyperglycemia induces anoikis and impairs angiogenesis by RGD and GFOGER motif modification. *Diabetes* **2006**, *55*, 1961–1969.
114. Chong, S.A.; Lee, W.; Arora, P.D.; Laschinger, C.; Young, E.W.; Simmons, C.A.; Manolson, M.; Sodek, J.; McCulloch, C.A. Methylglyoxal inhibits the binding step of collagen phagocytosis. *J. Biol. Chem.* **2007**, *282*, 8510–8520.
115. Nowotny, K.; Grune, T. Degradation of oxidized and glycoxidized collagen: Role of collagen cross-linking. *Arch. Biochem. Biophys.* **2014**, *542*, 56–64.
116. Yuen, A.; Laschinger, C.; Talior, I.; Lee, W.; Chan, M.; Birek, J.; Young, E.W.; Sivagurunathan, K.; Won, E.; Simmons, C.A.; *et al.* Methylglyoxal-modified collagen promotes myofibroblast differentiation. *Matrix Biol.* **2010**, *29*, 537–548.
117. Talior-Volodarsky, I.; Arora, P.D.; Wang, Y.; Zeltz, C.; Connelly, K.A.; Gullberg, D.; McCulloch, C.A. Glycated collagen induces alpha11 integrin expression through TGF-beta2 and Smad3. *J. Cell. Physiol.* **2015**, in press.
118. Talior-Volodarsky, I.; Connelly, K.A.; Arora, P.D.; Gullberg, D.; McCulloch, C.A. Alpha11 integrin stimulates myofibroblast differentiation in diabetic cardiomyopathy. *Cardiovasc. Res.* **2012**, *96*, 265–275.
119. Pozzi, A.; Zent, R.; Chetyrkin, S.; Borza, C.; Bulus, N.; Chuang, P.; Chen, D.; Hudson, B.; Voziyan, P. Modification of collagen IV by glucose or methylglyoxal alters distinct mesangial cell functions. *J. Am. Soc. Nephrol.* **2009** *20*, 2119–2125.

120. Sassi-Gaha, S.; Loughlin, D.T.; Kappler, F.; Schwartz, M.L.; Su, B.; Tobia, A.M.; Artlett, C.M. Two dicarbonyl compounds, 3-deoxyglucosone and methylglyoxal, differentially modulate dermal fibroblasts. *Matrix Biol.* **2010**, *29*, 127–134.
121. Charonis, A.S.; Tsilibary, E.C.; Saku, T.; Furthmayr, H. Inhibition of laminin self-assembly and interaction with type IV collagen by antibodies to the terminal domain of the long arm. *J. Cell Biol.* **1986**, *103*, 1689–1697.
122. Haitoglou, C.S.; Tsilibary, E.C.; Brownlee, M.; Charonis, A.S. Altered cellular interactions between endothelial cells and nonenzymatically glucosylated laminin/type IV collagen. *J. Biol. Chem.* **1992**, *267*, 12404–12407.
123. Cohen, M.P.; Ku, L. Inhibition of fibronectin binding to matrix components by nonenzymatic glycosylation. *Diabetes* **1984**, *33*, 970–974.
124. Liu, B.; Bhat, M.; Padival, A.K.; Smith, D.G.; Nagaraj, R.H. Effect of dicarbonyl modification of fibronectin on retinal capillary pericytes. *Investig. Ophthalmol. Vis. Sci.* **2004**, *45*, 1983–1995.
125. Duran-Jimenez, B.; Dobler, D.; Moffatt, S.; Rabbani, N.; Streuli, C.H.; Thornalley, P.J.; Tomlinson, D.R.; Gardiner, N.J. Advanced glycation end products in extracellular matrix proteins contribute to the failure of sensory nerve regeneration in diabetes. *Diabetes* **2009**, *58*, 2893–2903.
126. Collaboration, E.R.F. Diabetes mellitus, fasting blood glucose concentration, and risk of vascular disease: A collaborative meta-analysis of 102 prospective studies. *Lancet* **2010**, *375*, 2215–2222.
127. Kilhovd, B.K.; Juutilainen, A.; Lehto, S.; Rönnemaa, T.; Torjesen, P.A.; Hanssen, K.F.; Laakso, M. Increased serum levels of advanced glycation endproducts predict total, cardiovascular and coronary mortality in women with type 2 diabetes: A population-based 18 year follow-up study. *Diabetologia* **2007**, *50*, 1409–1417.
128. Meerwaldt, R.; Lutgers, H.L.; Links, T.P.; Graaff, R.; Baynes, J.W.; Gans, R.O.B.; Smit, A.J. Skin autofluorescence is a strong predictor of cardiac mortality in diabetes. *Diabetes Care* **2007**, *30*, 107–112.
129. Meerwaldt, R.; Graaff, R.; Oomen, P.; Links, T.; Jager, J.; Alderson, N.; Thorpe, S.; Baynes, J.; Gans, R.; Smit, A. Simple non-invasive assessment of advanced glycation endproduct accumulation. *Diabetologia* **2004**, *47*, 1324–1330.
130. Del Turco, S.; Basta, G. An update on advanced glycation endproducts and atherosclerosis. *BioFactors* **2012**, *38*, 266–274.
131. Kume, S.; Takeya, M.; Mori, T.; Araki, N.; Suzuki, H.; Horiuchi, S.; Kodama, T.; Miyauchi, Y.; Takahashi, K. Immunohistochemical and ultrastructural detection of advanced glycation end products in atherosclerotic lesions of human aorta with a novel specific monoclonal antibody. *Am. J. Pathol.* **1995**, *147*, 654–667.
132. Nerlich, A.G.; Schleicher, E.D. Nε-(carboxymethyl)lysine in atherosclerotic vascular lesions as a marker for local oxidative stress. *Atherosclerosis* **1999**, *144*, 41–47.
133. Zhou, Y.J.; Wang, J.H.; Zhang, J. Hepatocyte growth factor protects human endothelial cells against advanced glycation end products-induced apoposis. *Biochem. Biophys. Res. Communi.* **2006**, *344*, 658–666.

134. Li, Y.; Li, J.; Cui, L.; Lai, Y.; Yao, Y.; Zhang, Y.; Pang, X.; Wang, J.; Liu, X. Inhibitory effect of atorvastatin on AGE-induced HCAEC apoptosis by upregulating HSF-1 protein. *Int. J. Biol. Macromol.* **2013**, *57*, 259–264.

135. Chen, J.; Song, M.; Yu, S.; Gao, P.; Yu, Y.; Wang, H.; Huang, L. Advanced glycation endproducts alter functions and promote apoptosis in endothelial progenitor cells through receptor for advanced glycation endproducts mediate overpression of cell oxidant stress. *Mol. Cell. Biochem.* **2010**, *335*, 137–146.

136. Inagaki, Y.; Yamagishi, S.; Okamoto, T.; Takeuchi, M.; Amano, S. Pigment epithelium-derived factor prevents advanced glycation end products-induced monocyte chemoattractant protein-1 production in microvascular endothelial cells by suppressing intracellular reactive oxygen species generation. *Diabetologia* **2003**, *46*, 284–287.

137. Vlassara, H.; Fuh, H.; Donnelly, T.; Cybulsky, M. Advanced glycation endproducts promote adhesion molecule (VCAM-1, ICAM-1) expression and atheroma formation in normal rabbits. *Mol. Med.* **1995**, *1*, 447–456.

138. Schmidt, A.M.; Hori, O.; Chen, J.X.; Li, J.F.; Crandall, J.; Zhang, J.; Cao, R.; Yan, S.D.; Brett, J.; Stern, D. Advanced glycation endproducts interacting with their endothelial receptor induce expression of vascular cell adhesion molecule-1 (VCAM-1) in cultured human endothelial cells and in mice. A potential mechanism for the accelerated vasculopathy of diabetes. *J. Clin. Investig.* **1995**, *96*, 1395–1403.

139. Ishibashi, Y.; Matsui, T.; Takeuchi, M.; Yamagishi, S.-I. Glucagon-like peptide-1 (GLP-1) inhibits advanced glycation end product (AGE)-induced up-regulation of VCAM-1 mRNA levels in endothelial cells by suppressing AGE receptor (RAGE) expression. *Biochem. Biophys. Res. Commun.* **2010**, *391*, 1405–1408.

140. Ishibashi, Y.; Matsui, T.; Maeda, S.; Higashimoto, Y.; Yamagishi, S.-I. Advanced glycation end products evoke endothelial cell damage by stimulating soluble dipeptidyl peptidase-4 production and its interaction with mannose 6-phosphate/insulin-like growth factor II receptor. *Cardiovasc. Diabetol.* **2013**, *12*, 125–125.

141. Yamagishi, S.; Fujimori, H.; Yonekura, H.; Yamamoto, Y.; Yamamoto, H. Advanced glycation endproducts inhibit prostacyclin production and induce plasminogen activator inhibitor-1 in human microvascular endothelial cells. *Diabetologia* **1998**, *41*, 1435–1441.

142. Quehenberger, P.; Bierhaus, A.; Fasching, P.; Muellner, C.; Klevesath, M.; Hong, M.; Stier, G.; Sattler, M.; Schleicher, E.; Speiser, W.; *et al.* Endothelin 1 transcription is controlled by nuclear factor-κB in AGE-stimulated cultured endothelial cells. *Diabetes* **2000**, *49*, 1561–1570.

143. Zhang, L.; Zalewski, A.; Liu, Y.; Mazurek, T.; Cowan, S.; Martin, J.L.; Hofmann, S.M.; Vlassara, H.; Shi, Y. Diabetes-induced oxidative stress and low-grade inflammation in porcine coronary arteries. *Circulation* **2003**, *108*, 472–478.

144. Bucala, R.; Tracey, K.J.; Cerami, A. Advanced glycosylation products quench nitric oxide and mediate defective endothelium-dependent vasodilatation in experimental diabetes. *J. Clin. Investig.* **1991**, *87*, 432–438.

145. Xu, B.; Chibber, R.; Ruggiero, D.; Kohner, E.; Ritter, J.; Ferro, A. Impairment of vascular endothelial nitric oxide synthase activity by advanced glycation end products. *FASEB J.* **2003**, *17*, 1289–1291.

146. Chakravarthy, U.; Hayes, R.G.; Stitt, A.W.; McAuley, E.; Archer, D.B. Constitutive nitric oxide synthase expression in retinal vascular endothelial cells is suppressed by high glucose and advanced glycation end products. *Diabetes* **1998**, *47*, 945–952.
147. Barbato, J.E.; Tzeng, E. Nitric oxide and arterial disease. *J. Vasc. Surg.* **2004**, *40*, 187–193.
148. Yamagishi, S.; Amano, S.; Inagaki, Y.; Okamoto, T.; Koga, K.; Sasaki, N.; Yamamoto, H.; Takeuchi, M.; Makita, Z. Advanced glycation end products-induced apoptosis and overexpression of vascular endothelial growth factor in bovine retinal pericytes. *Biochem. Biophys. Res. Commun.* **2002**, *290*, 973–978.
149. Stitt, A.W.; Hughes, S.J.; Canning, P.; Lynch, O.; Cox, O.; Frizzell, N.; Thorpe, S.R.; Cotter, T.G.; Curtis, T.M.; Gardiner, T.A. Substrates modified by advanced glycation end-products cause dysfunction and death in retinal pericytes by reducing survival signals mediated by platelet-derived growth factor. *Diabetologia* **2004**, *47*, 1735–1746.
150. Antonetti, D. Eye vessels saved by rescuing their pericyte partners. *Nat. Med.* **2009**, *15*, 1248–1249.
151. Moore, T.C.; Moore, J.E.; Kaji, Y.; Frizzell, N.; Usui, T.; Poulaki, V.; Campbell, I.L.; Stitt, A.W.; Gardiner, T.A.; Archer, D.B.; *et al.* The role of advanced glycation end products in retinal microvascular leukostasis. *Investig. Ophthalmol. Vis. Sci.* **2003**, *44*, 4457–4464.
152. Lu, M.; Kuroki, M.; Amano, S.; Tolentino, M.; Keough, K.; Kim, I.; Bucala, R.; Adamis, A.P. Advanced glycation end products increase retinal vascular endothelial growth factor expression. *J. Clin. Investig.* **1998**, *101*, 1219–1224.
153. Stitt, A.W.; Bhaduri, T.; McMullen, C.B.; Gardiner, T.A.; Archer, D.B. Advanced glycation end products induce blood-retinal barrier dysfunction in normoglycemic rats. *Mol. Cell Biol. Res. Commun.* **2000**, *3*, 380–388.
154. Mamputu, J.C.; Renier, G. Advanced glycation end products increase, through a protein kinase C-dependent pathway, vascular endothelial growth factor expression in retinal endothelial cells. Inhibitory effect of gliclazide. *J. Diabetes Complicat.* **2002**, *16*, 284–293.
155. Yamagishi, S.; Inagaki, Y.; Okamoto, T.; Amano, S.; Koga, K.; Takeuchi, M.; Makita, Z. Advanced glycation end product-induced apoptosis and overexpression of vascular endothelial growth factor and monocyte chemoattractant protein-1 in human-cultured mesangial cells. *J. Biol. Chem.* **2002**, *277*, 20309–20315.
156. Ghayur, M.N.; Krepinsky, J.C.; Janssen, L.J. Contractility of the renal glomerulus and mesangial cells: Lingering doubts and strategies for the future. *Med. Hypotheses Res.* **2008**, *4*, 1–9.
157. Yamagishi, S.; Inagaki, Y.; Okamoto, T.; Amano, S.; Koga, K.; Takeuchi, M. Advanced glycation end products inhibit *de novo* protein synthesis and induce TGF-beta overexpression in proximal tubular cells. *Kidney Int.* **2003**, *63*, 464–473.
158. Throckmorton, D.C.; Brogden, A.P.; Min, B.; Rasmussen, H.; Kashgarian, M. PDGF and TGF-bold beta mediate collagen production by mesangial cells exposed to advanced glycosylation end products. *Kidney Int.* **1995**, *48*, 111–117.
159. Yang, C.W.; Vlassara, H.; Peten, E.P.; He, C.J.; Striker, G.E.; Striker, L.J. Advanced glycation end products up-regulate gene expression found in diabetic glomerular disease. *Proc. Natl. Acad. Sci. USA* **1994**, *91*, 9436–9440.
160. Fukami, K.; Ueda, S.; Yamagishi, S.; Kato, S.; Inagaki, Y.; Takeuchi, M.; Motomiya, Y.; Bucala, R.;

Iida, S.; Tamaki, K.; *et al.* AGEs activate mesangial TGF-beta-Smad signaling via an angiotensin II type I receptor interaction. *Kidney Int.* **2004**, *66*, 2137–2147.

161. Oldfield, M.D.; Bach, L.A.; Forbes, J.M.; Nikolic-Paterson, D.; McRobert, A.; Thallas, V.; Atkins, R.C.; Osicka, T.; Jerums, G.; Cooper, M.E. Advanced glycation end products cause epithelial-myofibroblast transdifferentiation via the receptor for advanced glycation end products (RAGE). *J. Clin. Investig.* **2001**, *108*, 1853–1863.
162. Burns, W.C.; Twigg, S.M.; Forbes, J.M.; Pete, J.; Tikellis, C.; Thallas-Bonke, V.; Thomas, M.C.; Cooper, M.E.; Kantharidis, P. Connective tissue growth factor plays an important role in advanced glycation end product-induced tubular epithelial-to-mesenchymal transition: Implications for diabetic renal disease. *J. Am. Soc. Nephrol.* **2006**, *17*, 2484–2494.
163. Roestenberg, P.; van Nieuwenhoven, F.A.; Wieten, L.; Boer, P.; Diekman, T.; Tiller, A.M.; Wiersinga, W.M.; Oliver, N.; Usinger, W.; Weitz, S., *et al.* Connective tissue growth factor is increased in plasma of type 1 diabetic patients with nephropathy. *Diabetes Care* **2004**, *27*, 1164–1170.
164. Chung, A.C.K.; Zhang, H.; Kong, Y.-Z.; Tan, J.-J.; Huang, X.R.; Kopp, J.B.; Lan, H.Y. Advanced glycation end-products induce tubular CTGF via TGF-β-independent Smad3 signaling. *J. Am. Soc. Nephrol.* **2010**, *21*, 249–260.
165. Thornalley, P.J. Dietary AGEs and ALEs and risk to human health by their interaction with the receptor for advanced glycation endproducts (RAGE)—An introduction. *Mol. Nutr. Food Res.* **2007**, *51*, 1107–1110.
166. Ramasamy, R.; Yan, S.F.; Schmidt, A.M. Arguing for the motion: Yes, rage is a receptor for advanced glycation endproducts. *Mol. Nutr. Food Res.* **2007**, *51*, 1111–1115.
167. Heizmann, C.W. The mechanism by which dietary AGEs are a risk to human health is via their interaction with RAGE: Arguing against the motion. *Mol. Nutr. Food Res.* **2007**, *51*, 1116–1119.

3

4-Hydroxy-nonenal—A Bioactive Lipid Peroxidation Product †

Rudolf J. Schaur [1,*], Werner Siems [2], Nikolaus Bresgen [3] and Peter M. Eckl [4]

[1] Institute of Molecular Biosciences, University of Graz, Heinrichstrasse 33a, 8010 Graz, Austria
[2] Institute for Medical Education, KortexMed GmbH, Hindenburgring 12a, 38667 Bad Harzburg, Germany; E-Mail: siems@kortexmed.de
[3] Division of Genetics, Department of Cell Biology, University of Salzburg, Hellbrunnerstasse 34, 5020 Salzburg, Austria; E-Mail: nikolaus.bresgen@sbg.ac.at
[4] Division of Genetics, Department of Cell Biology, University of Salzburg, Hellbrunnerstasse 34, 5020 Salzburg, Austria; E-Mail: peter.eckl@sbg.ac.at

† This review is dedicated to the late Professor Hermann Esterbauer, the discoverer of 4-hydroxy-nonenal.

* Author to whom correspondence should be addressed; E-Mail: rudolf.schaur@uni-graz.at;

Academic Editors: Michael Breitenbach and Peter Eckl

Abstract: This review on recent research advances of the lipid peroxidation product 4-hydroxy-nonenal (HNE) has four major topics: I. the formation of HNE in various organs and tissues, II. the diverse biochemical reactions with Michael adduct formation as the most prominent one, III. the endogenous targets of HNE, primarily peptides and proteins (here the mechanisms of covalent adduct formation are described and the (patho-) physiological consequences discussed), and IV. the metabolism of HNE leading to a great number of degradation products, some of which are excreted in urine and may serve as non-invasive biomarkers of oxidative stress.

Keywords: 4-hydroxy-nonenal; HNE; n-6 polyunsaturated fatty acids; Michael adducts; mercapturic acid

---

Table of Contents

Tables and Figures

Preface

Due to its diverse biochemical reactions and far-reaching biological effects 4-hydroxy-nonenal (HNE) is a fascinating compound, the discovery of which is hidden deep in the scientific literature. When Hermann Esterbauer isolated this aldehyde for the first time a quantitative analysis of its chemical elements was made resulting in values in between those for 4-hydroxy-octenal and 4-hydroxy-nonenal (Percent Composition: C 69.19%, H 10.32%, O 20.48%). Since these values were closer to those of 4-hydroxy-octenal Esterbauer and his supervisor Schauenstein published it with this chemical formula [1]. Only later the experimental error became evident. Thus, if one searches the literature for 4-hydroxy-nonenal, one would hardly find the paper describing its discovery as a product of lipid peroxidation.

Lipid peroxidation is a prominent feature of oxidative stress (for a review see the classical monograph of Halliwell and Gutteridge [2]), to which reference is made here. Additionally, a multi-author work on lipid peroxidation was edited by Catala [3]. Lipid peroxidation is a free radical amplification process yielding a great variety of bioactive products. One of the most extensively studied examples is HNE, which will be treated in this review. Since several reviews on HNE appeared previously [4–6], the focus of this contribution is directed on recent advances. In a recent review Fritz and Petersen [7] describe the generation and chemical reactivity of HNE and other lipid-derived aldehydes with a special attention to the homeostatic responses to electrophilic stress. Zarkovic *et al.* [8] describe the pathophysiological relevance of aldehydic protein modifications with special attention to HNE. A recent review of Perluigi *et al.* [9] describes the role of lipid peroxidation, particularly of HNE-induced protein modification, in neurodegenerative diseases. Furthermore, a review of Riahi *et al.* [10] addresses the origin and cellular functions of 4-hydroxyalkenals. Forman [11] reviewed the role of HNE and other α,β-unsaturated aldehydes as well as reactive oxygen species in signal transduction. New concepts and molecular mechanisms in cell signalling by reactive lipid species are discussed by Higdon *et al.* [12]. Similarly, Dwivedi *et al.* [13] as well as Ayala *et al.* [14] describe HNE as a signaling molecule. The chemistry and analysis of HNE was reviewed by Spickett [15]. LoPachin *et al.* [16] discussed the chemical attributes of HNE and acrolein that determine their toxicities.

A review of Catala [3] provide a synopsis of identified effects of HNE and other hydroxy-alkenals and oxidized phospholipids on cell signaling, from their intracellular production to their action as intracellular messengers, up to their influence on transcription factors and gene expression. Mattson [17] gave an overview of the roles of HNE in obesity, the metabolic syndrome, and associated vascular and neurodegenerative disorders. Finally, Guichardant and Lagarde [18] reviewed HNE as well as 4-hydroxy-hexenal and their degradation products as biomarkers of oxidative stress.

1. Lipid Peroxidation as a Free Radical Amplification Process

In this chapter a brief overview will be presented on the molecular mechanisms underlying lipid peroxidation. Lipid peroxidation was defined by Tappel [19] as the *oxidative deterioration of polyunsaturated lipids*. Poly-unsaturated fatty acids are those that contain two or more carbon-carbon double bonds. Initiation of lipid peroxidation can be caused by addition of a reactive species (RS) or, *more usually*, by hydrogen atom abstraction from a methylene ($-CH_2-$) group by a RS; in both cases, a carbon radical results. For example, the hydroxyl radical $^{\bullet}OH$ can react by addition (Equation (1)) or by $H^{\bullet}$ abstraction (Equation (2)).

$$-HC{=}CH- + {}^{\bullet}OH \rightarrow -HC(OH)-{}^{\bullet}CH- \tag{1}$$

$$-CH_2- + {}^{\bullet}OH \rightarrow -{}^{\bullet}CH- + H_2O \tag{2}$$

A double bond weakens the bond energy of the C-H bonds present on the next carbon atom (the allylic hydrogens), especially when there is a double bond on either side of the C-H bond (Figure 1), giving bis-allylic hydrogens.

Hydroxyl radicals readily initiate peroxidation of fatty acids, lipoproteins and membranes. In addition the hydroperoxyl radical $HOO^{\bullet}$ can abstract H from some polyunsaturated fatty acids (Equation (3))

$$HOO^{\bullet} + CH \rightarrow HOOH + C^{\bullet} \tag{3}$$

and it can stimulate peroxidation by reaction with preformed lipid hydroperoxides to generate peroxyl radicals [20] (Equation (4)).

$$HOO^{\bullet} + ROOH \rightarrow ROO^{\bullet} + H_2O_2 \tag{4}$$

The carbon radicals often stabilize by molecular rearrangement to form conjugated dienes (Figure 1).

Under aerobic conditions carbon radicals readily combine with O_2 to give a peroxyl radical (Equation (5)).

$$R^{\bullet} + O_2 \rightarrow ROO^{\bullet} \tag{5}$$

Peroxyl radicals can also abstract $H^{\bullet}$ from an adjacent fatty-acid side-chain (Equation (6)).

$$ROO^{\bullet} + CH \rightarrow ROOH + C^{\bullet} \quad (6)$$

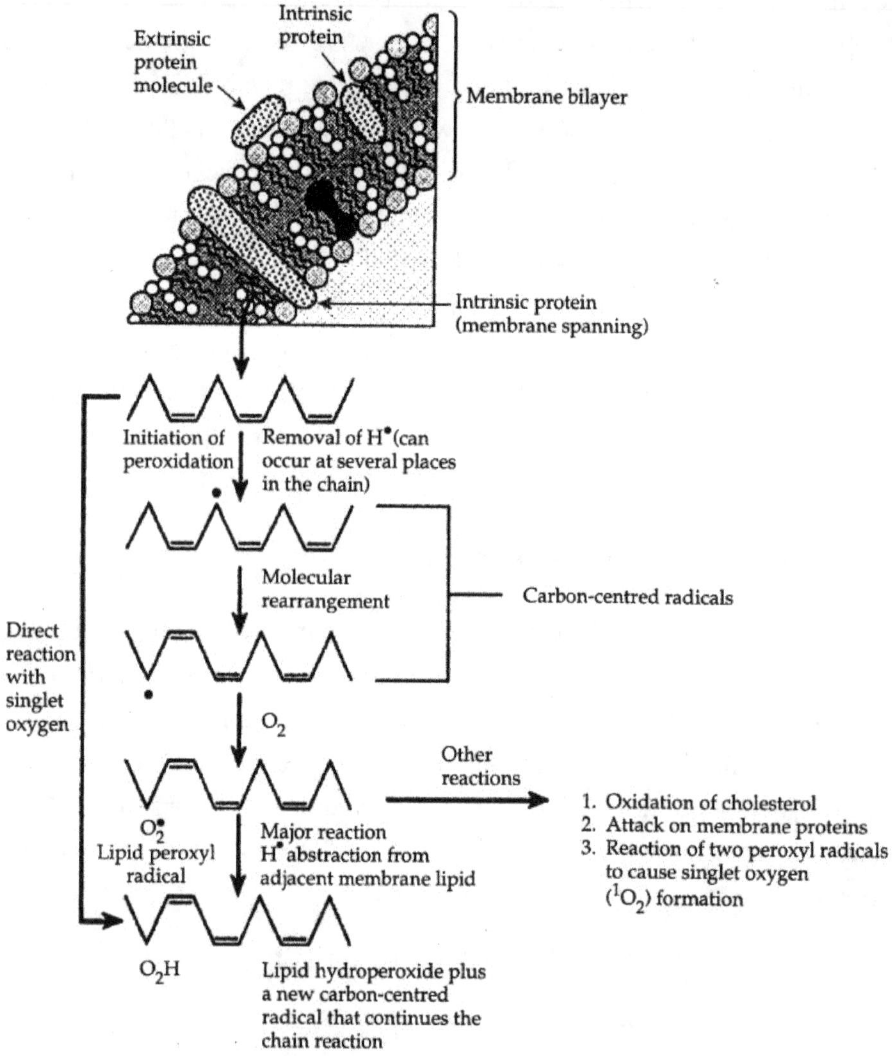

Figure 1. Idealized representation of the initiation and propagation reactions of lipid peroxidation. The peroxidation of fatty acids with three double bonds is shown. Reprinted from [2].

This is the propagation stage of lipid peroxidation. It forms new carbon radicals that can react with O_2 to form new peroxyl radicals, and so the chain reaction of lipid peroxidation continues (Figure 1). The $ROO^{\bullet}$ combines with the $H^{\bullet}$ that it abstracts to give a lipid hydroperoxide (ROOH). A single initiation event thus has the potential to generate multiple peroxide molecules by a chain reaction.

Cyclic and bicyclic peroxides can also form (Figure 2).

Peroxidized lipids are bioactive. They can have effects on cells similar to those of H_2O_2: low levels can stimulate proliferation, higher levels block proliferation and yet higher ones induce apoptosis and necrosis [21]. Such events occur in atherosclerosis and maybe in the gastrointestinal tract. Products of lipid peroxidation can also have more specific effects. For example, some mimic the actions of platelet activating factor (PAF = 1-*O*-alkyl-2-acetyl-sn-glycero-3-phospho-choline). Peroxidation of phosphatidyl-choline (lecithin) can generate fragments that bind to PAF receptors on target cells and exert PAF-like activity. Oxidized PAPC (1-palmitoyl-2-arachidonoyl-*sn*-glycero-3-phospho-choline) (ox-PAPC) can *stimulate* cytokine production in endothelial cells, but can *inhibit* activation of transcription factor NFκB by endotoxin and subsequent cytokine production, *i.e.*, it can be anti-inflammatory under some circumstances. It can also increase levels of haem-oxygenase, a cytoprotective enzyme in endothelial cells. Intraperitoneal injection of ox-PAPC allows mice that had been treated with a normally lethal dose of endotoxin, to survive [22]. Hence several specific lipid oxidation products act as signalling molecules and their actions are not always bad. It can be predicted that more such roles will be found.

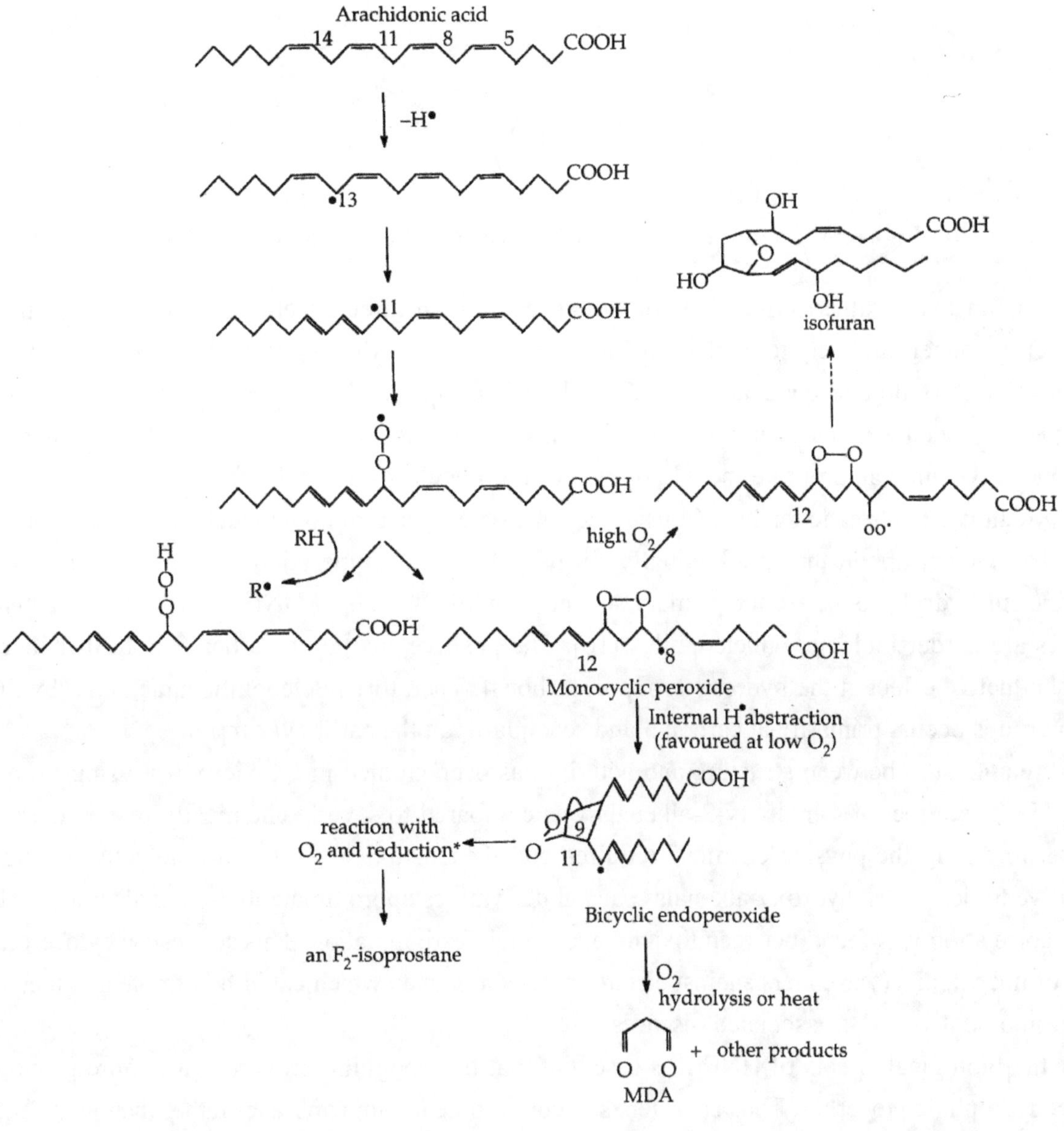

Figure 2. Formation of lipid hydroperoxides and cyclic peroxides from arachidonic acid. The chemistry is complex because abstraction of H• can occur at different sites; for simplicity we show it only for C13. The ROO• radicals formed can generate hydroperoxides or cyclize to monocyclic peroxides. This results when peroxyl radicals attack a double bond in the same chain. Monocyclic peroxides can form bicyclic structures that can give rise to isoprostanes, or react with O_2, eventually producing isofurans. Unstable peroxides can decompose to aldehydes such as malondialdehyde. Even products containing bicyclic endoperoxide and cyclic peroxide groups in the same molecule (dioxolane-isoprostanes)

can be generated [23]. O_2 concentration affects the relative levels of isoprostanes and other products. Reprinted from [2].

2. Structure, Properties and Generation of HNE

The chemical structure of 4-hydroxy-2-*trans*-nonenal (4-hydroxy-2E-nonenal, HNE) is shown in Figure 3.

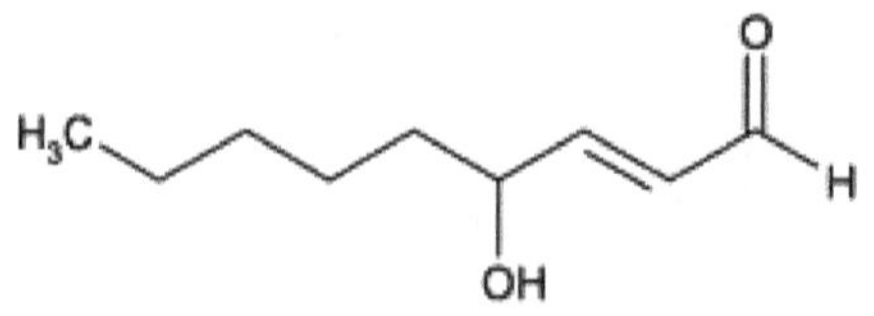

Figure 3. Chemical structure of 4-hydroxy-2-*trans*-nonenal (HNE).

HNE is an extraordinarily reactive compound containing three functional groups which in many cases act in concert and help to explain its high reactivity [24]. Most importantly there is a conjugated system of a C=C double bond and a C=O carbonyl group which provide a partial positive charge to carbon 3, because this structure contains mobile π-electrons. The carbonyl oxygen atom is electronegative and can promote the withdrawal of mobile electron density from the beta-carbon atom causing regional electron deficiency. Considering this type of electron polarizability, HNE is considered to be a *soft* electrophile that preferentially forms 1,4-Michael type adducts with *soft* nucleophiles. Cysteine sulfhydryl groups are the primary soft nucleophilic targets of HNE while lysine and histidine residues are harder biological nucleophiles [16]. The positive charge on carbon 3 is further increased by the inductive effect of the hydroxy group at carbon 4. Therefore nucleophilic attack e.g., by thiol or amino groups occurs primarily at carbon 3 and secondarily at the carbonyl carbon 1.

The relationship between structure and activity has been studied in a toxicity test using L6 muscle cells [25]. The effect of 4-hydroxy-2-alkenals was compared to several chemically related derivatives in order to clarify the physico-chemical requirement of their toxicity in L6 muscle cells. The rank of derivative toxicity was: hydroxy-alkenals > acetal derivatives approximate to 2-alkenals > alkanals and a high correlation was found between toxicity and protein carbonylation. This suggests a cyto-protective effect of nucleophilic scavengers against electrophilic compounds which could be of potential therapeutic benefit in oxidative stress associated diseases.

For the biological effects of HNE it is essential that the lipophilic properties are more pronounced than its hydrophilic properties. Thus HNE tends to concentrate in biomembranes rather than in the aqueous space of cells. HNE can be easily transferred from a membrane to both cytosol and the extracellular space. Studying the behaviour within membranes by using the time-dependent fluorescence shift method and molecular dynamics simulations demonstrated the stabilization of HNE in the carbonyl region of a 1-palmitoyl-2-oleoyl-sn-glycero-3-phosphocholine bilayer [26]. Thus, HNE is able to react with cell membrane proteins and lipids albeit stabilization in the membrane is moderate and HNE shuttling to either the extra- or intracellular space occurs in the microsecond range. These characteristics of the HNE—lipid membrane interaction provide a good explanation for the observed reactivity of HNE with proteins inside and outside the cell.

HNE shows chirality at carbon 4. This might be also biologically relevant. Although HNE is formed in tissues as a racemate, enantiospecific HNE effects have not yet been widely investigated. Different cellular responses have been reported for treatment with (R)-HNE, (S)-HNE, or racemic HNE. For instance, (S)-HNE and racemic HNE potently stimulate phosphorylation of Jun kinase and Akt while (R)-HNE is strongest in phosphorylating MAPK. Also, (S)-HNE shows a more pronounced cytotoxicity [27,28] which points at the relevance of HNE enantiomers in cellular responses to HNE. Furthermore, Guéraud *et al.* [29] have shown, that (R)- and (S)-HNE are enantioselectively metabolised in rats and d_{11}-4-hydroxy-2-(E)-nonenal is conjugated with glutathione in an enantio-selective mode, and exported from the liver [30].

Generally, HNE reacts with some sulfhydryl groups in proteins in a stereoselective manner. This has been characterized in detail by Wakita *et al.* [31]. Incubation of *N*-acetylcysteine with the S- and R-HNE entantiomer generated HNE-cystein adducts characterized as anomeric isomers by reverse-phase HPLC and NMR. The stereoselective formation of HNE-cyteine adducts was also demonstrated in the redox-regulatory protein thioredoxin, the active site at Cys^{32} showing a preference for R-HNE cysteine adducts. These findings provide insight into structural aspects of lipid peroxidation product/HNE—based sulfhydryl modification and the chemical characterization of protein S-associated aldehydes *in vitro* and *in vivo*. In this context it is worth to mention that an efficient, enantioselective synthesis of (R)- and (S)-HNE has been worked out recently by Komisarski *et al.* [32].

Of special relevance, HNE and its glutathione conjugates are able to regulate oxidative stress related transcription factors such as NFκB and AP-1 by addressing protein kinase cascade mediated stress signaling. This transcriptional activation leads to an upregulated expression of several genes involved in cell differentiation and cell death control. The mechanisms by which HNE and other lipid aldehydes transduce activation of NFκB signaling pathways have been described by Yadav *et al.* [33].

Regulatory effects of HNE can be based on purely metabolic mechanisms. Disruption of GST-based HNE conjugation, resulting in elevated HNE-levels, promotes obesity in mice and the nematode *Caenorhabditis elegans* suggesting a phylogenetically conserved, HNE-dependent mechanism involved in lipid metabolism [34,35]. In *C. elegans*, enhanced HNE-conjugation favours a lean phenotype while an inhibition of conjugation or oxidation of HNE yields fat accumulation. Together with the finding that synthetic HNE also stimulates lipid deposition in *C. elegans*, a causative role of HNE, acting on a metabolic basis in the development of an obese phenotype cannot be ruled out.

HNE is a degradation product of ω-6 polyunsaturated fatty acids, and the different mechanisms of HNE formation have been reviewed recently by Schneider *et al.* [36] and Spickett [15]. Notably, 4-hydroxy-hexenal (HHE), representing the corresponding hydroxy-alkenal derived from ω-3 polyunsaturated fatty acids shares several properties with HNE, but also shows important differences particularly with respect to targets of aldehyde adduct formation and detoxification pathways [37].

HNE formation may follow several radical-dependent oxidative routes involving the generation of hydroperoxides, alkoxyl radicals, epoxides, and fatty acyl cross-linking reactions. Cleavage of the oxidized fatty acyl chain results in HNE formation from the methyl end, and 9-oxo-nonanoic acid from the carboxylate or esterified end of the chain, although many other products are also possible.

Furthermore, the enzymatic production of HNE has been described by Wang *et al.* [38], reporting that *Escherichia faecalis*, a human intestinal commensal, triggers COX-2 based HNE formation in macrophages.

Finally, HNE is hepato- and nephrotoxic in mammalians [39]. Interestingly, high HNE concentrations (≥20 μM) are considered to contribute to cardiac ischemia-reperfusion injury, while lower, sublethal aldehyde concentrations (5 μM) serve cardioprotection by supporting cellular stress resistance [40]. At low concentrations, HNE stimulates the Nrf2 (NFE2-related factor 2) mediated upregulation of γ-glutamylcysteine ligase and the core subunit of the high-affinity cystine transporter Xc(-) yielding a 1.45-fold shift of intracellular GSH levels.

3. Major Reaction Mechanisms

Essentially, HNE exhibits a high reactivity towards thiol and amino groups which are responsible for most of the biochemical effects of HNE. An overview of the reactions of HNE with different biomolecules is shown in Figure 4.

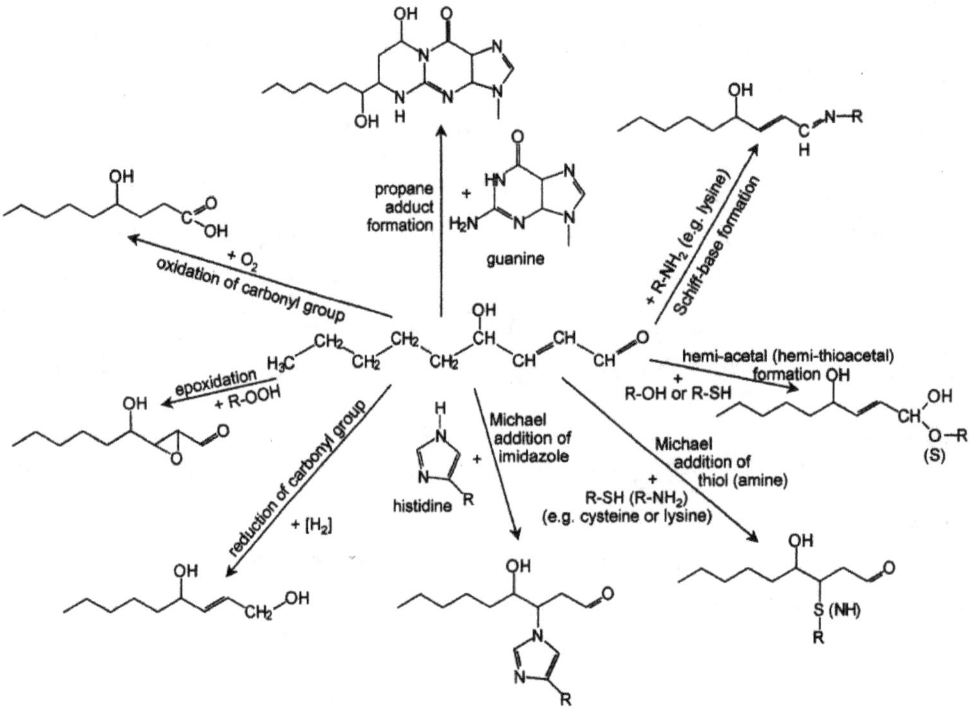

Figure 4. Overview of the reactions of 4-hydroxy-nonenal with different biomolecules. Explanations in the text. Reprinted from [24].

3.1. Reactions of the C=C Double Bond

3.1.1. Michael Additions

In Michael additions a nucleophile, e.g., cysteine or glutathione (GSH), is added to the C=C double bond (Figure 4). In some cases the reaction is reversible (Retro Michael cleavage). In case of GSH the reaction velocity is greatly enhanced by the enzyme glutathione-S-transferase (GST) as reviewed by Mannervik (2012) [41]. In particular the GSTA4-4 isoform strongly catalyzes the reaction between

HNE and GSH which is due to the high alkenal activity related to C-terminal interactions [42]. The elimination of an excess of HNE by GST activity may confer cytoprotection, however, this can change to the opposite in the presence of interacting factors such as the multidrug resistence protein-1 (MRP1). This is demonstrated by the HNE mediated cytotoxicity seen in MCF7 cells stably expressing MRP1 together with human GST-M1 (GSTM1) [43]. In these cells, the co-expression of GSTM1 and MRP1 leads to increased GSH depletion and formation of protein adducts upon HNE exposure which facilitated a cytotoxic outcome rather than protection.

GSTA4-4 was also used to address the question whether HNE metabolizing enzymes avoid HNE mediated inactivation [44]. Despite marked structural similarities between several GST isoforms (GSTA1-1, GSTA4-4, GSTP1-1) GST A4-4 shows the highest catalytic activity to metabolize HNE but is insensitive to covalent HNE adduct formation.

Similar to thiols amino compounds such as lysine, ethanol-amine, guanine and the imidazole group of histidine can also undergo Michael additions to HNE. For example, the histidine-containing dipeptide carnosine is a target of HNE [45].

As a recent example of Michael adduction of proteins by HNE, the modification of recombinant human phosphatase PTEN (Phosphatase and Tensin homolog) by HNE was studied by Shearn *et al.* [46] who identified several Michael adducts of HNE on Cys^{71}, Cys^{136}, Lys^{147}, Lys^{223}, Cys^{250}, Lys^{254}, Lys^{313}, Lys^{327} and Lys^{344} by mass analysis (LC/MS/MS). HNE adducts to Cys^{71} adjacent to the active center and Lys^{327} locating to the C2 domain of PTEN may lead to PTEN inhibition via steric hindrance of the active-site pocket or abrogation of PTEN functions associated with the C2 domain. Shearn *et al.* [46] also propose that Akt2 could be activated by the HNE mediated inhibiton of PTEN which suggests a potential novel mechanism of lipid accumulation initiated by an increased production of reactive aldehydes during chronic ethanol administration.

In rare cases Michael addition reactions of HNE as well as acrolein (ACR) have been observed with aromatic hydroxy groups of polyphenols. The formation of adducts with HNE or ACR was reported for naturally occurring flavan-3-ols, theaflavins, cyanomaclurin, dihydrochalcones and phloretin which showed the most efficient scavenging activity [47]. With respect to this, certain polyphenols could be of benefit in antagonizing "carbonyl stress" via quenching of α,β-unsaturated aldehydes. Finally, Michael addition of HNE to ascorbic acid has also been described, targeting an activated C-H bond followed by ring closure and hemiacetal formation (*i.e.*, addition of a hydroxyl group to a carbonyl group) [31,48].

Several methods are available at present to study Michael adduction by HNE. In particular, polyclonal and monoclonal antibodies (e.g., applied as ELISA) are intensively used to assess Michael-adduct formation of HNE with proteins [49]. Michael addition-type amino acid adducts of HNE can also be readily detected by HPLC methodology employing 2-aminopyridine (2-AP) as fluorescent probe [50]. Moreover the relation between protein adduct formation and gene expression is studied by the integration of protein adduction, gene expression, protein-DNA interaction, and protein-protein interaction data following a systemic approach [51].

3.1.2. Reduction

The C=C double bond can be reduced by an alkenal/one oxidoreductase with NAD(P)H as cofactor to yield 4-hydroxynonanal (Figure 4).This enzyme can metabolize HNE quite efficiently, which is demonstrated by the effect of NADP(H) overexpression seen in human embyonic kidney cells, efficiently protecting the cells from HNE challenge [52].

3.1.3. Epoxidation

As further reaction, epoxidation in presence of a hydroperoxide may introduce an oxirane ring into HNE (Figure 4). Albeit the exact underlying mechanism for this reaction remains to be defined, epoxides are formed by the incubation of HNE with hydrogen peroxide or 9- or 13-linoleic acid hydrogen peroxide at 37 °C for 24 h as well as in the presence of a fatty acid (linoleic acid, γ-linolenic acid, or arachidonic acid) and lipoxygenase [53].

3.2. Reactions of the Carbonyl Group

3.2.1. Acetal and Thio-Acetal Formation

The reaction of the carbonyl group of HNE with alcohols or thiols is a two step mechanism yielding (thio)acetals via (thio)hemi-actals (Figure 4). Since acetals are much more stable than HNE the reaction of HNE with methanol or ethanol is frequently used in synthetic chemistry for the storage of HNE. The free aldehyde can easily be released again in acidic medium. The hydroxyl group at carbon 4 of HNE can undergo a hemi-acetal formation only as a secondary reaction after a primary reaction such as a Michael addition which confers rotatory freedom to the C2-C3 bond.

3.2.2. Schiff-Base Formation

Schiff-base formation with primary amines, e.g., lysine, is a competitive reaction to the Michael addition of amines (Figure 4). This reaction contributes frequently to the crosslinking of proteins by HNE as was observed in mouse sciatic nerves. It has been shown that HNE adducts characterize axons of the mouse sciatic nerve from birth to senility. Using an antibody specific for a lysine-lysine cross-link, Perry *et al.* [54] examined axonal labeling in these nerves. As demonstrated by immunoblotting, HNE mediated intramolecular cross-linking is restricted to heavy and medium neurofilament subunits, the extent of HNE modification remaining relatively constant over the life span. Furthermore, Schiff-base formation may also have therapeutic implications. This is suggested by the reaction of the HNE carbonyl group with the hydrazine functional group of *phenelzine* underlying the HNE scavenging effect of this agent in traumatic brain injury [55].

3.2.3. Oxidation

The oxidation of the HNE carbonyl group producing 4-hydroxy-nonenoic acid is catalyzed by aldehyde dehydrogenase (ALDH), NAD^+ or $NADP^+$ acting as cofactor (Figure 4). Fournet *et al.*, 2013 critically reviewed the existing literature with resepct to the presumptive role of aldehyde dehydrogenases (ALDH1, ALDH2 and ALDH3) in regulating HNE and other endogenous apoptogenic aldehydes (methional, malondialdehyde) showing conflicting results concerning the contribution of different ALDH isoforms to HNE detoxification [56].

ALDH2 may play an important role in neuronal injury, the overexpression or activation in ALDH2 deficient rats providing neuroprotection via enhanced HNE removal [57]. Supportive to this, the neuroprotective effect of protein kinase Cε—a positive regulator of ALDH2 was abolished in ALDH2-knockdown rats. Notably, ALDH2 also contributes to myocardial protection from ischaemia/reperfusion injury which appers to be attributable to HNE detoxification as well as autophagy regulated by AMPK- and Akt-mTOR signalling during ischaemia and reperfusion [58].

Furthermore, the ALDH3 subfamily comprises a group of enzymes which oxidize medium-chain aliphatic and aromatic aldehydes (e.g., fatty and peroxidic aldehydes). ALDH3A1, a phase II drug-metabolizing enzyme is poorly expressed in the normal liver but present at high levels in the lung, stomach, cornea and keratinocytes, cells with a high ALDH3A1 content revealing an increased resistance to the cytostatic and cytotoxic effects of lipidic aldehydes [59]. Transfection studies demonstrated that ALDH3, but not class 1 ALDH, protects from severe GSH depletion and HNE mediated apoptosis in V79 fibroblasts and RAW264.7 macrophages which is associated with reduced levels of HNE protein-adducts [60]. Moreover, HNE serves as substrate for ALDH3 catalyzed oxidation suggesting that the oxidative decomposition of HNE by class 3 ALDH constitutes an important mechanism protecting from HNE toxicity [60]. Notably, this report also showed that the contributions of HNE functional groups to HNE toxicity follows the order aldehyde ≥ C2=C3 double bond >> C4-hydroxyl group. The role of ALDH3A1 in oxidative stress responses was further characterized in a rabbit corneal keratocyte cell line (TRK43) overexpressing ALDH3A1 which confirmed the protective role of ALDH3A1 towards the adverse effects of HNE which may occur under conditions of oxidative stress [61]. Additionally, Yoval-Sanchez *et al.*, 2012 reported that human ALDHs reveal different sensitivities to inactivation by LPO-derived aldehydes, ALDH2 most effciently oxidizing HNE, acrolein, and MDA [62]. Interestingly, while these compounds are also processed by ALDH1A1, ALDH3A1 specifically oxidizes HNE. Of note, ALDH2 is irreversibly inactivated by these LPO-metabolites which may target the ALDH aldehyde and NAD^+ binding sites. In contrast, ALDH3A1 tolerates concentrations of HNE or MDA of 20 mM. Hence, the pronounced resistance towards inactivation by aldehydic LPO metabolites emphasizes a particular role of ALDH3A1 (and also ALDH1A1) in cellular detoxification of LPO-derived aldehydes [62].

3.2.4. Reduction

The opposite reaction, reduction of the carbonyl group, can be catalyzed by alcohol dehydrogenase (ADH) and NADH (Figure 4). Of special relevance, amyloid beta peptide (Aβ) binding-alcohol dehydrogenase (ABAD) which is present in the brain metabolizes different substrates including β-estradiol as well as n-isopropanol in addition to its Aβ binding property. In the pathogenesis of Alzheimer disease (AD), the association of mitochondria derived oxidative stress with Aβ plays a considerable role and the aldehyde detoxification by ABAD has been suspected to play a particular role. Indeed, in SH-SY5Y neuroblastoma cells, representing a widely used *in vitro* model of neuronal function and differentiation, ABAD is able to decompose HNE and to antagonize HNE cytotoxicity wich is suppressed by Aβ [63]. Based on this, it is suggested that the impairment of ABAD mediated HNE-detoxification by Aβ contributes to the pathogenesis of AD.

A second enzyme, *aldose reductase*, preferentially reduces the glutathione conjugate of HNE. Aldose reductase protects against early atherosclerotic lesion formation in apolipoprotein E-null mice by removing HNE [64] and protects also against myocardial ischemia-reperfusion injury in mice [65].

Several studies indicate that a third group of enzymes, namely several members of the *aldo-keto reductase* (AKR) family are also involved in the metabolism of HNE [66–68]. For instance, AKR1C3 contributes to HNE reduction and thereby protects human SH-SY5Y cells from HNE cytotoxicity. Notably, AKR expression is under the control of Nrf2, a regulatorys effect which may be crucial to the cytoprotection conferred by Nrf2 towards the toxic effects of lipid peroxides. Similarly, AKR7A2 renders V79-4 Chinese hamster lung cells resitant towards HNE cytotoxicity, but fails to have the same effect on the toxicity of crotonaldehyde and acrolein [66]. Overexpression of AKR7A2 in V79 cells lowered HNE genotoxicity, improved the resistance to the redox-cycler menadione by lowering menadione-induced ROS generation and GSH depletion. With resepct to this, it cannot be ruled out that the participation of AKRs like AKR7A2 in cellular detoxifying pathways may also be of relevance to antioxidant defenses *in vivo.* This is also indicated by the activity of AKR1B10, another AKR isoform which detoxifies HNE and other dietary and lipid-derived α,β-unsaturated carbonyls at physiological levels [69]. In addition, retinals have been shown to be the best substrate of AKR1B10 [70]. Recombinant rat AKR1B10 (R1B10), derived from the corresponding cDNA isolated from rat brain reduces different aldehydes including HHE, HNE and 4-oxo-2-nonenal as well as α-dicarbonyls such as methylglyoxal and 3-deoxyglucosone, NADPH serving as preferred coenzyme [71]. 4-oxo-2-nonenal is most efficiently reduced into 4-oxo-2-nonenol by recombinant R1B10 which also lowers its cytotoxic properties. Expression of R1B10 (mRNA) is highest in rat brain and heart, low levels detectable also in other tissues and skin fibroblasts. Similar to AKR7A2, these findings suggest that R1B10 serves antioxidant defenses in rat tissues which also holds true for rat AKR1C15, which is upregulated by HNE and protects endothelial cells from HNE mediated cell damage [72].

In photoreceptor cells a fourth group of HNE reducing enzymes was detected: the *retinol dehydrogenases* RDH11 and RDH12, which are capable of reducing LPO products such as HNE. As demonstrated by Marchette *et al.* [73], expression of RDH11 and RDH12 in human embryonic kidney (HEK-293) cells suppressed HNE protein modification and apoptosis triggered by this aldehyde. Moreover, LPO and HNE formation is induced upon exposure of the mouse retina to bright light peroxidation which is accompanied by HNE protein modification in the inner segments of the photoreceptor containing RDH11 and RDH12. Interestingly, HNE adduct formation and light-induced photoreceptor apoptosis was suppressed only by RDH12 which suggests a substrate specifity of RDH12 for HNE. From the mechanistical aspect, the protection from the light-induced prooxidant and cytotoxic effects is considered to be based on the reduction of HNE by RDH12 to the nontoxic alcohol 1,4-dihydoxy-nonene. It is worth to note that the mutation of the RDH12 gene is causal to *early onset retinal dystrophy Leber congenital amaurosis* (LCA), a severe inherited eye-disease with a still incompletely understood pathogenesis.

Finally, the carbonyl group of HNE can also be reduced or oxidized by members of the *cytochrome P450* family. For example, the incubation of cytochrome P450 3A4 with HNE, O_2, and NADPH leads to the formation of several compounds including 1,4-dihydroxynonene (DHN) and 4-hydroxy-2-nonenoic acid (HNA) [74]. The reduction reaction is catalyzed also by other P450 isoforms especially (human) P450 2B6, revealing an activity similar to P450 3A4 and P450 1A2, P450 2J2, (mouse) P450 2c29, which are less active. In contrast, human P450 2E1 and rabbit P450 2B4 lack a reducing activity of the carbonyl group.

3.3. Reactions of the Hydroxy Group

The formation of cyclic hemi-acetals—as it is known for sugars—as secondary reactions has already been mentioned above. Similarly, the oxidation of HNE to 4-oxo-nonenal (Figure 4) has also been described as a secondary event only.

In addition, catabolic HNE degradation in the liver may also involve *phosphorylation* of the hydroxy group [75] as well as *ω-oxidation, i.e.*, the oxidation of the methyl group. ω-Oxidation, following oxidation of the carbonyl group is indicated by the presence of intermediates (4-hydroxynonanedioyl-CoA, 4-hydroxyheptanedioyl-CoA and 4-hydroxy-pentanedioyl-CoA) which account for a degradation of HNE from either end of the molecule [76].

4. Biophysical Effects

Besides biochemical effects, also biophysical changes of protein and lipid membrane conformation—e.g., loss of phospholipid asymmetry have to be taken into consideration. Using electron paramagnetic resonance spectroscopy in combination with protein-specific spin labeling (2,2,6,6-tetramethyl-4-maleimidopiperidin-1-oxyl), the induction of conformational changes of synaptosomal membrane proteins by physiologically relevant HNE concentrations (1 μM) was demonstrated [77].

HNE may also interfere with membrane fluidity, albeit results dealing with this aspect are conflicting. In gerbil synaptosomal membranes, fluidity of the bilayer is increased at higher HNE conentrations (*i.e.*, 50 μM) [77]. On the contrary, the fluidity of hepatic mitochondrial membranes in rats was decreased by HNE and to a lower extent also MDA supposedly as result of a direct interaction of HNE with membrane phospholipids [78]. Also, HNE directly inhibits the dopamine transporter and also decreases Na^+/K^+ ATPase activity in synaptosomes, the effect on Na^+/K^+ ATPase activity most likely resulting from HNE induced loss of membrane fluidity [79].

5. Biochemical Targets of HNE

The cellular reactions to HNE are strongly dose-dependent. According to recent concepts and supported by transcriptome analysis, low HNE concentrations provide protection from oxidative-stress mediated cell damage while increasing HNE doses promote genotoxic effects and initiate apoptosis or severely deregulates cell integrity [80].

The smallest reported biochemical target of HNE is hydrogen sulfide (H_2S) which has been recognized as neuromodulator and gasotransmitter. H_2S protects SH-SY5Y neuroblastoma cells from HNE cytotoxicity and HNE-related protein modification leading to the assumption that H_2S could represent an important antagonist to "carbonyl stress" arising from aldehydic LPO metabolites [81].

5.1. Reactions with Peptides and Proteins

From a quantitative perspective, proteins and peptides represent the most important class of biomolecules which are targeted by HNE. About 1% to 8% of the HNE formed in a cell are considered to modify proteins [82] and several investigations have been carried out to identify protein targets of HNE modification [83–85]. For example, more than 1500 proteins were identified as HNE targets in human colon carcinoma cells, 417 of these proteins showing a significant concentration dependent HNE-adduct formation [86]. Another study by Chavez *et al.* [85] using HNE treated human monocytic THP-1 cells demonstrated distinct, site-specific (His, Cys) HNE modifications of 16 proteins comprising cytoskeletal elements (tubulin α-1B chain, α-actinin-4, vimentin), proteins associated with stress responses and metabolic enzymes (D-3-phosphoglycerate dehydrogenase, aldolase A). Also, heat shock protein Hsp90 as well as the endoplasmic stress response protein, protein disulfide isomerase (PDI) may also become targets of HNE generated by alcohol-induced oxidative stress in rat liver [87]. Another illustrative *in vivo* example of selective targeting of enzymes involved in glycolysis, energy production as well as CO_2 hydration has been described for the ventilatory muscles of rats under the conditions of severe sepsis and strenuous muscle contractions [88]. Under these conditions, HNE adduction of enolase 3b, aldolase and triosephosphate isomerase 1, creatine kinase, carbonic anhydrase III, aconitase 2, dihydrolipoamide dehydrogenase, and electron transfer flavoprotein-beta is detectable. Together these studies account for different sensitivities to HNE protein modification between cellular proteins. In this context it has to be noted that HNE protein adduct formation affects the whole cell while prooxidant regimens such as H_2O_2 preferentially trigger cytosolic protein oxidation [89].

Enolase activity is incrementally declining at increasing HNE concentrations *in vitro* which demonstrates a dose-dependent negative effect of HNE-adduct formation on enzyme activity [88]. Indeed, in most instances protein function is impaired by HNE adduct formation. The aforementioned modification of Hsp70 and Hsp90 in alcoholic liver disease adversely interferes with the protein binding and ATPase activity of these stress response proteins. Moreover, HNE adducts negatively affect the activity of *protein disulfide isomerase* which is required for the proper establishment of disulfide bonds in protein folding. As a consequence, HNE modification will promote the accumulation of damaged or misfolded proteins which is detectable in the liver upon chronic alcohol ingestion [87,90,91].

It is well established that HNE forms adducts with three different side chains in proteins, namely Cys, His and Lys. The differential reactivity of these residues as well as of Arg and Glu for adduct formation with HNE in synthetic poly-aminoacid model compounds has been studied [92]. It turned out that Cys residues displayed by far the highest reactivity; the order of the molar HNE/AA ratio was Cys (0.6) >> His (1×10^{-3}) > Lys (3×10^{-4}) >> Arg (4×10^{-5}). No reaction of HNE was detected with Glu. Similar results were obtained for the keto-derivative of HNE, 4-oxononenal: (Cys >> His > Lys > Arg) with Arg being a target for 4-oxononenal but generally not for HNE [93]. However, one case has been described (cytochrome c), where Arg^{38} is also modified by HNE besides His^{33} and Lys^{87} residues [94]. The above mentioned order does not imply that Cys residues are always the preferential targets of HNE in proteins. Several other factors besides reactivity may determine the adduction site, such as polarity of the microenvironment and accessibility in the tertiary/quarternary structure. For example it has been shown for HNE- and 4-oxo-2-nonenal-treated myoglobin and apomyoglobin, that the latter more "open" protein structure resulted in more extensive modification [95]. Changes in the polarity of the microenvironment may explain, why the presence of Arg in a model peptide increases the reaction rate of Cys for adduct formation with HNE as well as 4-oxononenal by a factor of 5–6 compared to the Cys nucleophile alone [93]. The impact of accessibility and acidity has been shown for actin. Among several Cys residues only Cys^{374} is adducted by HNE because of significant accessible surface and substantial thiol acidity due to the particular microenvironment surrounding Cys^{374} [96].

Furthermore, the presence of other ligands bound to proteins can strongly influence adduct formation with HNE. The *in vitro* modification of Cys^{120} of epithelial fatty acid-binding protein by HNE is markedly potentiated by noncovalently bound fatty acids [97].

Adduct formation of HNE with proteins is a rather fast reaction. The half life for the covalent modification of the epithelial fatty acid-binding protein was reported to be $t_{1/2} < 60$ s *in vitro* [97]. Adduct formation of HNE with protein thiols can be reversible in the presence of low molecular thiols such as GSH. Physiological (e.g., 4 mM) concentrations of GSH were capable of removing the HNE adducts from PDI [98]. It is likely this mechanism serves as a protective mechanism against inactivation by HNE and other lipid peroxidation products.

Protein adducts of HNE are physiological constituents of mammalian organisms. Notably these adducts occur in blood, where they increase with age [99]. In an age-dependence study on HNE-adducted serum proteins in Fischer 344 rats albumin, transferrin and immunoglobulins were most prominently adducted by HNE [100]. In a similar study HNE-adducted proteins have also been determined in serum of aging Fischer 344 rats [101]. Among these, 16 are involved in blood coagulation, lipid transport, blood pressure regulation, and protease inhibition.

Recent evidence suggesting a role for protein modification by HNE in the pathogenesis of several diseases has been reviewed in [102]. The precise mechanisms are currently unknown but likely result from adduction of proteins involved in cellular homeostasis or biological signaling. The extent of HNE adduction to proteins can increase dramatically under pathophysiological conditions. e.g., using a heterotropic rat heart transplantation model the formation of protein adducts of HNE was studied as a response to cold storage and warm blood reperfusion in the recipient [103]. In hearts submitted to ischemia only, the adduct formation was low (<1% of the tissue area). However, transplantation and reperfusion in the recipient increased the amount of protein adducts to about 6% and it was assumed that this damage negatively affects longterm survival of the transplant.

In a murine model for alcoholic liver disease 414 protein targets for modification by reactive aldehydes (HNE, 4-oxononenal, acrolein, and MDA) were identified by a hydrazide method coupled to a highly sensitive 2-dimensional liquid chromatography tandem mass spectrometry (2D LC-MS/MS or MuDPIT) technique [104].

Also for individual proteins a high degree of HNE adduction has been reported in a condition of oxidative stress. In progressive supranuclear palsy, a neurodegenerative disorder, which may possibly be induced by oxidative stress, two-thirds of total glutathione peroxidase in the cerebrospinal fluid was found to be adducted with HNE. It was assumed that the antioxidant system might be unable to function effectively because of conjugation with HNE [105].

Adduction by HNE tends to make the targeted protein more succeptible to degradation by the proteasomal pathway which is responsible for most intracellular proteolysis. While under normal conditions the proteasomal system is able to remove the bulk of oxidatively damaged and modified proteins, under severe oxidative stress the accumulation of modified proteins occurs. This takes place either due to cross-linking of proteins or due to malfunctioning of the proteolytic machinery of the cell. The efficiency of the removal of HNE-modified proteins clearly depends on the dose of cellular exposure to HNE. Taking into consideration the permanent formation of HNE and its metabolism it has been shown that low levels of HNE-modification are sufficient to increase the proteolytic susceptibility of proteins [106]. Equine ADH treated with a 2-fold molar excess of HNE was degraded by a rabbit reticulocyte lysate system approximately 1.5-fold faster than the control, while treatment with concentrations up to 100-fold molar excess of HNE were inhibitory to degradation [107].

On the other hand, the proteasome itself is a target for HNE adduction causing inhibition of proteolysis [108,109]. In the liver of rats which were fed ethanol for one month HNE—produced presumably by P450 cytochrome 2E1 (CYP2E1)—formed an adduct with the Rpt4 subunit of the 26 S proteasome [110]. It was proposed that this adduct could impede the association of 19 S and 20 S subunits and thus accounts for the observed decrease of proteasomal activity. In accordance with these data a decline in trypsin-like (but not chymotrypsin-like) activity of the proteasome in HepG2 cells overexpressing CYP2E1 was associated with higher levels of HNE-modified proteins compared to control cells [111]. An altered balance between protein modification, ubiquitination, and degradation was proposed for these cells.

Modification of another subunit of the proteasome has also been reported leading to a decline of chymotrypsin-like activity. A catalytic site-specific inhibition of the α6/C2 subunit of the 20 S proteasome by HNE was observed in liver [112]. Furthermore, the degradation of proteins by a proteasome-independent pathway may also be triggered by adduct formation with HNE as was shown for glyceraldehyde-3-phosphate dehydrogenase in U937 cells [113].

The reaction of HNE with proteins is frequently associated with their covalent crosslinking leading to the formation of fluorophores. A major fluorophore has been identified as a lysine-derived dihydro-pyrrole derivative [114]. Finally, it should be mentioned that protein-HNE adducts also represent useful biomarkers of oxidative stress, lipid peroxidation and oxidative homeostasis [115].

In the next chapter several well studied examples of peptides and proteins are described as targets of HNE including synaptic proteins such as synaptosomal-associated protein 25 [116].

5.1.1. Substrates

5.1.1.1. Glutathione

Glutathione (GSH), whose concentration in mammalian liver is in the millimolar range, is likely to be the quantitatively most important target of HNE. Several glutathione-S-transferases (GST) such as GSTA4-4 and hGST5.8 are involved in the conjugation of HNE to GSH in humans. The isoenzyme GSTA4-4 with high specificity for HNE has been found in human liver mitochondria. The specific mitochondrial GST activities toward HNE exceeded that observed in liver cytosol. These observations are suggestive of a role of GST in protecting against mitochondrial injury during the secondary phase of oxidative stress, or modulation of HNE-mediated mitochondrial signaling pathways [117].

The conjugation of HNE with glutathione is reversible. Both the spontaneous and the GST-catalyzed retro-Michael cleavages of HNE-GSH are the source of HNE in the urine of rats [118]. Strategies such as irrigation with gamma-glutamyl-cysteine to increase glutathione levels have been suggested as effective means for normalizing pathophysiological HNE levels after traumatic injuries [119].

GSTs (in humans hGSTA4-4 and hGST5.8) and the transporter 76 kDa Ral-binding GTPase activating protein (RLIP76) regulate intracellular concentrations of HNE through a coordinated action in various human cell lines and erythrocytes and catalyzes the ATP-dependent transport of the conjugate into the extracellular space, *i.e.*, a mild stress caused by heat, UV-A, or hydrogen peroxide with no apparent effect on the cells in culture caused a rapid, transient induction of hGST5.8 and RLIP76. These stress preconditioned cells acquired the ability to metabolize and exclude HNE at an accelerated pace and acquired relative resistance to apoptosis by UV and oxidative stress as compared to unconditioned control cells [120].

Patrick *et al.*, 2005 demonstrated that depletion of HNE in hGSTA4-4-transfected adherent HLE B-3 cells results in profound changes in gene expression, phenotypic transformation and immortalization [121]. At the genetic level this transfection caused changes in the expression of genes involved in cell adhesion, cell cycle control, proliferation, cell growth, and apoptosis, which is consistent with the phenotypic changes of the transformed cells. For example, the expression of p53, p21, p16, fibronectin 1, laminin gamma 1, connexin 43, Fas, integrinα6, TGFα, and c-jun was down-regulated, while the expression of protein kinase C beta II, c-myc, cyclin-dependent kinase 2, extracellular signal regulated kinase and transforming growth factor β1 was up-regulated [121].

The consequences of severe oxidative stress have also been studied in GSTA4-4 deficient mice. The acute hepatotoxicity of CCl_4 in GSTA4-4$^{-/-}$ mice was compared with wild type GSTA4-4$^{+/+}$ mice [122]. The results indicated that GSTA4-4 is an important component during the early stages (1–6 h) of cellular defense against oxidative stress and lipid peroxidation, although it is not effective in protecting against the degree of overall cell injury.

In addition chronic hepatotoxicity was studied in mGSTA4 null mice. Mice homozygous for the disrupted mGSTA4 allele were viable and appeared normal except for lower litter size, higher fat content in bones, and greater susceptibility to bacterial infection. The null mice had a lower survival time than wild-type controls when chronically treated with relatively low doses of paraquat. These mice had a reduced ability to conjugate HNE, and had an increased steady-state level of HNE in tissues [123].

The genotoxicity of HNE is highly dependent on cellular GSH status and those GSTs that contribute toward HNE conjugation, including hGSTA4-4. Depletion of GSH in HT29 cells, derived from a human colorectal adenocarcinoma, by BSO (L-buthionine-S,R-sulfoximine) decreased cellular GSH levels by 77% without significant changes in cell viability. Associated with this decrease was a 2-fold higher level of HNE-induced DNA damage as measured by the comet assay [124].

The glutathione adduct of HNE has signaling properties of its own [125]. For instance, it enhances the peritoneal leukocyte infiltration and stimulates the formation of proinflammatory lipid mediators. Moreover, the reduced form of the glutathione conjugate of HNE (GS-DHN) elicits strong mitogenic signaling in smooth muscle cells. Also, GSH-HNE is not only a product of adipocyte oxidative stress but also an activator of macrophage inflammation [126] and is pro inflammatory *in vivo* in mice [127].

5.1.1.2. Carnosine

Carnosine is a dipeptide (beta-alanyl-L-histidine) present in mammalian tissue and in particular at high concentration in skeletal muscle. Growing evidence indicates that it acts as quencher of reactive and cytotoxic carbonyl species including HNE [45,128]. The reaction mechanism was found to be an intra-molecular Michael addition. Two reaction products of carnosine were identified, in a pH-dependent equilibrium: (a) the Michael adduct, stabilized as a 5-member cyclic hemi-acetal and (b) an imine macrocyclic derivative. The adduction of carnosine to HNE thus appears to start with the formation of a reversible α,β-unsaturated imine, followed by ring closure through an intra-molecular Michael addition [128]. By trapping HNE in stable covalent adducts, carnosine can inhibit HNE-induced protein cross-linking.

The biological role of carnosine as a quencher of α,β-unsaturated aldehydes was verified by detecting carnosine-HNE adducts in oxidized rat skeletal muscle homogenates [129]. In a therapeutic approach it was shown that dietary carnosine prevents early atherosclerotic lesion formation in apolipoprotein E-null mice [130]. The facile Michael adduct formation of carnosine with 4-hydroxyalkenal species was explored and a sensitive, facile, shotgun lipidomics-based method was developed for quantification of these compounds directly from organic solvent lipid extracts of biological samples [131].

A perspective for the design of novel derivatives, active as exogenous agents able to detoxify carbonyl compounds has been proposed [132]. Histidyl-containing carnosine analogues bearing hydrazide or 1,2-diol moieties have been synthesized, some of which have demonstrated higher aldehyde-sequestering efficiency than carnosine and were also efficient in protecting SH-SY5Y neuroblastoma cells and rat hippocampal neurons from HNE-mediated death. The cytoprotective efficacy of these compounds suggests their potential use as therapeutic agents for disorders that involve excessive membrane lipid peroxidation and HNE-mediated neuronal toxicity [133]. Vistoli *et al.*, 2009 described a set of aryl derivatives that are characterized by high stability in human plasma and a quenching activity toward HNE, as a model of reactive carbonyl species, up to threefold greater than D-carnosine [134].

The synthesis and the physicochemical and biological characterization of a series of carnosine amides bearing alkyl substituents on the amido group endowed with different lipophilicity have been described by Bertinaria *et al.*, 2011 [135]. These products can be considered as metabolically stable analogues of carnosine and are worth of additional investigation as potential neuroprotective agents.

5.1.1.3. Thioredoxin

Thioredoxins (TRXs) are scavengers of intracellular ROS and participate in the anti-oxidant system of the retina. Disruption of these systems leads to dysfunction of retinal pigment epithelial (RPE) cells, which then accelerates the development of age-related macular degeneration (AMD). The role of TRXs in the protection against HNE-induced oxidative stress was studied in RPE cells by assessing the effect of TRX overexpression on cell viability, morphology, NFκB expression, and mitochondrial membrane potential [136]. It was found that overexpression of TRXs reduced cell death caused by HNE: HNE caused perinuclear NFκB accumulation, which was absent in TRX-overexpressing cells. Moreover, overexpression of TRXs prevented depolarization of mitochondrial membranes. TRX2 was more effective than TRX1 in maintaining the membrane potential. The difference in the protective effects of these TRXs against oxidative stress may be due to their expression profile. TRX2 was expressed in the mitochondria, while TRX1 was expressed in the cytoplasm. It was therefore suggested that the effect of TRXs on mitochondria may be a key to prevent oxidative stress in RPE cells [137].

5.1.1.4. Cytochrome c

The nature of cytochrome c modification by HNE was investigated in an *in vitro* study [138]. The overwhelming reaction observed is Michael addition by Lys side-chains in addition to the modification of His^{33}. While the Lys-HNE adducts were generally observed to be reversible, the HNE-His^{33} was found to be stable with half of the formed adduct surviving the denaturation and proteolysis protocols used to generate proteolytic peptides for LC-ESI-MS/MS.

5.1.2. Enzymes

When the amino acid targets of HNE are in the active site of an enzyme, the activity is frequently diminished. It was therefore proposed that HNE for example may disrupt the active site of lipoxygenase-1 by forming a Michael adduct with one or more of the three histidines that ligate the iron active site [139]. However, the inactivation of enzymes by HNE need not necessarily be due to the modification of the catalytic center but may be due to the selective modification of amino acids primarily located on the surface of the enzyme as has been shown for glyceraldehyde-3-phosphate dehydrogenase (GAPDH). HNE is a potent inhibitor of sulfhydryl enzymes such as the glycolytic enzyme GAPDH. It has therefore been suggested that HNE exerts an inhibitory effect on the enzyme due to the modification of Cys^{149} at the catalytic site generating the HNE-cysteine Michael addition-type adduct [140]. To test this hypothesis, ESI-MS of tryptic peptides was carried out and identified five peptides, which contained the HNE adducts at His^{164}, Cys^{244}, Cys^{281}, His^{327}, and Lys^{331} and revealed that both His^{164} and Cys^{281} were very rapidly modified within 5 min, followed by Cys^{244} at 15 min and His^{327} and Lys^{331} at 30 min. However, a modification of the catalytic center, Cys^{149}, by HNE was not observed [141].

As mentioned, the inhibition of enzymes is in some cases reversible in the presence of thiol-containing compounds such as glutathione [142,143] or acetylcysteine. For example, cysteine restores the activity of the pyruvate dehydrogenase complex [144] but in most instances irreversible destruction takes place which may also be site-specific [145]. Profound consequences are to be expected when HNE forms adducts both with the substrate and the enzyme of a biochemical reaction as is the case of the TRX/TrxR system. TrxR is most sensitive to inhibition by HNE. This seleno-enzyme regulates redox-sensitive proteins involved in inflammation and carcinogenesis, including ribonucleotide reductase, p53, NFκB, and others. HNE and also its precursor, the 15-lipoxygenase-1 product, 15(S)-hydroperoxy-5,8,11-*cis*-13-*trans*-eicosatetraenoic acid, inhibit TrxR with IC_{50} = 1 μM and 13 μM, respectively, in colo-rectal cancer cells [146]. Besides inhibition of TrxR irreversible inactivation of TRX occurs. A pair of Cys residues (Cys^{32}/Cys^{35}) in Trx and the selenocysteine (Sec) and Cys residues (Cys^{496}/Sec^{497}) in the active site of TrxR are primary targets for HNE modification [147]. Because the TRX system is one of the core regulatory enzymes of cellular functions, it is assumed that inhibition of both TrxR and TRX by HNE provides a possibly novel mechanism for the explanation of its cytotoxic effect and signaling activity, as well as the further damage indirectly caused under oxidative stress conditions.

However, HNE is also an inducer of TrxR [148]. It has been shown to exert an adaptive cytoprotective effect at low concentrations through induction of TrxR 1 via transcriptional activation of Nrf2. As a consequence pretreatment with HNE at sublethal concentrations significantly protected PC12 cells rat adrenal pheochromocytoma line, against the subsequent oxidative cell death induced by H_2O_2 and 6-hydroxydopamine. Since HNE also exhibited adaptive protection in human arterial endothelial cells far reaching conclusions were drawn from these findings. It was suggested that this may be a general effect of HNE and may lead to a reappraisal of the eventual role of ROS and LPO in organisms [149]. It has to be determined whether these complex effects of HNE—inhibitor on the one side, inducer on the other—can be explained by tissue-specific responses.

In a few instances activation rather than inhibition of enzymes may occur. Interestingly, in the case of a protease HNE is promoting reactions via four separate mechanisms under pathophysiological conditions: activation and induction of the enzyme, modification of the substrate and inhibition of expression of the substrate [150]. During the destruction of cartilage HNE is prominently produced in osteoarthritic (OA) synovial cells. *In vitro*, HNE binding to matrix metalloproteinase 13 (MMP-13) activated this enzyme, and modification of type II collagen (Col II) by HNE accelerated its degradation by MMP-13. The authors concluded that the increased level of HNE in OA cartilage and the ability of HNE to induce transcriptional and posttranslational modifications of Col II and MMP-13 suggest that HNE could play a role in OA [150].

In the following section several examples of enzymes from different enzyme classes and other proteins with pathophysiological significance are presented as targets of HNE. An overview over the enzymes identified to form adducts with HNE has been presented previously [4].

5.1.2.1. Oxidoreductases

5.1.2.1.1. Lactate Dehydrogenase

Lactate dehydrogenase (LDH) activity can regenerate NADH. Mass spectrometric examination revealed that HNE binds with cysteine and histidine residues of LDH at pH 5.6 and 7.4, and covalent binding of HNE decreases NADH formation [151].

5.1.2.1.2. Glyceraldehyde-3-Phosphate Dehydrogenase (GAPDH)

Schlisser *et al.*, 2009 conducted an investigation on organogenesis in mice with the goal to determine whether exposure of CD1 mice to the model teratogen hydroxyurea (HU) on gestation day nine generates region-specific HNE-protein adducts in the embryo, and to identify the proteins targeted [152]. They found out that the formation of HNE-protein adducts was elevated in the caudal region of control embryos and that HU exposure further increased HNE-protein adduct formation in this area. Interestingly, among the eight identified HNE-modified proteins, GAPDH appears to play a specific role since it was shown that HNE adducts reduced GAPDH enzymatic activity by 20%. The authors therefore concluded that GAPDH may play a role in stress response during development.

5.1.2.2. Transferases

5.1.2.2.1. Glutathione-S-Transferase (GST)

HNE is not only a substrate for GSTs, but also a modifier under pathophysiological conditions such as Alzheimer's disease (AD), a neurodegenerative disorder characterized pathologically by intracellular inclusions including neurofibrillary tangles and senile plaques. Oxidative stress is associated with the pathogenesis of the disease leading to oxidative modifications of cellular macromulecules such as lipids and proteins [153,154], and an increased formation of HNE [155]. Since the alpha class of GST can detoxify HNE and may interact with the multidrug resistance protein-1 (MRP1) to export the glutathione conjugate of HNE to confer cellular protection. Sultana and Butterfield, 2004 investigated oxidative modifications of GST and MRP1 in AD brain [156]. They found that HNE is covalently bound to GST and MRP1 proteins in excess in AD brain and suggested that HNE may be an important mediator of oxidative stress-induced impairment of this detoxifying system which thus may play a role in promoting neuronal cell death.

Besides modification HNE can induce the expression of GST via Nrf2. Nrf2 regulates a battery of antioxidative and phase II drug metabolizing/detoxifying genes via binding to the antioxidant response elements (ARE). Nrf2-ARE signaling plays a central role in protecting cells from a wide spectrum of reactive toxic species including reactive oxygen/nitrogen species (RONS). The role of Nrf2 in regulating the HNE induced gene expression of antioxidant and detoxifying enzymes was studied in HeLa cells [157]. When HeLa cells were treated with HNE, Nrf2 rapidly translocated into the nucleus and increased Nrf2 protein in the nuclear fraction causing a dose-dependent transcriptional activity of ARE followed by the expression of GST A4, AKR1C1 and heme oxygenase-1 (HO-1). Moreover, this induction was attenuated by knocking down Nrf2, highlighting the role of Nrf2 in mediating HNE induced expression of antioxidant and detoxifying genes.

GSTA4-4 has been postulated to be a rigid template that is preorganized for HNE metabolism. However, the combination of high substrate chemoselectivity and low substrate stereoselectivity is intriguing [158]. GSTA4-4 must metabolize both enantiomers of HNE to efficiently detoxify the biologically formed mixture and achieves this by an ideal location of the active site residue Arg15 enabling the interaction with the 4-hydroxyl group of either HNE enantiomer [158].

5.1.2.2.2. Liver Kinase B1 (LKB1)

LKB1 is characterized as a serine/threonine kinase that phosphorylates and activates the AMP-activated protein kinase (AMPK) and 12 other AMPK-related kinases. HNE treatment (10 μM for 1 h) of human embryonic kidney 293T (HEK293T) cells expressing LKB1 resulted in the formation of HNE-LKB1 adducts and decreased LKB1 kinase activity but had no effect on the association of LKB1 with its adaptor proteins sterile-20-related adaptor and mouse protein 25. Mutation of LKB1 Lys^{97} residue reduced HNE adduct formation and attenuated the effect of HNE on LKB1 activity. It was therefore concluded that adduction of LKB1 Lys^{97} mediates the inhibitory effect of HNE [159].

5.1.2.2.3. 5'-AMP-Activated Protein Kinase (AMPK)

AMPK is a major regulator of β-oxidation, which is altered in ALD. In an *in vitro* cellular model, AMPK was identified as a direct target of HNE adduction resulting in inhibition of H_2O_2- and 5-aminoimidazole-4-carboxyamide ribonucleoside (AICAR)-induced downstream signaling [160]. Using a murine model of ALD the authors were able to demonstrate that treatment with high concentrations of ethanol resulted in an increase in phosphorylated as well as carbonylated AMPKα. Mass spectrometry thereafter identified Michael addition adducts of HNE on Cys^{130}, Cys^{174}, Cys^{227}, and Cys^{304} on rAMPKα and Cys^{225} on rAMPKβ which by molecular modeling analysis of the adducted sites revealed an inhibition of AMPK by steric hindrance of the active site pocket and inhibition of hydrogen peroxide induced oxidation.

5.1.2.2.4. ZAK Kinase (Sterile Alpha Motif and Leucine Zipper Containing Kinase AZK)

ZAK kinase is a member of the MAPKKK family of signal transduction molecules. It contains an N-terminal kinase catalytic domain, followed by a leucine zipper motif and a sterile-alpha motif (SAM). It mediates gamma radiation signaling leading to cell cycle arrest and the activity of this protein plays a role in cell cycle checkpoint regulation. The protein also has pro-apoptotic activity.

ZAK kinase is adducted by HNE on a conserved, active site-proximal cysteine and the resulting enzyme inhibition generates a negative feedback mechanism that can suppress the activation of JNK pathways normally induced by oxidative stress [161].

5.1.2.2.5. Serine/Threonine-Protein Kinase AKT2 (Proteinkinase B2)

The effects of HNE on insulin-dependent stimulation of the AKT2 pathway have been evaluated by Shearn *et al.*, 2011 [162]. The regulatory effect of AKT2 is mediated through serine and/or threonine phosphorylation of a range of downstream substrates. Following HNE treatment, the level of downstream phosphorylation of Akt substrates such as glycogen synthase kinase-3-β (GSK3β) was significantly decreased in the hepatocellular carcinoma cell line HepG2, and this effect was shown to be mediated by Michael addition adducts of HNE with His^{196} and Cys^{311} of rAkt2 suggesting inhibition of GSK3β peptide binding in the Akt2 substrate binding pocket. This inhibition of Akt by HNE provides a novel mechanism for increased insulin resistance in ALD [162].

5.1.2.3. Hydrolases

5.1.2.3.1. ATP Synthase

Oxidative stress has been proposed as a mechanism for impaired beta-cell function in type 2 diabetes. Pancreatic islets from humans with type 2 diabetes were used to study the occurrence of HNE adducts in these cells [163]. The major HNE adduct was a 52-kDa protein seen with four different antibodies (two antibodies against HNE and two other antibodies generated against reactive small aliphatic compounds) that was also seen in islets of nondiabetic humans, rat islets, and insulinoma cells and in mitochondria of various rat tissues. It was identified as the β-chain of the mitochondrial F-ATP synthase, an enzyme responsible for 95% of ATP formed in tissues. The ATP synthase β-subunit has also been identified as a major target in isolated rat liver mitochondria upon HNE exposure [164].

5.1.2.3.2. Phosphatase and Tensin Homolog Deleted on Chromosome 10 (PTEN)

The tumor suppressor PTEN is a key regulator of Akt or protein kinase B (Akt/PKB) activation in hepatocytes, and plays a role in the etiology of alcoholic liver disease. PTEN negatively regulates AKT activation via its lipid phosphatase activity. PTEN is a phosphatidylinositol 3-phosphatase catalyzing the removal of the 3-position phosphate from phosphatidylinositol-3,4,5-trisphosphate [PtdIns(3,4,5)P-3] to produce phosphatidylinositol-4,5-bisphosphate.

Treatment of both the human hepatocellular carcinoma cell line (HepG2) and primary hepatocytes with subcytotoxic concentrations of HNE resulted in the activation of Akt within 30 min as demonstrated by increased phosphorylation of residues Ser^{473} and Thr^{308} [165]. This increased phosphorylation is accompanied by a 6-fold increase in total PtdIns(3,4,5)P-3 and an increased immunostaining at the plasma membrane after HNE treatment. However, PTEN lipid phosphatase activity is decreased due to the formation of a single HNE adduct with the active site cysteine in PTEN (Cys^{124}).

PTEN has also been used to demonstrate a new methodology to interrogate effects of reactive electrophiles on specific target proteins in cells [166]. In this study a target specific electrophile delivery platform was introduced that ultimately paves the way to interrogate effects of reactive electrophiles on specific target proteins in cells. This new methodology was demonstrated by photoinducible targeted delivery of HNE to the PTEN and Keap1. Covalent conjugation of the HNE-precursor to HaloTag fused to the target proteins enables directed HNE delivery upon photoactivation and allows a more precise determination of the pathophysiological consequences of HNE-induced protein modifications [166]. HaloTag again was designed as a modular protein tagging system that allows different functionalities to be linked onto a single genetic fusion, either in solution, in living cells, or in chemically fixed cells [167]. The protein tag (HaloTag) is a modified haloalkane dehalogenase designed to covalently bind to synthetic ligands (HaloTag ligands). The synthetic ligands comprise a chloroalkane linker attached to a variety of useful molecules, such as fluorescent dyes, affinity handles, or solid surfaces. Covalent bond formation between the protein tag and the chloroalkane linker is highly specific, occurs rapidly under physiological conditions, and is essentially irreversible. The utility of this system for cellular imaging and protein immobilization was demonstrated by analyzing multiple molecular processes associated with NFκB-mediated cellular physiology, including imaging of subcellular protein translocation and capture of protein-protein and protein-DNA complexes [167].

5.1.2.3.3. Sirtuin 3 (SIRT3)

SIRT3 is a mitochondrial class III histone deacetylase, which is inhibited by HNE via a thiol-specific modification [168]. HNE covalently modifies *rSIRT3 at* Cys^{280} altering the conformation of the zinc-binding domain and thus allosterically inhibits SIRT3 activity.

5.1.2.3.4. Cathepsins

Since reduced lysosomal capacity may contribute to lipofuscinogenesis and progressive dysfunction of the retinal pigment epithelium (RPE) during the pathogenesis of age-related macular degeneration Krohne *et al.*, 2010 investigated the effects of HNE and MDA on both isolated lysosomes from primary human retinal pigment epithelial cells (RPE) and on cultured RPE cells. Both HNE and MDA inhibited the cysteine proteases cathepsin B and L at a concentration of 1 μM by 88%–94% and this effect was due to HNE and MDA adducts in the active center of these proteases [169].

5.1.2.3.5. Neprilysin (NEP)

As one of the dominant amyloid-β peptide (Aβ) proteases, NEP plays a crucial role in maintaining a physiological balance between Aβ production and catabolism. Wang *et al.*, 2003 and 2009 showed that NEP is modified by HNE adducts, resulting in decreased activity in the brain of AD patients and cultured SH-SY5Y and H4 APP695wt cells [170,171]. The inactivation of NEP by HNE-adduction was associated with, at least partly, reduced Aβ cleavage and enhanced Aβ accumulation [171]. The HNE-induced modification and inactivation can be prevented by the carbonyl-scavenger *N*-acetylcysteine [172].

5.1.2.4. Lyases

5.1.2.4.1. Mitochondrial Aconitase (ACO2)

ACO2 is inactivated by HNE. Liu *et al.*, 2013 related the inactivation of ACO2 by HNE to structural features and determined the HNE addition reaction rates using the iTRAQ approach (isobaric Tags for Relative and Absolute Quantitation) [173]. The most reactive sites were Cys^{358}, Cys^{421}, and Cys^{424}, the three iron-sulfur cluster-coordinating cysteines, Cys^{99}, the closest non-ligated cysteine to the cluster, and Cys^{565}, which is located in the cleft leading to the active site.

Not only aconitase but also other enzymes of the citric acid cycle are targeted by HNE. Zhao *et al.*, 2014 identified several HNE-modified mitochondrial proteins in mouse heart mitochondria after doxorubicin (DOX) treatment [174]. The majority of the identified proteins were related to mitochondrial energy metabolism. These included proteins in the citric acid cycle and electron transport chain.

5.1.2.4.2. α-Enolase

The formation of HNE adducts with the multifunctional protein α-enolase was demonstrated by MS and confirmed by immunoblotting experiments, in HL-60 human leukaemic cells [175]. HNE caused a dose- and time-dependent reduction of the binding of plasminogen to α-enolase. As a consequence, HNE reduced adhesion of HL-60 cells to HUVECs (human umbilical vein endothelial cells) suggesting a potential role for HNE in the control of tumour growth and invasion.

Enolase is among the six enzymes which are excessively adducted by HNE in early AD (EAD) inferior parietal lobule (IPL) compared to age-related control brain [176]. These proteins play roles in antioxidant defense (manganese superoxide dismutase), neuronal communication and neurite outgrowth (dihydropyriminidase-related protein 2), and energy metabolism (α-enolase, malate dehydrogenase, triosephosphate isomerase, and F1 ATPase, α-subunit. These results are consistent with the hypothesis that LPO is an early event in the progression of AD.

5.1.2.5. Isomerases

5.1.2.5.1. Protein Disulfide Isomerase (PDI)

PDI is an abundant endoplasmic reticulum (ER)-resident chaperone and oxidoreductase that catalyzes formation and rearrangement (isomerization) of disulfide bonds, thereby participating in protein folding. PDI modification by HNE or oxidized LDL (oxLDL) inhibits its enzymatic activity and potentiates both ER stress (increased mRNA expression of the stress-regulated transcription factors CHOP (C/EBP homologous protein) and apoptosis by oxLDL [177].

5.1.2.5.2. Peptidyl-Prolyl *Cis/Trans*-Isomerase A1 (Pin1)

Pin1, an enzyme that catalyzes the conversion of the peptide bond of pSer/pThr-Pro moieties in signaling proteins from *cis* to *trans*, is highly susceptible to HNE modification. This was demonstrated by Aluise *et al.*, 2013 upon incubation of purified Pin1 with HNE leading to Michael adducts at the active site residues His^{157} and Cys^{113} [178]. Additionally the authors confirmed Cys^{113} adducts in the Pin1 active site in MDA-MB-231 breast cancer cells treated with tagged HNE (8-alkynyl-HNE) and showed that knockdown of Pin1 in MDA-MB-231 cells partially protected the cells from HNE-induced toxicity.

5.1.2.6. Ligases: Glutamine Synthetase

Glutamine synthetase is among the proteins modified by HNE in the retina as shown by Tanito *et al.*, 2006 [179]. By analyzing HNE-modified proteins increased after intense white light exposure nine proteins including voltage-dependent anion channel, enolase 1α, aldolase C, crystallins αA and βB3, heterogeneous nuclear ribonucleoprotein A2/B1, albumin, and glutamine synthetase were identified. The results indicate that HNE modifications of retinal proteins are specific to a particular set of proteins rather than random events on abundant proteins.

5.1.3. Carriers

5.1.3.1. Albumin

Toyokuni *et al.*, 2000 investigated the interaction of HNE with human serum albumin and found that it is rapidly quenched by human serum albumin (HSA) due to the covalent adduction to different accessible nucleophilic residues of the protein [180]. The molecular characterization of the covalent modifications led to the identification of eight Michael adducts and three Schiff bases involving nine nucleophilic sites. Cys^{34}, His^{146}, and Lys^{199} were found to be the most reactive HNE adduction sites. The latter adducted peptides were proposed as useful biomarkers of oxidative and carbonylation damage in humans [181]. In another MS investigation only those modified peptides were considered which were supported by high mass accuracy Orbitrap precursor ion measurements (high confidence hits) [182]. With HNE:HSA ratios of 1:1 and 10:1, 3 and 15 addition sites, respectively, were identified. This investigation confirmed previous work that Cys^{34}, the only free cysteine, is the most reactive residue in HSA.

HNE-modified HSA is highly immunogenic eliciting high titre immunogen specific antibodies as was shown for systemic lupus erythematosus (SLE), a chronic autoimmune disease, which is primarily characterized by increased levels of autoantibodies, predominantly against *double stranded DNA* [183]. A preferential binding of SLE autoantibodies to HNE-modified HSA as compared to native HSA or native DNA was shown and indicates a potential role of HNE-modified HSA in SLE etiopathogenesis.

5.1.3.2. Hemoglobin and Myoglobin

While hemoglobin (Hb) can be modified by HNE in cell-free experiments, HNE-Hb adducts were not detected in erythrocytes under oxidative stress induced by gamma-irradiation [184]. In myoglobin (Mb) HNE forms adducts with His residues and destabilizes Mb redox state, affecting meat colour [185]. LC-ESI-MS/MS of chicken Mb reacted with HNE identified covalent adduction of His^{64} and His^{93} at pH 7.4.

5.1.3.3. Liver and Adipocyte Fatty Acid-Binding Protein (FABP)

FABPs have been characterized as facilitating the intracellular solubilization and transport of long-chain fatty acyl carboxylates via noncovalent interactions. Sites of HNE adduction on mouse liver recombinant FABP (L-FABP) were mapped on apo (Lys^{57} and Cys^{69}) and holo (Lys^{6}, Lys^{31}, His^{43}, Lys^{46}, Lys^{57} and Cys^{69}) L-FABP [186]. These modifications revealed minor conformational changes in global protein structure of apo and holo L-FABP by molecular modeling simulations, but apparent differences were observed within the internal binding pocket.

The FABP of adipocytes is the first protein for which the crystal structure of its HNE adduct was determined. Recent work has shown that the adipocyte FABP is covalently modified by HNE *in vivo* on Cys^{117}. To evaluate HNE binding and modification, the crystal structures of adipocyte FABP covalently and noncovalently bound to HNE have been solved to 1.9 Å and 2.3 Å resolution, respectively [187]. While HNE in the noncovalently modified protein is coordinated similarly to a carboxylate of a fatty acid, the covalent form shows a novel coordination through a water molecule at the polar end of the lipid. Other defining features between the two structures with HNE and previously solved structures of the protein include a peptide flip between residues Ala^{36} and Lys^{37} and the rotation of the side chain of Phe^{57} into its closed conformation [187].

5.1.3.4. Apolipoprotein B-100 (ApoB)

The adduct formation of HNE and 4-oxo-nonenal (ONE) resp. with ApoB demonstrate that reactive aldehydes generated by LPO can differ in their biological effects, and that these differences can be mechanistically explained by the structures of the protein adducts formed [188]. In a study to understand the pathobiology of reactive lipid aldehydes the effects of HNE and ONE on the transport and secretion of very low-density lipoprotein was studied. Physiologically relevant concentrations of HNE and ONE rapidly disrupted cellular microtubules of HepG2 cells in a concentration-dependent manner. ONE reduced ApoB secretion while HNE did not significantly impair secretion. Both HNE and ONE formed adducts with ApoB protein, but HNE adducts were detectable as mono-adducts, while ONE adducts were present as protein-protein cross-links [188]. Interestingly another apolipoprotein, namely apolipoprotein A-I, the major HDL protein, is not modified by HNE *in vitro*, while MDA is adducted to lysine residues under the same conditions [189].

5.1.3.5. β-Lactoglobulin

No clear function has been identified for β-lactoglobulin, although it binds to several hydrophobic molecules, suggesting its role as a carrier. The strong suggestion is that the molecule exists primarily as a food source. β-Lactoglobulin was used as a model protein to study adduct formation *in vitro* during peroxidation of linolenic acid (LA) in the presence of Fe(II), and ascorbate [190]. By mass spectrometry the authors identified adducts including HNE-His Michael adducts, ONE-Lys 4-ketoamide, ONE-Lys pyrrolinone, and a Cys/His-ONE-Lys pyrrole cross-link. However, reversibly formed adducts, such as the HNE-Lys Schiff base were not present at detectable levels.

5.1.4. Transporters and Channels

5.1.4.1. Glutamate Transport Protein

HNE can inhibit glutamate transport *in vitro* and *in vivo*. Numerous *in vitro* and cell culture experiments indicate that oxidative damage decreases astrocyte glutamate transport activity. The hypothesis that LPO products impair glutamate and glucose transport *in vivo* has been tested by Ou *et al.*, 2002 [191]. Their findings indicate that LPO products that irreversibly modify protein lysyl residues cause a two- to sixfold elevation in extracellular glutamate in striatum and cerebral cortex of rats undergoing microdialysis and that LPO product-evoked extracellular glutamate appeared to be derived from nonneuronal sources. These results support the hypothesis that oxidative damage leads to inhibition of glutamate transport and thereby contributes to the progression of neurodegenerative diseases. Mattson and Chan [192] showed that in AD Aβ can induce membrane LPO and the formation of HNE. Aβ thereby impairs the function of glutamate and glucose transporters, membrane ion-motive ATPases, and can enhance calcium influx through voltage-dependent and ligand-gated calcium channels. In sporadic amyotrophic lateral sclerosis (ALS) patients increased modification of proteins by HNE has been reported in the lumbar spinal cord compared to neurologically normal controls. Biochemical analysis revealed that one of the proteins modified by HNE was the astrocytic glutamate transporter EAAT2 [193]. The authors conclude that the function of proteins modified by HNE can be severely compromised leading to impairment of glutamate transport, and excitotoxic motor neuron degeneration in ALS.

For a homolog of HNE, 4-hydroxyhexenal, similarly the impairment of glutamate transport in astrocytes has been demonstrated [194].

5.1.4.2. α-Synuclein (α-Syn)

α-Syn is a small protein that is abundant in various regions of the brain [195]. Although the physiological function of α-Syn is not well understood, its function is likely to involve presynaptic vesicle pool size and neurotransmitter release [196,197], and vesicle recycling [198]. One of the pathological hallmarks of Parkinson disease (PD) is the presence of intracellular inclusions called Lewy bodies that consist of aggregates of α-Syn (for review see [199]). Xiang *et al.*, 2013 [200] investigated posttranslational modifications (PTMs) of α-Syn caused by oxidative stress, including modification by HNE (HNE-α-Syn), nitration (N-α-Syn), and oxidation (O-α-Syn), which have been reported to promote oligomerization of α-Syn and found that modification of α-Syn by HNE increased dopaminergic toxicity by increasing the interaction of extracellular α-Syn with neurons. In another study [201], Bae *et al.*, 2013 investigated the oligomerization of recombinant human α-Syn via HNE adduct formation at the lysine and histidine residues and found that (a) HNE-induced α-Syn oligomers are distinct from amyloid fibrils at both conformation and ultrastructure levels; (b) the HNE-induced oligomers are capable of seeding the amyloidogenesis of monomeric α-Syn under *in vitro* conditions; (c) after treatment with HNE both the translocation of α-Syn into vesicles and the release of this protein from cells were increased. The HNE- or ONE-induced α-Syn oligomers have been characterized by Nasstrom *et al.*, 2011 [202]. Both oligomers are rich in beta-sheet structure and have a molecular weight of about 2000 kDa. Atomic force microscopy analysis revealed that neither oligomer type polymerized into amyloid-like fibrils despite prolonged incubation. According to findings of Shibata *et al.*, 2010 both ONE- and HNE-induced α-Syn oligomers were cytotoxic when added exogenously to a neuroblastoma cell line, but HNE-induced α-Syn oligomers were taken up by the cells to a significantly higher degree [203].

5.1.4.3. Sarco/Endoplasmic Reticulum Ca^{2+}-ATPase (SERCA1a)

HNE has a dual effect on Ca^{2+} transport through sarcoplasmic reticulum membranes from rabbit fast-twitch skeletal muscle as reported by Hortigon-Vinagre *et al.*, 2011 [204]. The authors showed that exposure of the membranes to HNE resulted in inhibition of the maximal ATPase activity and Ca^{2+} transport ability of SERCA1a, the Ca^{2+} pump in these membranes, reduced ATP binding and phosphoenzyme formation from ATP, whereas Ca^{2+} binding to the high-affinity sites was altered to a lower extent, and that HNE reacted with Lys^{515} within the nucleotide binding pocket of SERCA1a.

5.1.4.4. Transient Receptor Potential Vanilloid 1 (TRPV1)

TRPV1 is a nociceptive, Ca^{2+}-selective ion channel activated by capsaicin, heat, and protons. The function of TRPV1 is detection and regulation of body temperature. In addition, TRPV1 provides sensing of scalding heat and pain (nociception). It has been reported that activation of TRPV1 expressed in esophageal mucosa is involved in gastroesophageal reflux disease (GERD) or in nonerosive GERD symptoms [205]. Capsaicin has been shown to significantly increase the production of HNE-modified proteins in Het1A cells, and IL-8 production in capsaicin-stimulated Het1A cells was enhanced by synthetic HNE treatment. By analyzing protein modifications it turned out that TRPV1 was modified by HNE [206]. It was therefore concluded that TRPV1 functions in chemokine production in esophageal epithelial cells, and this function may be regulated by HNE via posttranslational modification of TRPV1.

5.1.4.5. Dopamine Transporter

LoPachin *et al.*, 2009 characterized the synaptosomal toxicity of HNE and evaluated the role of putative nucleophilic amino acid targets and were able to show that HNE exposure of striatal synaptosomes inhibited (3)H-dopamine membrane transport and vesicular storage which was paralleled by decreases in synaptosomal sulfhydryl content [206]. Furthermore they demonstrated that the sulfhydryl thiolate state was the HNE target, that the rate of adduct formation was pH-dependent and that *N*-acetyl-L-cysteine, but not *N*-acetyl-L-lysine or β-alanyl-L-histidine, reduced *in vitro* HNE neurotoxicity.

5.1.5. Receptors

5.1.5.1. Platelet-Derived Growth Factor Receptor-β (PDGFR-β)

The PDGFR-β signaling pathway regulates smooth muscle cell (SMC) migration and proliferation and plays a role in the vascular wall response to injury. HNE and oxidized low-density lipoprotein (oxLDL) induce a dual effect on PDGFR-β signaling [207]. Short-term incubation of SMC with HNE or oxLDL induces PDGFR-β pathway activation via the formation of PDGFR-adducts and oxidative stress increase. In contrast, long-term incubation triggers a desensitization of PDGFR to its own agonist, leading to a progressive inhibition of PDGF-β-mediated signalling and proliferation, resulting from decreased PDGF binding and inhibition of PDGFR-β tyrosine kinase activity. PDGFR inhibition was associated with increased formation of HNE- and acrolein- PDGFR-adducts. PDGFR-β adducts were detected in aortae of apolipoprotein-deficient mice, in hyper-cholesterolemic rabbits and in human carotid plaques. The aldehyde scavengers dinitrophenylhydrazine and hydralazine prevented both HNE- and oxLDL-induced structural modification and PDGFR-β signaling dysfunction in cells and *in vivo* [207]. In conclusion, PDGFR-β acts as a sensor for both oxidative stress and oxidized lipids in atherosclerotic areas and its progressive inhibition by HNE or acrolein may contribute to defective SMC proliferation, and decrease the stability of a vulnerable plaque.

5.1.5.2. Lectin-Like Oxidized Low-Density Lipoprotein Receptor-1 (LOX-1)

LOX-1 is an endothelial scavenger receptor that is important for the uptake of oxLDL. However, the precise structural motifs of oxLDL that are recognized by LOX-1 are unknown. By investigating the ability of BSA modified by lipid peroxidation to compete with AcLDL (acetylated low-density lipoprotein), Kumano-Kuramochi *et al.*, 2012 showed that HNE-modified proteins most potently inhibited the uptake of AcLD, that both the modification of BSA and the oxidation of LDL resulted in the formation of HNE-histidine Michael adducts, that the HNE-histidine adduct inhibited the uptake of AcLDL in a dose-dependent manner, and that the HNE-histidine adduct stimulated the formation of ROS and activated extracellular-signal-regulated kinase 1/2 (ERK 1/2) and NFκB thus initiating endothelial dysfunction and leading to atherosclerosis [208].

5.1.5.3. Toll-Like Receptor 4 (TLR4)

Toll-like receptors (TLRs) detect invading microbial pathogens and initiate immune responses as part of host defense mechanisms. They also respond to host-derived substances released from injured cells and tissues to ensure wound healing and tissue homeostasis. Dysregulation of TLRs increases the risk of chronic inflammatory diseases and immune disorders. Inflammatory events are often accompanied by oxidative stress which generates lipid peroxidation products such as HNE. Therefore, Kim *et al.*, 2009 investigated whether HNE affects TLR activation and found that HNE blocked LPS (a TLR4 agonist)-induced activation of NFκB and IRF3 as well as expression of IFN-β, IP-10, RANTES and TNFα [209]. Furthermore, they were able to demonstrate that HNE suppressed both ligand-induced and ligand-independent receptor dimerization and that HNE formed adducts with cysteine residues of synthetic peptides derived from TLR4 suggesting that the reactivity of HNE with sulfhydryl moieties is implicated in the inhibition of TLR4 activation. Since inhibition of TLR4 activation by HNE also resulted in down-regulation of the phagocytic activity of macrophages it can further be concluded that HNE blocks TLR4-mediated macrophage activation and phagocytic functions.

5.1.6. Cytoskeletal Proteins

5.1.6.1. Tau Proteins

Tau proteins are microtubule-associated proteins found in neurons in the brain. These proteins interact with tubulin to stabilize microtubules and promote tubulin assembly into axonal microtubules. Tau has two ways of controlling microtubule stability: isoforms and phosphorylation. The formation of tau-containing neurofibrillary tangles is a major feature of AD and other neurodegenerative diseases. Fibers are correlated with disease severity and they have been implicated as playing a direct role in disease pathophysiology. Biochemical findings show that tau oxidative modifications are regulated by phosphorylation and that tau found in neurofibrillary tangles is oxidatively modified, suggesting that it is closely linked to the biology, not toxicity, of AD [210]. Several studies support a link between tau protein phosphorylation and adduction of tau by reactive carbonyls. The phosphorylation-dependent adduction of tau by carbonyl products resulting from LPO creates the neurofibrillary tangle-related antigen, Alz50. To determine whether epitopes of carbonyl-modified tau are major conformational changes associated with neurofibrillary tangle formation, seven distinct antibodies raised against neurofibrillary tangles have been examined that recognize unique epitopes of tau in AD. Consistently, all seven antibodies recognize tau more strongly (4- to 34-fold) after treatment of normal tau with HNE, but only when tau is in the phosphorylated state [210]. *In vitro* HNE as well as several quinones facilitated the phosphorylated tubulin binding of tau into fibrillar polymers which have a morphology very similar to that of paired helical filaments present in the brains of AD patients [211]. These findings not only support the idea that oxidative stress is involved in neurofibrillary tangle formation occurring in brains of AD patients, but also show that HNE modifications of tau promote and contribute to the generation of the major conformational properties defining neurofibrillary tangles [210].

5.1.6.2. Ankyrin

Ankyrin is an adaptor protein in the membrane cytoskeleton of erythrocytes. The modification of ankyrin is pathologically relevant in malaria [212], however, the molecular basis for the prevalence of blood group 0 in regions where malaria is endemic remains unclear [212]. Mendez *et al.*, 2012 [212] therefore investigated whether there are differences in carbonylated membrane proteins between the different blood groups and found that group 0 blood showed a reduced protein oxidation pattern compared to groups A, B and AB. By examining HNE modified proteins, ankyrins were found to be differentially carbonylated in group 0 upon malaria infection as compared to A and B groups.

5.1.6.3. Spectrins

Both and spectrin of the cytoskeleton of erythrocytes (RBC) are targets of HNE [213]. Spectrin strengthens the RBC membrane through its direct association with membrane lipids and through protein-protein interactions. Spectrin loss on the other hand reduces membrane stability and results in various types of hereditary spherocytosis. However, less is known about acquired spectrin damage. Arashiki *et al.*, 2010 were able to show that α- and β-spectrin in human RBC are the primary targets of HNE, *i.e.*, the level of HNE adducts in spectrin (particularly α-spectrin) was increased following HNE treatment of RBC membrane ghosts [213]. In contrast, ghost preparation in the presence of MgATP reduced HNE adduct formation, with preferential β-spectrin modification and increased cross-linking of the HNE-modified spectrins resulted in selective HNE-spectrin adduct formation. These findings were interpreted in terms of preferential HNE adduction in spectrin at the interface between the skeletal proteins and lipid bilayer. A combined protocol of immune-detection, peptide enrichment, mass spectrometry, and *de novo* protein sequencing was applied to study the adduction sites in human β-spectrin [214]. HNE-lysine and HNE-histidine Michael adducts were detected in β-spectrin under physiological conditions.

5.1.7. Chaperones: Heat Shock Proteins 70 and 90

In a rat model of chronic alcohol-induced oxidative stress [91] Carbone *et al.*, 2004 investigated the modification of both cytosolic Hsp72, the inducible variant of Hsp70, and Hsp90 by HNE. By applying mass sprectrometrical analysis they were able to show that Hsp72 treated with 10 and 100 μM HNE caused adduct formation at Cys^{267} in the ATPase domain of the chaperone. Cys^{572} was found to be a site for HNE modification in Hsp90 [90]. HNE adducted Hsp70 has been identified in neuronal necrotic cell death [215] and in ALS an age-related, fatal motor neuron degenerative disease [216]. Although multiple mechanisms contribute to the pathogenesis of motor neuron injury in ALS, it is likely that oxidative stress plays a significant role in the amplification, and possibly the initiation, of the disease. HNE levels are increased in spinal cord motor neurons of ALS patients, indicating that LPO is associated with the motor neuron degeneration in ALS. Three significantly HNE-modified proteins were found in the spinal cord of G93A-SOD1 transgenic mice: besides Hsp70 dihydropyrimidinase-related protein 2, and possibly α-enolase.

HNE-modified Hsp70 is much more vulnerable to calpain cleavage [217]. The calpain-cathepsin cascade is mediating necrotic neuronal death from simpler organisms to primates. The main event of this cascade is calpain-mediated lysosomal rupture and the resultant release of lysosomal cathepsins into the cytoplasm. However, the *in vivo* substrate of calpain for inducing lysosomal destabilization still remains completely unknown. Recent data obtained with post-ischemic hippocampal CA1 tissues and glaucoma-suffered retina from primates suggests that Hsp70.1 might be the *in vivo* substrate of activated μ-calpain at the lysosomal membrane of neurons. Hsp70.1 is known to stabilize lysosomal membrane by recycling damaged proteins and protect cells from oxidative stresses [217]. Sahara *et al.*,

2010 studied the molecular interaction between activated μ-calpain and the lysosomal Hsp70.1 in monkey hippocampal CA1 neurons after ischemia-reperfusion insult and found that Hsp70.1 in CA1 tissue is an *in vivo* substrate of activated μ-calpain and that Hsp70.1 carbonylated by HNE or hydrogen peroxide is much more vulnerable to calpain cleavage [217]. These data further suggest that Hsp70.1 can become a target of the carbonylation by HNE, and Hsp70.1 is a modulator of calpain-mediated lysosomal rupture/permeabilization after ischemia-reperfusion injury.

The adduction of Hsp90 by HNE has been studied by protein-selective capture [218] in RKO colorectal cancer cells and identified His^{450} and His^{490} adducts of Hsp90α. Five other histidine residues were also adducted on Hsp90β: His^{171}, His^{442}, His^{458}, His^{625} and His^{632}.

Both Hsp70 and Hsp90 are together with PDI, liver FABP and the protein-serine/threonine kinases ERK1 and ERK2 targets of HNE in early stages of ALD [219]. The adduction leads to inhibition of these enzymes and it was concluded that inhibition of Hsp70, Hsp90 and PDI function could be involved in initiation of the early phases of ER stress contributing to stimulation and accumulation of hepatic lipids. Likewise, impairment of L-FABP activity could also disrupt lipid transport contributing to steatosis. The modification and inhibition of Erk1/2 by HNE may in addition contribute to the decreased hepatocellular proliferation associated with ALD [219].

The electrophilic adduction of Hsp70 and Hsp90 by HNE appears to mediate in part the activation of heat shock factor 1 (HSF1), which is normally maintained in an inactive cytosolic complex [220]. This is one of the mechanisms, by which cellular responses to HNE are elicited, as was revealed by a systems analysis approach of protein modification and cellular responses induced by electrophile stress aiming to define the chemistry of protein modification and its biological consequences using lipid-derived α,β-unsaturated aldehydes such as HNE as model electrophiles. In this global approach, two large data sets were analyzed: one represented the identity of proteins modified over a wide range of electrophile concentrations, and the second comprised changes in gene expression observed under similar conditions. Informatics tools showed theoretical connections based primarily on transcription factors hypothetically shared between the two data sets, downstream of adducted proteins and upstream of affected genes [220]. This technique identified a multitude of HNE targets such as several transcription factors as potential mediators of the cellular response to HNE-adducted proteins. Among these, HSF1 was confirmed as a sensitive and robust effector of HNE-induced changes in gene expression. Among the genes induced by HSF1, Bcl-2- associated athanogene 3 (BAG3) is notable for its actions in promoting cell survival through stabilization of antiapoptotic Bcl-2 proteins, thus having a critical role in mediating cellular protection against electrophile-induced death [220].

The ER homolog of HSP70 is glucose-regulated protein 78 (GRP78). It plays a critical role in the cellular response to ER stress by serving as a chaperone assisting protein folding and by regulating the signaling of the unfolded protein response (UPR). GRP78 revealed a marked propensity for Lys and His adduction by HNE within the ATPase domain and a relative paucity of adduct formation within the peptide-binding domain [221]. Consistent with these findings, a concomitant dose-dependent decrease in ATP-binding and ATPase activity was observed without any discernible impairment of chaperone function.

5.1.8. Uncoupling Proteins 2 and 3 (UCP2 and UCP3)

Most studies with HNE were focused on mitochondrial UCP2 and UCP3. Mitochondria are potent producers of cellular superoxide from complexes I and III of the electron transport chain, and mitochondrial superoxide production is a major cause of the cellular oxidative damage that may underlie degradative diseases and aging. This superoxide production is very sensitive to the proton motive force, so it can be strongly decreased by mild uncoupling of oxidative phosphorylation. Superoxide and the downstream LPO products including hydroxyalkenals such as HNE, are potent activators of proton conductance by mitochondrial uncoupling proteins such as UCP2 and UCP3, although the mechanism of activation has yet to be established. It was therefore suggested that superoxide releases iron from aconitase, leading to LPO and the release of molecules such as HNE that covalently modify and activate the proton conductance of UCPs and other proteins such as the adenine nucleotide translocase [222]. UCP2 and UCP3 do not catalyse proton leak in the absence of such acute activation. They can also catalyse export of fatty acid and other anions, although the relationship of anion transport to proton transport remains controversial [223]. The notion that HNE plays a role in the function of UCPs to minimize superoxide formation is supported by a study on plant mitochondria. HNE and a structurally related compound, trans-retinal, stimulated proton conductance in potato mitochondria, that is inhibitable by GTP, a characteristic of UCP. Proof that the effect of HNE and trans-retinal are mediated by UCPs was provided by examination of proton conductance in transgenic plants overexpressing UCP. An increase in UCP content resulted in a modest but significant decrease in the rate of superoxide production [224]. A hypothesis has been forwarded for the main function of uncoupling proteins: to cause mild uncoupling to diminish mitochondrial superoxide production, hence protecting against disease and oxidative damage at the expense of a small loss of energy. This simple feedback loop would constitute a self-limiting cycle to protect against excessive superoxide production. A more recently evolved role of UCP2 is perhaps as part of a signaling pathway to regulate insulin secretion in pancreatic beta cells [222,225,226].

Recently it was also shown that fatty acids are key players in HNE-mediated activation of UCP1 and UCP2 [227]. The molecular mechanisms of HNE action were investigated by evaluating the separate contributions of lipid and protein phases of the membrane and by comparing UCP1 and UCP2, which were reconstituted in planar lipid bilayers. Via this approach it was demonstrated that HNE does not directly activate either UCP1 or UCP2. Instead, it strongly potentiates the membrane conductance increase (G(m)) mediated by different long-chain fatty acids in UCP-containing and in UCP-free membranes. This G(m) increase is concentration-dependent and exhibits a typical saturation kinetics. In addition, for the amoeba *Acanthamoeba castellani* it was shown that the HNE induced UCP-mediated mitochondrial uncoupling is GTP-sensitive [228].

HNE is able to increase the proton conductance of the inner mitochondrial membrane through effects not only on uncoupling proteins but also the adenine nucleotide translocase (ANT). To clarify the relative contribution of the two carriers to these effects mitochondria were isolated from skeletal muscle and heart of wild-type and Ucp3 knockout (Ucp3KO) mice [229]. To increase UCP3 expression, some mice were i.p. injected with LPS which did not change basal proton conductance. HNE increased the proton conductance of skeletal muscle and heart mitochondria. In skeletal muscle, this increase was lower in Ucp3KO mice and higher in LPS-treated wild-type mice, and was partially abolished by GDP (UCPs inhibitor) and completely abolished by carboxyatractylate (ANT inhibitor) or addition of both inhibitors. GDP had no effect on HNE-induced conductance in heart mitochondria, but carboxyatractylate or administration of both inhibitors had a partial effect. GDP-mediated inhibition of HNE-activated proton conductance in skeletal muscle mitochondria was not observed in Ucp3KO mice, indicating that GDP is specific for UCP3, at least in muscle. It was therefore concluded that, in skeletal muscle, HNE-induced increase in proton conductance is mediated by UCP3 (30%) and ANT, whereas in the heart the increase is mediated by ANT and other carriers, possibly including UCP3 [229].

5.1.9. Growth Factors: Platelet-Derived Growth Factor (PDGF)

Jin *et al.*, 2013 investigated breast cancer-associated changes of HNE, GSH, nitrotyrosine and halotyrosine adducts in 27 secreted proteins, for a total of 108 candidate biomarkers in plasma samples for both cases and benign controls and identified HNE-modified PDGF and GSH-modified ceruloplasmin which were significantly altered in samples from cancer patients [230]. The authors concluded that these changes may reflect redox changes in breast cancer, and that HNE-modified PDGF may be useful as plasma biomarker in breast cancer.

5.1.10. Peptide Hormones

5.1.10.1. Insulin

Human insulin is a target of both HNE and 4-hydroxy-hexenal (HHE) [231]. This was demonstrated by incubation of insulin in the presence of HNE or HHE and adduct analysis. The formation of covalent adducts on insulin was analyzed by mass spectrometry analysis and revealed one to five Michael adducts on insulin. Furthermore the authors report that glucose uptake in 3T3-L1 and L6C5 cells was significantly reduced after treatment with adducted insulin compared to native insulin, indicating that the formation of HNE- and HHE-Michael adducts significantly disrupted the biological activity of insulin.

5.1.10.2. Angiotensin II

Another peptide hormone adducted by HNE is angiotensin II [232]. The octapeptide angiotensin II is the primary active hormone of the renin/angiotensin system (RAS) and has been implicated in various cardiovascular diseases. Numerous structure activity relationship studies have identified Asp^1, Arg^2, and His^6 of Ang II to be critical for its biological activity and receptor binding [232]. Subsequently, it has been elaborated that the oxidative modifications on the *N*-terminus of Ang II disrupt interactions with Ang II type 1 receptor and aminopeptidase A (which cleaves the N-terminal Asp residue of Ang II to generate Ang III), which could affect the regulation of cardiovascular function [232].

5.1.11. Extracellular Matrix Proteins: Collagen

HNE forms adducts with type II collagen (Col II). El-Bikai *et al.*, 2010 addressed the question whether interactions between human osteoarthritic chondrocytes and HNE-modified Col II affect cell phenotype and functions and to determine the protective role of carnosine (CAR) treatment in preventing these effects [233]. Their major findings were that Col II modification by HNE at a molar ratio (MR) approximately 1:20, strongly induced ICAM-1, α1β1-Integrin and metallo-proteinase-13 (MMP-13) expression as well as extracellular signal-regulated kinases 1 and 2 (ERK1/2) and NFκB-p65 phosphorylation without having an impact on cell adhesion and viability or Col II expression. In contrast, Col II modification with HNE at MR approximately 1:200, altered chondrocyte adhesion by evoking cell death and caspase-3 activity, and the inhibition of α1β1-integrin and Col II expression as well as ERK1/2 and NFκB-p65 phosphorylation. At the same time a release of PGE_2, expression of COX-2 and p38 MAPK phosphorylation were observed. All these effects were prevented by CAR, an HNE-trapping drug (see chapter 5.1.1.2., p.20). The authors concluded that HNE-binding to Col II results in multiple abnormalities of chondrocyte phenotype and function, suggesting its contribution in osteoarthritis development [233].

5.1.12. Histones: Histone-H2A

Histones are DNA associated nucleoproteins, which adopt different structures under oxidative stress. A study was undertaken to test the role of HNE-modified histone-H2A (HNE-H2A) in *systemic lupus erythematosus* (SLE) [234]. The data revealed that HNE-mediated LPO in histone-H2A caused alteration in histidine, lysine and cysteine residues. In addition, protein carbonyl contents were also high in HNE-H2A. The specificity of autoantibodies from SLE patients were analyzed towards HNE-H2A and their results were compared with sex- and age-matched controls. SLE autoantibodies showed preferential binding to HNE-H2A in comparison with histone-H2A. In addition, HNE-H2A was also detected in SLE peripheral blood mononuclear cells suggesting a likely role of HNE-H2A in the initiation/progression of SLE [234].

5.2. Reactions with Lipids

The phospholipids phosphatidyl-ethanolamine (PE) and—to a lesser extent—phosphatidyl-serine, which contain a primary amino group, can undergo both Michael addition and Schiff-base formation with HNE [235]. The Schiff-base of PE partly undergoes cyclization to a pyrrole derivative, and PEs modified by HNE and other aldehydes induce monocyte adhesion to cultured endothelial cells.

LPO generates a large family of aldehyde-modified PEs (al-PEs), many of which have the potential to drive inflammation [236], *i.e.*, gamma-ketoaldehydes (isolevuglandins, IsoLGs) form inflammatory mediators by modifying the ethanolamine headgroup of PEs. When the ability of al-PEs was tested to induce THP-1 monocyte adhesion to cultured endothelial cells, it was found that PEs modified by HNE, ONE and MDA induced adhesion with potencies similar to those of PEs modified by IsoLGs, while PEs modified by acrolein or by glucose were only partial agonists for adhesion [236].

5.3. Reactions with Cofactors and Vitamins

5.3.1. Vitamin C (Ascorbic Acid)

Of interest is the observation that HNE, like other LPO-derived electrophiles, can undergo ascorbylation. Vitamin C can form a Michael-type conjugate with HNE *in vitro* [237,238]. The reaction has been described as a Michael addition of an activated C-H group of ascorbic acid to the C=C double bond of HNE followed by cyclisation to produce a hemiacetal [48]. Ascorbic acid is able to promote the detoxification and elimination of HNE in human monocytic THP-1 leukemia cells [239]. In this process adduct formation of ascorbic acid with HNE is only a minor pathway. It was therefore suggested that the protective effect of ascorbate against HNE cytotoxicity is largely via modulation of multidrug resistant protein (MRP)-mediated transport of GSH-HNE conjugate metabolites.

5.3.2. Pyridoxamine

Also the cofactor pyridoxamine is an HNE target. In search for new scavengers of carbonyl compounds, the amino group of pyridoxamine was demonstrated to react faster with HNE than the amino group of *N*-acetyl lysine, but much slower than a series of 4-ketoaldehydes with oxopentanal as model compound [240].

5.3.3. Lipoic Acid

The reduced form of lipoic acid contains two thiol groups which can be adducted by HNE. Both lipoic acid and its redox enzyme lipoamide dehydrogenase (LADH) are targets of HNE in AD brain. The AD brain shows increased levels of LPO products, including HNE. The levels of the HNE-modified lipoic acid in brain of subjects with AD and age-matched controls were therefore measured by Hardas *et al.*, 2013 [241]. Lipoic acid is a key co-factor for a number of proteins including pyruvate dehydrogenase and α-ketoglutarate dehydrogenase. Decreased levels of HNE-lipoic acid in the AD brain were observed in this study compared to that of age-matched controls. It was further demonstrated that both LADH levels and activities were significantly reduced in AD brain compared to age-matched control. The authors conclude that these data are consistent with a two-hit hypothesis of AD: oxidative stress leads to LPO that, in turn, causes oxidative dysfunction of key energy-related complexes in mitochondria, triggering neurodegeneration, and they suggest that lipoic acid supplementation could be a potential treatment for the observed loss of cellular energetics in AD to potentiate the antioxidant defense system to prevent or delay oxidative stress in and progression of this disorder [241].

5.4. Reactions with Nucleic Acids

The current knowledge on DNA damage induced by endogenously produced reactive aldehydes such as HNE, MDA, acrolein, crotonaldehyde and methylglyoxal in relation to the pathophysiology of human diseases has been reviewed by Voulgaridou *et al.*, 2011 [242]. Minko *et al.*, 2009 [243] gave an overview of the chemistry and biology of DNA containing 1,N^2-deoxyguanosine adducts of the α,β-unsaturated aldehydes HNE, acrolein, and crotonaldehyde.

HNE is genotoxic, e.g., for hepatocytes [244] and cerebral endothelial cells [245]. HNE in a concentration of 0.1 μM causes elevated levels of sister chromatid exchanges in primary hepatocytes, 1 μM causes elevated levels of chromosomal aberrations and at 10 μM the formation of micronuclei is observed. HNE is thought to contribute to the low level of DNA adducts which are abundant both in untreated rodent and human genomes. Two pathways are considered to be responsible for the mutagenicity of HNE: one route is the formation of an adduct by direct interaction with the guanosine moiety of

DNA (Figure 4). The reaction of HNE with DNA gives four diastereomeric 1,N^2-γ-hydroxypropano adducts of deoxyguanosine; background levels of these adducts have been detected in animal tissue. Stereospecific synthesis of these four adducts at the nucleoside level have been accomplished. In addition, a versatile strategy for their site-specific incorporation into oligonucleotides has been developed [246].

Reaction of HNE with calf thymus DNA resulted in a pair of diastereomeric adducts. Background levels of the 1,N^2-propane adduct of guanine (HNE-dG) in rat tissues are in the range of 18–158 adducts per 10^9 nucleotides with relatively high levels in the liver [247]. The guanine adducts can form interstrand cross-links. When placed opposite dC in the 5'-CpG-3' sequence, the (6S,8R,11S) diastereomer of the HNE-dG forms a N^2-dG:N^2-dG interstrand cross-link [246,248]. Its structure was refined by Huang *et al.*, 2010 and 2011 showing that the crosslink is a N^2-dG:N^2-dG carbinolamine [249,250]. These cross-linked -enal adducts are likely to contribute to the genotoxic effects of both HNE and acrolein. These cross-linked adducts occur at levels that are similar to 1%–2% of the levels of the monomeric 1,N^2-dG adducts in calf thymus DNA treated with either -enal.

Besides HNE also acrolein can form a cyclic adduct of deoxyguanosine (Acr-dG) in DNA. *In vivo* levels of Acr-dG in DNA are at least two orders of magnitude higher than those of the HNE adduct HNE-dG [251,252]. In addition to the facile reaction with acrolein, the higher levels of the acrolein adduct *in vivo* are due to a lower rate of repair.

The consequences of cellular formation of LPO products, particularly HNE, in terms of genomic stability have been reviewed by Winczura *et al.*, 2012 [253]. Elimination from DNA of LPO-induced lesions is executed by several repair systems: base excision repair (BER), direct reversal by AlkB family proteins, nucleotide excision repair (NER) and recombination. NER and recombination are involved in repair of HNE adducts to DNA bases in Escherichia coli [254]. According to the authors Acr-dG in a plasmid DNA is repaired by NER proteins, but it is repaired at a much slower rate than HNE-dG in human colon cell extracts, and the slow repair of Acr-dG is likely due to poor recognition/excision of the lesions in DNA. Using a plasmid DNA containing both adducts it was shown that the repair of Acr-dG is significantly inhibited by HNE-dG. In contrast, the repair of HNE-dG is not much affected by Acr-dG [251].

Interestingly, the 1,*N*-2-propane adduct of guanine in DNA can be bypassed during replication by Sulfolobus solfataricus P2 DNA polymerase IV [255].

The role of damage-specific DNA polymerases in mutagenesis induced by HNE was studied in M13mp18 phage by testing survival and mutation spectra in the lacZα gene [256]. The phage was grown in uvrA(−) Escherichia coli strains, carrying one, two or all three SOS DNA polymerases. When Pol IV was the only DNA SOS polymerase in the bacterial host, survival of HNE-treated M13 DNA was similar to, but mutation frequency was lower than in the strain containing all SOS DNA polymerases. When only Pol II or Pol V were present in host bacteria, phage survival decreased dramatically. Simultaneously, mutation frequency was substantially increased, but exclusively in the strain carrying only Pol V suggesting that induction of mutations by HNE is mainly dependent on Pol V [256].

The second route for the mutagenicity of HNE putatively involves the oxidation of HNE to its epoxide. The resulting etheno adducts have been found in human lung tissue [257]. The reaction mechanism was studied by Lee *et al.*, 2005 [258]. Analysis of the reaction between 4-hydroperoxy-2-nonenal (HPNE) and 2'-deoxyguanosine (dGuo) revealed the formation of 1,N^2-etheno-dGuo as well as heptanone-etheno-dGuo and trace amounts of dihydroxyheptane-etheno-dGuo. On the basis of these findings, it was concluded that HPNE, a primary product of LPO, is a major precursor to the formation of 1,N^2-etheno-dGuo, and it was proposed that 1,N^2-etheno-dGuo arises from the reaction of dGuo and HPNE via the intermediate formation of a cyclic hydroxy-ethano-epoxide derivative. In support of this assumption cytochrome P4502E1 appears to be responsible for the formation of HNE-etheno-DNA adducts in alcoholic liver disease [259].

Miscoding etheno-modified DNA adducts including 1,N(6)-etheno-2'-deoxyadenosine (ε-dA) are excreted in urine, following elimination from tissue DNA. An ultrasensitive and specific immunoprecipitation/HPLC-fluorescence detection method was developed for quantifying ε-dA excreted in urine [260]. Levels in urine of Thai and European liver disease-free subjects were found to be in the range of 3–6 fmol ε-dA/μM creatinine. Subjects with inflammatory cancer-prone liver diseases caused by viral infection or alcohol abuse excreted massively increased and highly variable ε-dA-levels. Based on this pilot study it was concluded that (i) high urinary ε-dA-levels, reflecting massive LPO-derived DNA damage *in vivo* may contribute to the development of hepatocellular carcinoma (HCC); (ii) ε-dA-measurements in urine and target tissues should thus be further explored as a putative risk marker to follow malignant progression of inflammatory liver diseases in affected patients and that (iii) etheno adducts may serve as biomarkers to assess the efficacy of (chemo-) preventive and therapeutic interventions [260].

6. Formation of HNE in Mammalian Cells and Tissues

High HNE levels or increases of HNE are generally an indication for accelerated LPO. However, both the high reactivity of HNE and the rapid metabolism of HNE will overlap the formation rate of HNE, *i.e.*, even if the rates of LPO and of HNE formation are very high, the topical increasing HNE level does not fully reflect the increase of LPO and HNE formation rates due to the overlapping of formation and degradation. The dynamics of HNE turnover, which is not easy to analyze—e.g., with tracer kinetic measurements—is much higher than the simple increase rate of HNE concentration. Thus, when significant increases of HNE levels are found in organelles, cells, tissues, organs, or in whole organisms, one may assume that the total increase in LPO rate and in HNE formation rate is extremely high.

6.1. HNE Formation in Cellular and Organ Systems

The oxidative stress during posthypoxic reoxygenation of cell suspensions or postischemic reperfusion of organs was demonstrated by an increase of HNE levels—often paralleled by an enhancement of MDA or thiobarbituric acid-reactive substances (TBA-RS), decrease of GSH and increase of disulfides, too. Those changes were found in hepatocytes [261], renal tubular cells [262], heart [263–265], and small intestine [266]. In earlier studies, HNE increases were observed in hepatocyte suspensions under oxidative stress by CCl_4 or ADP-iron [267]. An age-dependent increase of HNE in rats with traumatic brain injury was found in ipsilateral hippocampus one and seven days post injury [268]. Furthermore an increase of HNE was also found in hippocampus/parahippocampal gyrus, superior and middle temporal gyrus and cerebellum in mild cognitive impairment compared to age-matched control subjects [269]. This demonstrates that LPO occurs early in the pathogenesis of Alzheimer's disease.

Similar changes in Alzheimer's disease and other neurodegenerative diseases were described by Negre-Salvayre *et al.*, 2010 [270].

Table 1 presents a list of cellular, tissue or organ concentrations of HNE in humans and other species. Additionally, Figure 5 demonstrates human blood plasma concentrations of HNE in dependence on age of the blood donors.

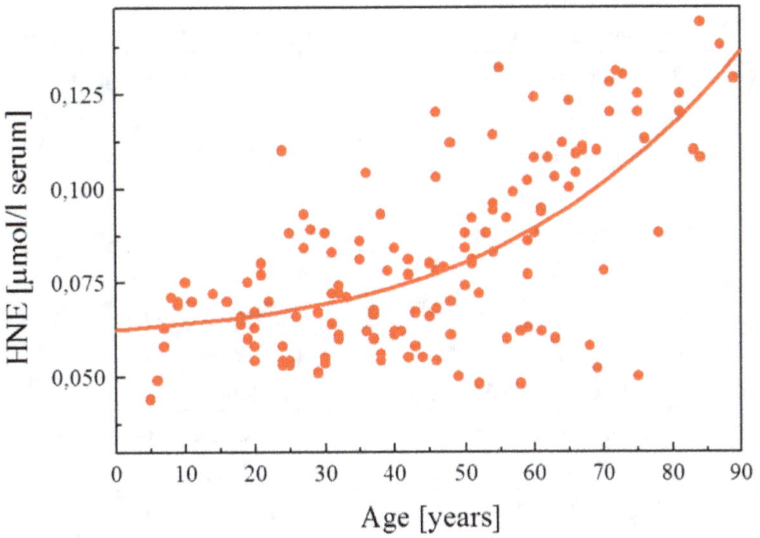

Figure 5. HNE plasma concentration in dependence on age of the blood donor (5 to 90 years).

HNE formation was also measured in perfused organs. Changes of perfusion conditions lead to changes in HNE formation rate. The tissue concentration of HNE only partly represents the real HNE formation rate. The dynamics of HNE formation rate was of particular interest in the postischemic reperfusion syndrome. The real flux rate of intracellular HNE formation during normoxic perfusion, during ischemia and during postischemic reperfusion was estimated on the basis of tissue HNE concentration, HNE utilization rate at defined HNE tissue concentration and HNE washout by perfusate circulation in rat small intestine [266]. Such estimations led to intestinal HNE formation rates of about 20 nmol/g w.w./min at normoxia, 20 to 40 nmol/g w.w./min at ischemia, 100 nmol/g w.w./min at 10 min of postischemic reperfusion, and 20 nmol/g w.w./min at 60 min of reperfusion). These data showed that the highest increase of HNE formation occured during the early reperfusion phase following ischemia.

A significant increase of HNE concentration was found in human renal tubular cells [262] and in rat hepatocytes [261,271] during postanoxic reoxygenation in comparison with preanoxic control values. In parallel to the HNE increase decreases of energy rich nucleotides, in particular of ATP, of GSH and cell viability as well as electron-microscopic morphological alterations of cells were observed. In human monocytes, HNE increased after phagocytosis of the malarial pigment hemozoin reaching 4 pmol/10^6 cells at 2 h (approximate intracellular concentration (AIE) 8 µM), 23 pmol/10^6 cells at 5 h (AIE 46 µM), and 7.9 pmol/10^6 cells (AIE 16 µM) at 12. A moderate increase in HNE, approximately 2 pmol/10^6 cells (AIE 4 µM), was also observed after phagocytosis of anti-D IgG-opsonized erythrocytes. HNE in unfed controls was approximately 0.5 pmol/10^6 cells (AIE 1 µM) [272]. The values which were measured in monocytes fed with hemozoin up to 50 µM were the highest intracellular concentrations which were ever measured. In some additional studies formation and metabolism of both HNE as well as MDA were compared [14].

Table 1. HNE concentrations in cells, tissues and organs.

Cells, Tissue	HNE-Concentration	Comment	Reference
Kidney tubular cells (human)	102 pmol/10^6 cells	Praeanoxic control	[262]
	307 pmol/10^6 cells	30 min reoxygenation after 60 min anoxia	
Hepatocytes (rat)	13 ± 5 pmol/10^6 cells	Freshly prepared	[267]
	27 ± 18 pmol/10^6 cells	CCl_4-treated for 1 h	
	40 ± 14 pmol/10^6 cells	ADP-Fe^{3+}-treated for 1 h	
	0.43 pmol/5.2 × 10^6 cells	Freshly prepared	[271]
	1.42 pmol/5.2 × 10^6 cells	15 min reoxygenation after 60 min anoxia	
Monocytes (human)	0.5 pmol/10^6 cells (1 µM) *	Basic value	[272]
	4 pmol/10^6 cells (8 µM) *	Hemozoin fed at 2 h	
	23 pmol/10^6 cells (46 µM) *	Hemozoin fed at 5 h	
	7.9 pmol/10^6 cells (16 µM) *	Hemozoin fed at 12 h	
Small intestine (rat)	0.68 nmol/g w.w.	perfusion, normoxia	[266]
	3.02 nmol/g w.w.	reperfusion after 60 min ischemia	
Blood plasma (human)	0.074 ± 0.028 µM	194 healthy woman and men	[273]
	0.069 ± 0.015 µM	18 to 29 years aged	
	0.070 ± 0.014 µM	30 to 39 years aged	
	0.072 ± 0.020 µM	40 to 49 years aged	
	0.083 ± 0.020 µM	50 to 59 years aged	
	0.096 ± 0.022 µM	60 to 69 years aged	
	0.107 ± 0.027 µM	70 to 84 years aged	
	106.3 ± 65.8 ng/mL	*O*-pentafluorobenzyl oxime	[274]
Umbilical cord plasma (human)	0.3 µM	Full-term healthy neonates	[275]
	0.5 µM	Term neonates with acidosis	
	0.6 µM	Term neonates with asphyxia	
	0.1 µM	Preterm neonates (healthy)	
	0.4 µM	Preterm neonates (asphyxia)	

* The approximate intracellular HNE concentration is given in parentheses.

6.2. HNE in the Whole Healthy Organism

Increased HNE levels in blood serum and tissues were measured in various diseases in children and adults such as reperfusion syndrome, inflammatory diseases, rheumatological diseases (e.g., rheumatoid arthritis), lymphedema, AIDS, diabetes, and cystic fibrosis. Increased serum levels of HNE, MDA or other LPO products have also been demonstrated in several systemic autoimmune diseases in adults such as rheumatoid arthritis and systemic lupus erythematosus (SLE). Children suffering from SLE have significantly higher HNE levels (146 ± 14 nmol/L) compared to controls (61 ± 10 nmol/L) while children with the focal type of scleroderma (SCL) had HNE levels similar to the control group [276].

In general, basic HNE concentrations of blood serum (for adults and children except neonates) were found to be in the range of 0.05 and 0.15 µM [273]. These values were measured by means of the DTNB method and were in very good agreement with the plasma HNE concentration of 106.3 ± 65.8 ng/mL measured by means of another assay, the *O*-pentafluorobenzyl oxime method [274]. In about 200 healthy humans of ages ranging from 18 to 84 years plasma HNE was measured together with various oxidative stress parameters, giving a mean value of 0.074 μM [273]. Accelerated oxidation during ageing was demonstrated by increases of HNE, MDA, GSSG and by the slight decrease of erythrocytic GSH with age. In the group aged up to 30 years the mean HNE was 68.9 ± 15.0, whereas in the group aged more than 70 years the corresponding value was 107.4 ± 27.3 nmol/L [273]. HNE levels were measured in umbilical arterial cord blood samples from healthy, acidotic, and asphyctic neonates with a gestational age ranging from 26 to 41 weeks [275]. HNE in umbilical cord plasma of full-term healthy neonates amounted to about 0.3 μM. After perinatal complications, HNE increased significantly in the group of term newborns with acidosis to about 0.5 μM and in term neonates suffering from asphyxia to about 0.6 μM ($p < 0.001$ in comparison with the control group of healthy term newborns). In healthy preterm neonates with a gestational age from 34 to 36 weeks the mean HNE level was 0.1 μM. HNE increased up to 0.4 μM in preterm newborns after perinatal asphyxia ($p < 0.001$) [275].

The presence of protein-bound HNE *in vivo* has been assessed in various human tissues, including the human aorta with atherosclerotic lesions [277], nigral neurons in Parkinson disease [278], renal cell carcinomas [279], amyloid deposits in systemic amyloidosis [280], in brain from subjects with mild cognitive impairment [281], and trophoblast cells of preeclamptic placenta [282]. Furthermore, the demonstration of HNE protein adducts in human lung cells after ozone exposure are consistent with a potential role for HNE in the toxic effects of ozone in these cells [283].

Moreover, endogenous cyclic DNA adducts derived from HNE were associated with neurodegenerative diseases such as Alzheimer's disease. A sensitive and selective capillary liquid chromatography nanoelectrospray isotope dilution mass spectrometric method was developed to identify and quantify endogenous cyclic DNA adducts derived from HNE with 2'-deoxyguanosine (HNE-dG) in human brain tissues [284].

6.3. Influence of Nutrition

Parallel to endogenous formation of HNE it can also be consumed with the diet. To assess human exposure to 4-hydroxy-2-alkenals in the diet, these lipid peroxidation products were monitored in vegetable oils, fish and shellfish [285] and the Korean daily exposure to 4-hydroxy-2-alkenals, excluding consumption from fried food, was calculated to be 4.3 μg/day (2.7 μg HNE and 1.6 μg HHE). Additionally, 4-hydroxy-2-alkenal intake from fried foods was calculated and amounted to more than 11.8 μg/day. Thus the combined exposure would be therefore, 16.1 μg/day corresponding to 0.3 μg/kg body weight/day for a 60 kg Korean adult [285]. It was also found that nutrition and various food constituents may have influence not only on HNE formation or intake, but also on HNE metabolism. Of course, this involves the sulfhydryl content of the food such as S-allyl-L-cysteine of garlic [286], but also other components. Dietary fibers are fermented by the gut flora to yield short chain fatty acids (SCFAs), which inhibit the growth of tumor cells, induce glutathione S-transferases, and protect cells from the genotoxic activity of HNE [287].

Beneficial effects of the probiotic VSL#3 (a medical food that delivers a high concentration of beneficial live bacteria) on parameters of liver dysfunction in chronic liver diseases were evaluated by Loguercio *et al.*, 2005 [288]. These effects were studied in non-alcoholic fatty liver disease, alcoholic liver cirrhosis, HCV-positive patients with chronic hepatitis without or with liver cirrhosis. Treatment with VSL#3 improved levels of HNE and MDA except in HCV-positive patients. Furthermore, routine liver damage tests and plasma S-NO levels were improved in all groups of patients. It was concluded that manipulation of the intestinal flora should be taken into consideration as possible adjunctive therapy in some types of chronic liver disease [288].

It was further shown that mean levels of HNE and MDA in rat brain tissues were decreased in rats fed soy-protein as a dietary antioxidant [289].

Data on the effect of green tea on LPO products formation and parameters of the antioxidative system of liver, blood serum and central nervous tissue of healthy young rats drinking green tea for five weeks were collected by Skrzydlewska *et al.*, 2002 [290]. The bioactive ingredients of green tea extract caused an increase in the activity of glutathione peroxidase and glutathione reductase and in the content of reduced glutathione as well as a marked decrease in lipid hydroperoxides, HNE and MDA in the liver while minor changes of the measured parameters were observed in the blood serum, but LPO products, particularly MDA, were significantly reduced. In the central nervous tissue the activity of superoxide dismutase and glutathione was found to be peroxidase while the activity of glutathione reductase and catalase were increased, and the level of lipid hydroperoxides, HNE and MDA significantly was found to be decreased after drinking green tea.

Beside of food uptake the uptake of environmental toxins may increase HNE concentrations in human tissues. Among the several converging factors leading to Parkinson disease, epidemiological studies indicate a correlation between Parkinson disease with living in a rural area and/or exposure to agricultural pesticides. In this context Leiphon and Picklo, 2007 demonstrated that mitochondrial aldehyde dehydrogenases are sensitive targets of pesticide inactivation and that pesticides such as maneb and benomyl can decrease the detoxification of HNE to HNA and therefore contribute to an increase of HNE in rat brain mitochondria [291].

7. Metabolism of HNE

HNE is rapidly metabolized in cells. The velocity and the pattern of HNE metabolism were studied in various cell types, in subcellular organelles and in whole organisms. In all cases the overall metabolic rate of exogenously added HNE was so high, that already within a few minutes equilibrium concentrations in the nanomolar range were obtained (Figure 6). In Table 2 the maximal velocity of HNE total degradation in different biological systems is compared. From these observations it is obvious, that HNE—even at very high lipid peroxidation rates—hardly can accumulate to high levels in biological systems. The highest level ever measured to our knowledge was 6.5 μM (besides the 50 μM in monocytes fed with hemozoin). This value of 6.5 μM was recorded in experiments with rat intestine during the postischemic reperfusion period [266].

In all experimental models investigated a variety of HNE metabolites was identified and quantified. However, one may differentiate these metabolites into primary and secondary intermediates. The primary metabolites are undergoing further metabolic conversion leading to secondary intermediates. Some of the secondary intermediates are stable end products. Possibly one may use such stable HNE products—such as mercapturic acids excreted in urine—as biomarkers for oxidative stress, specifically for LPO, also in medical diagnostics. Additionally, the HNE metabolism is an important component of the antioxidative defense system of cells because of the cytotoxic, mutagenic, and even carcinogenic effects of aldehydic products of LPO, namely of HNE.

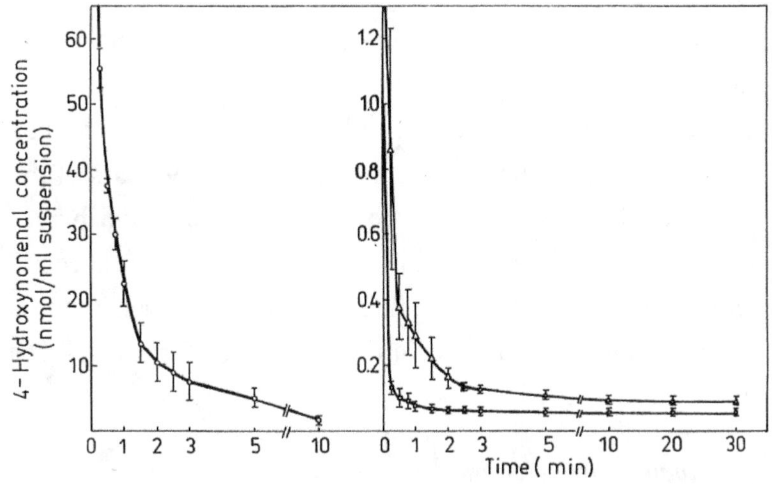

Figure 6. Degradation/metabolism of 4-HNE in rat hepatocytes. The method was the so-called *direct measurement of HNE by HPLC*. The remaining HNE shown in the graphs is the total of not metabolized HNE within the suspension, *i.e.*, intracellular plus extracellular ones at the indicated time-points. The differenciation between the intracellular and extracellular compartments of HNE is not necesssary for balancing the HNE degradation/metabolism. The experimental conditions were: 37 °C, pH 7.4, 1 × 10^6 hepatocytes/mL suspension; HNE was added at 100 μM (o), 5 μM (Δ) or 1 μM (□). Values are the mean ± S.D.

Table 2. Maximal velocity of total HNE degradation in cells, subcellular organelles, and perfused organs.

Biological System	Maximal Rate of HNE Catabolism	Reference
Kidney cortex mitochondria, rat	112.2 nmol/mg protein/min	[292]
Hepatocytes, rat	28.4 nmol/mg w.w./min	[293]
Ascites tumor cells	9 nmol/mg w.w./min	[264]
Thymocytes, mouse	27.7 nmol/mg w.w./min	[294]
Synovial fibroblasts, rabbit	27.3 nmol/10^6 cells/min	[295]
Perfused kidney, rat	160–190 nmol/g w.w./min	[296]
Perfused intestine, rat	22 nmol/g w.w./min	[266]
Perfused heart, rat	50 nmol/g w.w./min	[264]

The velocity of HNE metabolism was often measured by tracer kinetic methods. Mostly [2-3H]-HNE was used [266,292,293,297,298]. Sometimes only the so-called "direct HNE measurement" (free or unbound HNE) by HPLC was used detecting the release of added HNE [263,292]. In other studies, namely those investigating specific parts of HNE catabolism, glutathione consumption after adding different HNE amounts was analysed [299]. HNE added to hepatocyte suspensions (10^6 cells/mL) in concentrations of 1, 5, or even 100 μM was almost completely metabolized within three minutes [300–303] (Figure 6).

Originally it was assumed that HNE metabolism is especially fast in hepatocytes [300,304–306] because the first studies on cellular HNE metabolism demonstrated that compared with liver, all other tissues possess only a low capacity to metabolize HNE, ranging from 0% (fat pads) to 10% (kidney) of the activity present in liver [300]. However, later on also in other cells types a rapid HNE metabolism was found, in some cell types such as thymocytes even in a similar range as in hepatocytes. Furthermore, the first findings on the metabolism of HNE by isolated hepatocytes and by liver homogenate cytosolic fractions suggested that hepatocytes and rat liver cytosol, respectively convert HNE enzymatically mainly to the corresponding alcohol, dihydroxynonene (DHN; non-2-ene-1,4-diol) via alcohol dehydrogenase. It was therefore postulated that the rapid conversion of the cytotoxic HNE and other reactive aldehydes to alcohols, which are probably less toxic, could play a role in the defense system of the liver against toxic substances arising from free radical-induced LPO [300]. This view was expanded later after the identification of a multitude of HNE products with quantitative majorities of the GSH-adduct(s) and/or hydroxynonenoic acid (HNA) and its further degradation products rather than a majority of the alcohol DHN [293].

7.1. HNE Metabolism in Mammalian Cells and Organs

HNE metabolism was studied in hepatocytes [293,300,302,306], hepatoma cells [307–310], ascites tumor cells [264], mucosal cells [297], synovial fibroblasts [295], thymocytes [294], vascular smooth muscle cells [311] and also in organs such as heart [264,312] and kidney [296,298]. The rapid HNE degradation found and the intracellular metabolism implicate that HNE can rapidly enter cells. The velocity of HNE entry into the cells was measured by tracer kinetics in various cell types [295]. In ascites tumor cells, the total rate of HNE metabolism by cells in the proliferating phase (early phase) and cells in the stationary phase (late phase or resting phase of tumor growth) was almost the same [264]. The formation of glutathione conjugates following the addition of HNE was higher in early phase cells when compared with cells in the late phase of tumor growth. This observation was in accordance with the increased consumption of the reduced form of glutathione (GSH). The glutathione transferase activity in tumor cells of both proliferation stages was equal, but the substrate GSH had a higher level in cells of the proliferating phase than in the cells of the resting phase. The result was a higher glutathione-adduct formation in cells of the proliferating phase. The reductive pathway of HNE metabolism was high in proliferating phase tumor cells. In contrast, the oxidative pathway was overwhelming in resting phase tumor cells. This fact corresponded with the different energetic and redox state of both phases. The cells of the resting phase showed a higher rate of uncoupling than cells during the proliferating phase, in accordance with a lower $NADH/NAD^+$ and lactate/pyruvate ratio of resting cells. The consequence of the high demand of reduced NADH by partially uncoupled mitochondria for ATP production is a lower capacity of resting cells to use NADH for HNE reduction (DHN or GSH-DHN formation). In support, addition of NADH to the homogenates of rat hepatoma cells (MH1C1) resulted in a 1.5-fold increase of aldehyde consumption [309].

In hepatoma cells, such as MH1C1 and HTC, the metabolic pathway for HNE removal is markedly different from hepatocytes. The results of several investigations indicate that the liver biochemical pathways involved in HNE metabolism undergo a complex variety of changes during neoplastic transformation, so that hepatoma cells with different degrees of deviation show different metabolic patterns with respect to HNE. The analysis of aldehyde enzymatic activities for example showed that MH1C1 cells have decreased levels of all enzymes involved in the catabolism of HNE except for NAD and NADP aldehyde dehydrogenase in the cytosol, compared to hepatocytes and to HTC cells. HNE can, however, be metabolized through a sufficiently high activity of glutathione S-transferase. NAD and NADP dependent aldehyde dehydrogenases are of great importance in HTC cells for HNE metabolism. In HTC cells the metabolic efficiency is due mainly to oxidative enzymes [310].

Comparing Chinese hamster fibroblast control cells (HA-1) with Chinese hamster fibroblast H_2O_2-resistant cell lines (OC5 and OC14) the H_2O_2-resistant cells were found to be significantly more resistant than HA-1 cells to the cytotoxicity of HNE, as determined by clonogenic cell survival. This finding was related to a significant 2–3-fold increase in the HNE degradation rate of OC5 and OC14 cells, *i.e.*, the HNE removal from culture media containing 72 μM HNE when compared with HA-1 cells [313]. Thus, the resistance appeared to be related to increased cellular HNE metabolism.

7.2. HNE Metabolism in Subcellular Organelles

HNE metabolism was studied in several mitochondrial suspensions. By analyzing the decrease of the HNE concentration in the first 30 s after treatment with 100 μM HNE, Ullrich *et al.*, 1994 calculated the maximal HNE detoxification rate in rat kidney cortex mitochondria to be 112.2 nmol/mg protein/min [292]. After addition of 1 μM HNE the degradation of HNE approached a steady state after 2 min. The steady state for the remaining HNE level has been adjusted to about 0.1 μM, a range which is also known as the steady state value in human blood plasma.

HNE is metabolized by renal mitochondria to 4-hydroxynonenoic acid (HNA), 1,4-dihydroxynonene (DHN), and the GSH-HNE-adduct (GSH-HNE). GSH-HNE and HNA reached the maximum value about 2 min after HNE addition to mitochondria, and thereafter these metabolites slowly decreased. Almost 10% of the radioactivity was bound to proteins. The protein bound radioactive HNE was found predominantly bound to the proteins of the intermembrane space. Lower amounts were bound to the proteins of the inner membrane and to the proteins of the matrix. The nmol HNE value per mg protein was found to be very high in the intermembrane space and almost similar for outer and inner membrane and matrix proteins [292]. In rat brain mitochondria it was found, that *in situ* mitochondrial HNE detoxification is affected by decrements in NAD^+ availability and complex I activity [314]. In investigations of HNE metabolism during diethyl-nitrosamine-induced carcinogenesis in rat liver it was established that the NAD- and NADP-dependent aldehyde dehydrogenases of the cytosolic fraction and the NADP-dependent aldehyde dehydrogenase of the microsomes show higher values in nodules and hepatoma than in normal liver [315]. Both aldehyde dehydrogenase and glutathione S-transferase activities of rat liver mitochondria are reduced in aged animals. Mitochondrial HNE oxidation by aldehyde dehydrogenase and glutathione conjugation of HNE decline at 18 and 24 months, respectively, while these enzyme activities were found to be well-preserved in dietary restriction animals throughout their life span. These findings indicate that the prevention of the age-associated decrease in HNE detoxification by dietary restriction may be an important mechanism underlying enhanced aldehyde elimination, thus minimizing the functional deterioration observed in mitochondria of old animals [316].

After exposure to racemic HNE, rat brain mitochondria metabolized HNE enantioselectively with a higher rate of (R)-HNE metabolism [317]. By using purified enantiomers of HNE the authors were able to demonstrate that this enantioselective metabolism of HNE was the result of higher rates of enzymatic oxidation of (R)-HNE by aldehyde dehydrogenases compared to (S)-HNE. Conjugation of

HNE to glutathione was a minor metabolic pathway and was not enantioselective. These studies demonstrate that the chirality of HNE affects its mitochondrial metabolism.

7.3. HNE Metabolism in Whole Animals and Interorgan Relationships

The aim of many *in vivo* metabolic studies was the characterisation of end-products of HNE, especially in urine, in order to develop specific and non-invasive biomarkers of lipid peroxidation. When HNE is administered intravenously, it is mainly excreted into urine and bile as conjugated metabolites, in a proportion that is dependent on the administration route [118,318]. However, biliary metabolites undergo an enterohepatic cycle that limits the final excretion of faecal metabolites. Only a very low amount of metabolites is found to be bound to macromolecules.

According to Alary *et al.*, [118,318] the main urinary metabolites are represented by two groups of compounds. The first group derives from (a) the mercapturic acid formation from 1,4 dihydroxynonene-glutathione (DHN-GSH); (b) the lactone of 4-hydroxynonanoic-GSH (HNA-lactone-GSH) and (c) HNA-GSH. The second group of metabolites is derived from the ω-hydroxylation of HNA or HNA-lactone by cytochromes P450 4A, followed eventually, in the case of ω-oxidized-HNA-lactone, by conjugation with GSH and subsequent mercapturic acid formation. Biliary metabolites are GSH or mercapturic acid conjugates of DHN, HNE and HNA [118,318]. Stereochemical aspects of HNE metabolism are also discussed by Alary *et al.* [118,318]. Experiments with injection of tritium-labeled HNE [319] and also of tritium-labeled 4-hydroxyhexenal (HHE) [320] into the vein of rats have shown that part of the radioactivity was excreted in the urine as mercapturic acid conjugates. From these experiments and studies with perfused kidneys [296] it may be concluded that the excretion of mercapturic acid conjugates as well as the excretion of DHN is the main route of HNE-product disposal from the organism. The most important studies on urinary excretion of HNE metabolites were carried out by the group of Alary, Cravedy, and Gueraud. Following intravenous administration of [^{3}H]HNE into rats, the majority of the dose appeared in urine (67.1% after 48 h) [319]. The radio-HPLC metabolic profile showed that no unchanged parent compound was detected in urine whereas at least four metabolites were present, most of them corresponding to mercapturic acid conjugates.

Two major pathways were involved in the biotransformation of HNE *in vivo*: the reduction/oxidation of the aldehyde group, and the conjugation to endogenous glutathione leading to mercapturic acid conjugates in urine. These end products were isolated by HPLC and identified by mass spectrometry as HNE mercapturic acid, 1,4-dihydroxynonene mercapturic acid, 4-hydroxynonenoic mercapturic acid, and the corresponding lactone.

7.4. Primary HNE Intermediates—Enzymatic Reactions and Quantitative Results

Already in 1991 in the classical HNE-review of Esterbauer *et al.* [304] it was mentioned: the main enzymes involved in HNE metabolism are glutathione transferases, aldehyde dehydrogenases, and alcohol dehydrogenases. These are the enzymes forming the so-called primary HNE metabolites. The primary metabolites of HNE are therefore, the HNE-GSH adduct, the corresponding carboxylic acid of HNE, *i.e.*, HNA, and the corresponding alcohol, *i.e.*, DHN. Figures 7–9 provides examples for the analysis of HNE and HNA by HPLC (Figure 7), 1,4-DHN by HPLC plus MS with fluorimetric detection (Figure 8), and formation of HNA, HNE-GSH and DHN in hepatocytes (Figure 9). As displayed in

Figure 9, after addition of a single dose of HNE to a hepatocyte suspension all three primary HNE metabolites increased very rapidly within the first minute of the experiment and reached a maximum value—in case of 100 μM HNE—after about 2 min, and thereafter slowly decreased. At all time points the GSH-HNE conjugate was the main product of hepatocytes followed by HNA. The concentration of DHN was considerably lower. The relative proportion of the three metabolites in hepatocytes was approximately GSH-HNE:HNA:DHN = 6:4:1, a pattern which may be quite different in other cell types. The formation of the GSH-HNE conjugate always leads to a decrease of intracellular GSH concentration. The sum of the three primary metabolites, GSH-HNE + HNA + DHN, in hepatocytes after 3 min of incubation ranged between 60% and 65% of the added HNE.

Taking into account the small fraction of the non-metabolized HNE, it follows that about 33% to 36% of the HNE was converted into other metabolites. This fraction involves secondary metabolites and HNE-modified proteins and peptides. In hepatocytes, this fraction was below 10% within the very first initial phase of about 30 s of incubation after addition of exogenous HNE. With increasing duration of the incubation this fraction steadily increased—while the sum of GSH-HNE, HNA, and DHN decreased again—and reached a value of nearly 50% after 30 min. These findings strongly suggest that the primary metabolites GSH-HNE conjugate, HNA, and DHN are further metabolized to secondary products [293].

As mentioned, the metabolism of HNE involves multiple pathways, including conjugation with glutathione catalyzed by glutathione S-transferases (GST) [321,322], oxidation of the aldehyde functional group to form HNA catalyzed by aldehyde dehydrogenases [323,324], and reduction by alcohol dehydrogenase to 1,4-dihydroxynonene [300]. The HNE-metabolizing aldehyde dehydrogenase isoenzymes are present in the hepatic cytosol, mitochondria, and probably also in microsomes [303,325]. The HNE-metabolizing NADH-dependent alcohol dehydrogenase is localized mainly in the hepatic cytosol [303]. The glutathione transferases are ubiquitous enzymes, being particularly rich in hepatocytes and probably other liver cells. Even though a number of rat GST isozymes show catalytic activity toward HNE, rat liver glutathione transferase 8-8 has been demonstrated to be most effective in catalyzing the conjugation of this compound to GSH, and it has been suggested that this isozyme may be involved in the detoxification of HNE and related endogenous electrophiles [303,304,321,322].

A human acidic glutathione-S-transferase, hGST 5.8, was isolated from liver, heart, bladder, pancreas, and brain [326], and it was shown that this enzyme has an about 20-fold higher specific activity for HNE than for 1-chloro-2,4-dinitrobenzene, and expressed glutathione peroxidase activity toward phospholipid hydroperoxides. The hGST 5.8 was absent in lung, kidney, skeletal muscle, colon and erythrocytes [326]. Berhane *et al.*, 1994 characterized the detoxification of base propenals (degradation products of DNA) and other unsaturated aldehyde products of free radical reactions and lipid peroxidation by human glutathione transferases [327]. They found that in general, GST A1-1 and GST M1-1, in contrast to GST P1-1, were more active with 4-hydroxyalkenals than with base propenals. HNE readily reacts with GSH also in a nonenzymatic reaction. The GSH transferase-catalyzed reaction can, however, proceed about 300 to 600 times faster [304].

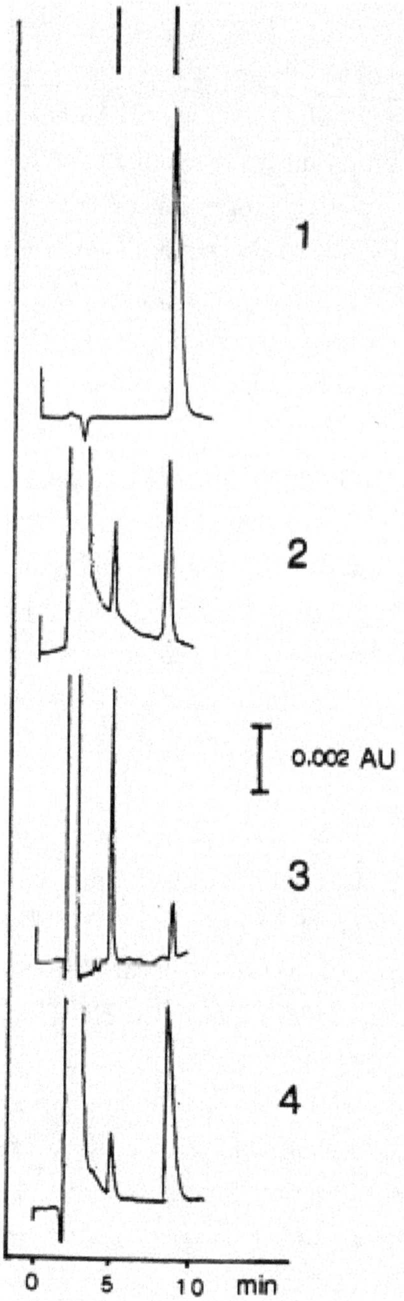

Figure 7. Identification of HNE and 4-hydroxynonenoic acid (HNA) by isocratic HPLC separation. Conditions: the eluent was 42% acetonitrile/water v/v 1 mL/min; columns were Spherisorb S5 ODS2 250 × 4.5 mm; detection of both substances at 223 nm; authentic standard of 4-HNA was produced in the system of HNE, dehydrogenase (NAD^+) and aldehyde dehydrogenase; (**1**) HNE standard solution 30 μM; (**2**) product of oxidation of HNE: HNA 15 μM; (**3**) mixture from cellular incubation mit CH_2Cl_2 extract of HNE oxidation; (**4**) mixture formed in rat liver cells (10^6 cells/mL) after addition of 100 μM HNE, the first peak representing HNA, the second peak representing HNE; for exact quantification of HNA the comparison of the HNA-peak with NADP formation in experiments on transition of HNE by means of aldehyde dehydrogenase was carried out.

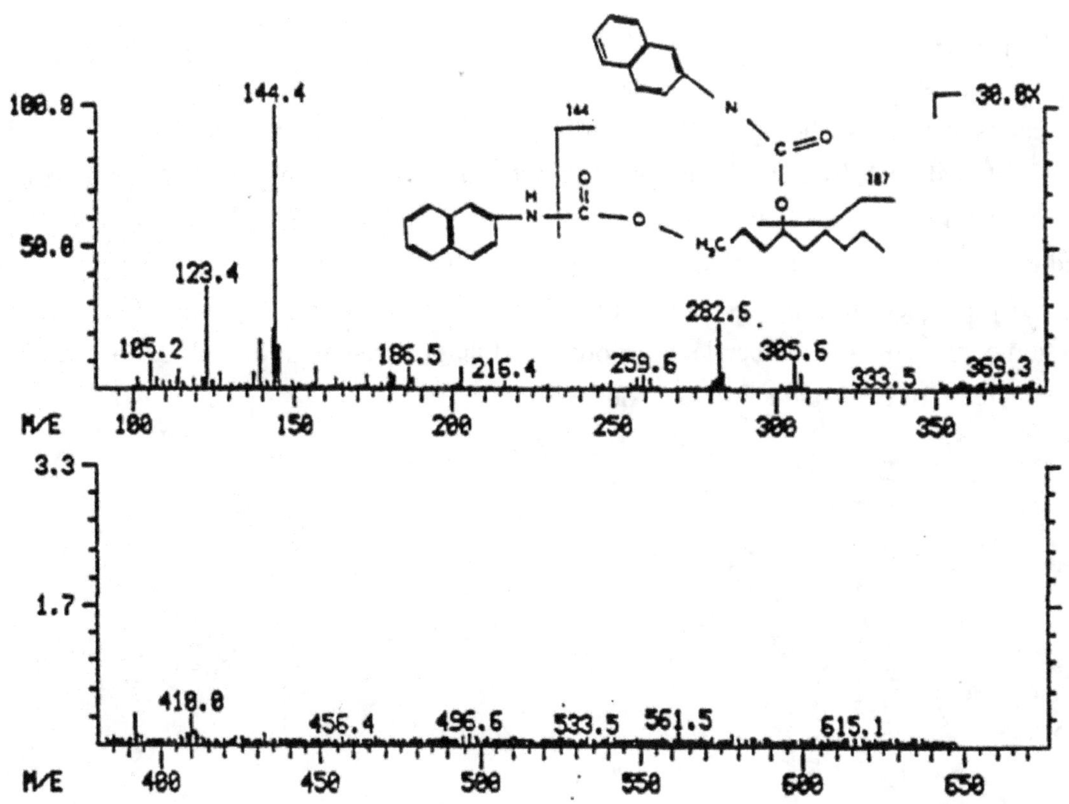

Figure 8. Mass spectrum of dihydroxynonene urethane (HPLC plus MS) with fluorimetric detection. The preparation of the reduced HNE product 1,4-DHN was carried out in the following manner: reduction of HNE with sodium boronate; extraction with dichloromethane; drying by means of sodium sulphate; purification by TLC separation, detection with 10% phosphomolybdenic acid dissolved in ethanol, extraction from TLC plates with CH_2Cl_2. Identification of 1,4-DHN: Siliylation; GC (85 °C, with 10 °/min up to 150 °C, 50 m SE30 column—i.d. 0.32 mm, FID); formation of urethane derivative (1,4-dinaphthyl-2,3-nonenylurethane); (305, 350 nm). Preparation of radioactively labelled 1,4-DHN: Reduction of radioactively labelled HNE with sodium boronate; extraction with dichloromethane; purification by means of TLC and elution on Bond Elut C18 columns (35% methanol/water, v/v). Chemical ionization (CH_4). Peaks: m/e 144 Ar-NH_3^+, m/e 187 Ar-NH-COOH, m/e 123 (M + 1)-(2 × 187); (M + 1) = 497 m/e [301].

GST can be induced by HNE. To investigate the effect of LPO products on the expression of GSTs Fukuda *et al.* [328] exposed normal rat liver epithelial cells (RL34) to aldehydic compounds and found out that GST activity was induced by α,β-unsaturated aldehydes, such as acrolein (1.3-fold), crotonaldehyde (1.4-fold), 4-hydroxyhexenal (HHE) (1.4-fold), and HNE 1.7-fold. The induction of GST activity by HNE was time-dependent, reaching a plateau after 16 h. The authors concluded that the induction of GSTs by HNE may represent an important cellular defense mechanism against oxidative injury [328]. Immunoblot analysis applying polyclonal antibodies against GST isozymes demonstrated that GST-P, a well-known tumor marker, was significantly induced 16 h after HNE treatment [328]. In early human fibrous atherosclerotic plaques, immunohistochemical studies demonstrated marked induction of hGST A4-4 in endothelial cells overlying plaque, and in proliferating plaque vascular smooth muscle cells [329]. The authors conclude that endothelial cell mGST A4-4 can play a key role in protecting blood vessels against oxidative stress and, thus, is likely to be a critical defense mechanism against oxidants that act as atherogens. In rat aortic smooth muscle A10 cells the induction of GSH and glutathione S-transferase by 3H-1,2-dithiole-3-thione (D3T, a chemoprotective chemical known to induce detoxification enzymes), and protective effects of D3T, *i.e.*, elevated cellular defenses against HNE-mediated toxicity were characterized [330]. As results the authors report that incubation of A10 cells with D3T resulted in a marked concentration-dependent induction of both GSH and GST, that this induction exhibited a time-dependent response, and that pretreatment of A10 cells with D3T led to a dramatic decrease of HNE-induced cytotoxicity.

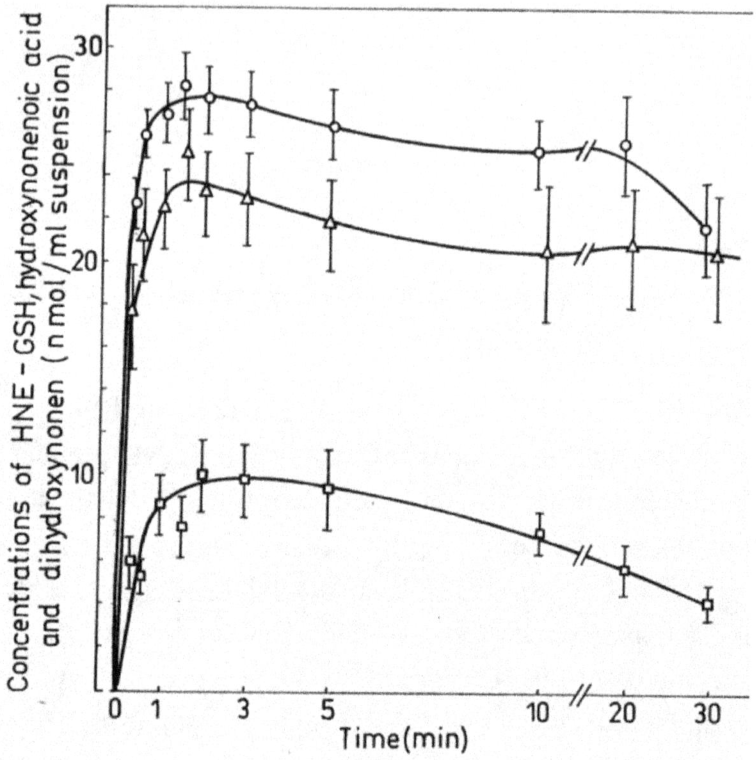

Figure 9. HNE metabolites. Formation of the so-called primary HNE metabolites HNA (Δ), HNE-GSH (o) and DHN (□) in rat hepatocyte suspensions after loading the cell suspensions with 100 μM tritium-labelled HNE. The glutathione (GSH) adduct of HNE was quantified as follows: Measurement as isoindol derivative following the reaction of the adduct with o-phthaldialdehyde in the presence of mercaptoethanol; HPLC separation of the isoindol derivative by isocratic HPLC with 65% methanol/0.1 M sodium acetate buffer pH 6.5 (v/v) in the presence of an ion-pair reagent (tetraethyl-ammoniumhydroxide/TAA); fluorimetric detection at 343 and 445 nm; columns: ODS 5 μm, 250 mm × 4.6 mm i.d.; purification of the extracts before HPLC separation on bond Elut C18 columns (methanol, water, sodium acetate buffer, hexane, methanol elutions) [293].

A redox activation of aldose reductase (AR) in the ischemic heart was shown by Kaiserova *et al.*, 2006 [331]. Their data indicate that reactive oxygen species formed in the ischemic heart activate AR by modifying its cysteine residues to sulfenic acids. To investigate the role of AR in the late phase of ischemic preconditioning, Shinmura *et al.*, 2002 treated conscious rabbits with six cycles of 4-min coronary occlusion/4-min reperfusion [332]. 24 h later they observed s a marked increase in AR protein and activity and in the myocardial content of sorbitol, a unique product of AR catalysis. AR was demonstrated to be an essential mediator of late preconditioning. Furthermore, the data suggest that myocardial ischemia/reperfusion injury is due in part to the generation of LPO products and that late preconditioning diminishes this source of injury by upregulating aldose reductase [332].

Aldose reductase (AR), an NADPH-dependent aldo-keto reductase, which catalyzes the reduction of glucose to sorbitol as part of the polyol pathway, is able to catalyze the reduction of a very broad range of aldehydes. HNE is an excellent substrate (Km = 22 μM, Kcat/Km = $4.6 \times 10^6\ M^{-1} \cdot min^{-1}$) [332]. Therefore, it was concluded that the extrahepatic reductive metabolism of HNE involves AR [332]. ARs, present in human kidney and other tissues play an important role in the reductive metabolism of endogenous aldehydes such as HNE [333]. The kidney contributes via this pathway to a rapid urinary excretion of HNE products from the organism. It is worth to mention that AR is the only major HNE oxidoreductase of the human heart. AR has been purified from bovine retina, too. It has an apparent molecular weight of 32 kilodaltons and shares immunological and kinetic properties with the much studied aldose reductases purified from various sources. Retinal aldose reductase displays a Km of approximately 40 μM for HNE and 4-hydroxyhexenal. Thus, AR was suggested to constitute a major detoxification route for these toxic aldehydes also in the retina [334]. Inhibition of aldose reductase by fidarestat suppresses several oxidative stress-induced inflammatory disorders. Fidarestat has already undergone phase III clinical trial for diabetic neuropathy and was found to be safe, though clinically not very effective [335]. It was also shown, that HNE and MDA induce the expression of the aldehyde reductase gene, which could be prevented by the carbonyl scavenger *N*-acetyl-L-cysteine (NAC) [336].

In other experiments effects of HNE addition on cellular glutathione levels were analyzed [299]. The authors studied the dose-dependent cytotoxicity of HNE after treatment of cultured fibroblasts at doses up to 50 nmol/10^6 cells for 3 h. In peroxide resistant OC14 and oxygen resistant O2R95 cells variants of HA1 cells, HNE treatment resulted in a statistically significant consumption of GSH at all doses tested. Cellular GSH levels in HA1 and O2R95 cells fell—starting from about 25 nmol GSH/mg cell protein—below 5 nmol GSH/mg cell protein at HNE doses of 25 and 40 nmol/10^6 cells, respectively. That argues for a very high rate of GSH-HNE adduct formation. In OC14 cells, GSH consumption—starting from about 35 nmol/mg protein—could not be detected until HNE doses exceeded 50 nmol/10^6 cells and did not fall below 5 nmol GSH/mg cell protein at any dose tested. In OC14 cells obviously the GSH-HNE adduct formation rate is very low [299]. This cannot be due to the capacity of glutathione-S-transferase which was even higher in OC14 cells compared to HA1 and O2R95 cells and stayed stable during the 3 h of HNE treatment in all three cell types, but rather to the enzymatic properties which obviously differ from one cell type to the other. These experiments also support the high stability of glutathione-S-transferase in these cultured fibroblast cell lines.

Table 3 displays the formation of primary HNE metabolites for different cells and tissues after the addition of 100 μM HNE to the biological system.

Table 3. Primary HNE metabolites in different cells and tissues after the addition of 100 μM HNE to the biological system.

Biological System	DHN	GSH-HNE	HNA	HNE-P	Reference
Kidney mitochondria (rat)	9.3 ± 1.8	9.8 ± 1.2	5.9 ± 0.7	8.4 ± 1.4	[292]
Hepatocytes (rat)	8.1 ± 2.1	27.5 ± 2.5	25.3 ± 5.6	2.0 ± 0.6	[293]
Enterocytes (rat)	5.4 ± 0.6	11.0 ± 0.5	4.2 ± 0.6	1.3 ± 0.2	[297]
Thymocytes (mouse)	6.9	10.2	6.8	2.8	[294]
Synovial fibroblasts (rabbit)	0.3 ± 0.1	29.1 ± 1.9	1.0 ± 0.3	4.3 ± 0.5	[295]
Kidney venous effluent (rat)	24 ± 5	14 ± 4	5 ± 1	n.d.	[296]
Ehrlich ascites cells, proliferating	2.0	10.5	3.0	7.2	[264]
Ehrlich ascites cells, resting	1.5.	8.0	3.8	1.5	

HNE-P, HNE-protein; n.d., not determined. Given values represent the % of the initial HNE concentration (mean; mean ± S.D.).

7.5. Secondary HNE Intermediates—Enzymatic Reactions and Quantitative Results

It is known from many studies that HNE can bind to thiol-groups of proteins. In general it is well established that HNE forms adducts with the following side chains in proteins: Cys, His and Lys. From tracer-kinetic experiments it was estimated that between 1% and 8% of exogenously added or endogenously formed HNE reacts with proteins to form HNE-protein conjugates. With that share of the totally added or formed HNE the proteins and peptides are the most important group of biomolecules targeted by HNE. In experiments studying the metabolism of exogenously added HNE always a dynamic pattern was seen: the binding of labelled HNE in the HNE-protein conjugates-pool reached a maximum—in hepatocytes 3 min after HNE addition—followed by a decline, underlining the fact that aldehyde-modified proteins underlie a removal of oxidatively damaged and modified proteins by the proteasomal system [337–339]. Nevertheless, the formation of HNE-protein and HNE-peptide compounds does not represent a metabolic pathway of HNE degradation. The chemical reactions leading to HNE modifications of proteins, peptides, and further biomolecules rather reflect this part of HNE "removal" which leads to damaging or signal effects by HNE. The real metabolic reactions of HNE lead to the primary and secondary HNE metabolites, which finally contribute to the detoxification and harmless removal of HNE.

Which are the real secondary metabolic products of HNE? The pool of secondary HNE intermediates contains degradation products of HNA, which is degraded via different products of β-oxidation to acetyl-CoA and, finally to carbon dioxide and water. This involves the reaction steps of the tricarboxylic acid cycle and of the respiratory chain. In experiments with ^{3}H-labeled HNE the formation of ^{3}H-labeled water could be shown in hepatocytes [293]. The rate of formation of radioactive water was slow, but the time curve showed not a transient maximum but a steady increase. In hepatocytes treated for 10 min with 100 μM HNE, 14% of the radioactivity appeared in the water peak. The formation of 3H-labeled water could be almost completely inhibited in presence of 4-pentenoic acid, an inhibitor of β-oxidation. In the presence of 4-pentenoic acid only traces of radioactive water were formed.

Furthermore, in the presence of this inhibitor there was a significantly higher accumulation of 4-hydroxynonenoic acid (HNA) in comparison with control experiments without inhibitor. In the presence of 4-pentenoic acid HNA became the major HNE metabolite. During the first 60 s of incubation, the inhibitor had almost no effect on the rate of the conversion of HNE to HNA. Thereafter, however, the time profile was markedly different. In the presence of the inhibitor, HNA steadily increased, reaching 45 nmol/mL after 3 min, whereas in the absence of the inhibitor the concentration of the carboxylic acid slowly decreased again. These experiments indicate that HNA is not an end product of HNE metabolism, but is further metabolized [293].

The oxidative pathway includes—when HNA is degraded—also tricarboxylic acid cycle products, such as aconitate, citrate, malate, fumarate, *etc.* In primary cultures of rabbit synovial fibroblasts 3 min after addition of 100 μM labelled HNE the shares of radioactivity of tricarboxylic acid cycle intermediates and of water were 7.4% and 34.1%, respectively [295]. The β-oxidation of HNA including the formation of tritiated water was also demonstrated in liver slices [340]. The identification of 4-hydroxy-9-carboxy-2-nonenoic acid and 4,9-dihydroxynonanoic acid demonstrated that ω-oxidation significantly contributes to the biotransformation of HNE in liver slices [340]. Alary *et al.*, 1998 demonstrated the existence of further secondary products generated from HNA. These are the ω-oxidation product of HNA, namely 9-hydroxy-HNA, and the subsequent oxidation product of 9-OH-HNA, 9-carboxy-HNA [341]. Furthermore, 9-hydroxy-DHN was identified as ω-hydroxylated product of DHN [341].

The sequence of β- and α-oxidations in the pathway from HNA-CoA to the citric acid cycle was described in detail by Li *et al.*, 2013 [342]. They showed that after the formation of HNA-CoA a β-oxidation takes place leading to the formation of 2-hydroxy-heptanoyl-CoA. This intermediate is converted by α-oxidation to hexanoyl-CoA, and by three additional β-oxidations to acetyl-CoA. HNA-CoA can alternatively be phosphorylated to give 4-phospho-nonanoyl-CoA. The latter is hydrolyzed to 3-hydroxy nonanoyl-CoA, which is converted by three β-oxidations to propionyl-CoA, then to methylmalonyl-CoA and finally to succinyl-CoA as intermediate of the citric acid cycle [342].

Another group of secondary intermediates which was identified and quantified are the GSH-DHN and GSH-HNA conjugates as additional GSH-conjugates besides GSH-HNE [118]. Mercapturic acids were demonstrated to be the final products of this part of HNE metabolism (see also Chapter 6.2.). Different further secondary HNE products were postulated, such as 1-hydroxy-4-oxononene and others. In astrocytic biotransformation of HNE two novel metabolites of HNE were discovered, gamma-nonalactone and the potent pyrrole forming compound, 4-oxo-nonanal [343].

Compared to other organs, liver and kidney seem to be the most important organs for the elimination of the final products of metabolism. The importance of the kidney in the formation of HNE-mercapturic acid conjugates was demonstrated by Alary *et al.*, 1995 [319] and Grune *et al.*, 1997 [296]. HNE mercapturic acid, 1,4-dihydroxynonene mercapturic acid, 4-hydroxynonenoic mercapturic acid, and the corresponding lactone as HNE intermediates in urinary excretion were already mentioned above [319]. In studies with perfused kidneys it was shown, that all metabolites of HNE are also found in the urine. HNA, the HNE-mercapturic acid conjugates, and DHN had a higher concentration in urine than in the venous effluent of the kidney [296]. When 100 μM HNE was used for perfusion the highest HNE metabolite concentrations were found for DHN with 53 nmol/mL in urine and 24 nmol/mL in the venous effluent. The GSH-HNE conjugate and the products of the TCA-cycle never reached concentrations above 5 nmol/mL in urine. In the venous effluent their concentrations were about 15–17 nmol/mL which was 4–6-fold higher than in urine. HNE-mercapturic acids seem to be stable products of HNE metabolism which are excreted from the organism. 1,4-dihydroxynonene mercapturic acid (DHN-MA), was shown by Alary *et al.*, 1998 to be the major urinary metabolite of HNE administered to the rat [344]. It was characterized and determined to be a normal constituent of both rat and human urine. DHN-MA was excreted as a mixture of at least two stereoisomers. The 24-h urinary excretion of this compound was about 10 ng and 5 μg for rat and human, respectively. Following i.v. administration of HNE and [4-^{3}H]HNE to rats, besides DHN-MA 15 polar urinary metabolites accounting for about 50% of the urinary radioactivity were separated by HPLC by the same group [341]. Among them, eight major compounds and tritiated water were quantified. Most of polar metabolites were found to originate from ω-oxidation of 4-hydroxy-2-nonenoic acid (HNA) to 9-hydroxy-HNA, its mercapturic acid conjugate, and two diastereoisomers of the corresponding lactone. 9-Hydroxy-HNA is a substrate for alcohol and aldehyde dehydrogenases leading to the formation and excretion of 9-carboxy-HNA and of the corresponding lactone mercapturic acid conjugate. The 9-hydoxy-HNA and the 9-carboxy-HNA were already mentioned above. The alcohol dehydrogenase metabolite 1,4-dihydroxy-2-nonene (DHN) was to a lower extent ω-hydroxylated, leading to 9-hydroxy-DHN which was found to be excreted as a mercapturic acid conjugate [341].

To characterize the metabolic pathways involved in the formation of urinary HNE-MA conjugates in the rat, the metabolism of HNE-thioethers HNE-GSH, HNE-MA, and HNE-Cys by rat liver and kidney cytosolic fractions was investigated by Alary *et al.*, 2003 [318]. The results showed that HNE-GSH is a good substrate for cytosolic incubations whereas HNE-MA and HNE-Cys are poorly metabolized. About 80% of the urinary MA conjugates originate from the primary and major HNE metabolite, namely, the hemiacetalized HNE-GSH. The direct reduction of HNE-GSH by a cytosolic aldo-keto reductase (NADPH) leads to DHN-GSH and subsequently to DHN-MA. The direct oxidation of HNE-GSH by aldehyde dehydrogenase (NAD^+) leads to 4-hydroxynonenoic-lactone-GSH, the partial hydrolysis of which occurs at physiological pH and accounts for the corresponding HNA-GSH. Both the spontaneous- and the glutathione S-transferases-catalyzed retro-Michael cleavages of HNE-GSH and HNA-lactone-GSH are the source of HNE and HNA-lactone, respectively. HNA-lactone with both lipophilic and electrophilic properties is available for microsomal ω-hydroxylation by cytochrome P450 4A enzymes and conjugation with thiol groups, and therefore is the most likely candidate for the formation of ω-hydroxylated HNE-MA conjugates excreted in rat urine [318].

HNE biotransformation products were also detected in the bile of rats following a single intravenous administration of [4-(3)H]HNE. Five metabolites were present in the bile, two of them corresponded to GSH-HNE and to GSH-DHN adducts. Two further metabolites were identified as DHN and HNA-lactone mercapturic acid conjugates, whereas the fifth metabolite in the bile remained unidentified [345]. The kinetics of excretion of biliary metabolites demonstrated a rapid metabolism of HNE in rats: Within 4 h of injection, the bile accounted for 19.5% of the injected radioactivity, whereas only 3% was found in the feces within 48 h. All rat recipients were found to have measurable levels of HNE metabolites in bile, confirming that HNE undergoes enterohepatic recirculation, and the extent of recycling was about 8% of the total dose [345]. In this context it has to be mentioned that the multidrug resistance-associated protein2 (MRP2) plays an important role in the biliary excretion of glutathione conjugates of HNE [346].

The metabolism of the two enantiomers of HNE, the (R)- and the (S)-enantiomer of its racemic mixture, was the aim of a study by Gueraud *et al.*, 2005 [29]. The pathways were investigated in male rats after intravenous administration of the corresponding radiolabeled compounds and compared with the racemic mixture. Although the difference in the excretion rates was not statistically significant, the metabolic urinary pattern showed qualitative and quantitative differences. The level of 3-mercapturic acid-1,4-dihydroxynonene, which is considered as the major urinary metabolite of HNE, was significantly lower in the case of (S)-HNE injected rats. In order to clarify the intermediate pathways the authors performed *in vitro* studies using rat liver cytosolic incubations and HNE-glutathione conjugates as substrate and found large differences in the reduction and retro-Michael conversion steps of the metabolism between the conjugates originating from the two enantiomers [29]. Studies on the enantioselective metabolism of HNE by brain mitochondria were already described above [317] (see Chapter 7.2).

Determination of HNE Metabolic Products and Pathways

For the determination of HNE intermediates HPLC, HPLC-MS, and TLC were used, sometimes applying radioactively labelled compounds. In tracer kinetic experiments using tritium labelled HNE the HNE metabolites were separated by means of different methods (TLC or HPLC), and the formation of the products was simply calculated using the radioactivity of the spot or peak and dividing it by the specific radioactivity of the added HNE which can be taken as constant taking into account the overshoot of added radioactive HNE in comparison with the "low" formation of new HNE. After addition of labelled HNE aliquots of suspensions were taken. The HNE metabolizing reactions were stopped by the addition of e.g., 0.5 mL suspension to an equal volume of acetonitrile/acetic acid mixture (96:4; v/v). After centrifugation two aliquots of the supernatant were eluted on TLC plates. The elution with hexane/diethylether (3:7; v/v) allows the separation of HNE, HNA, and DHN. The elution with butanol/acetic acid/water (4:1:1; v/v/v) allows the determination of GSH-adducts of HNE [294]. For quantification of HNE utilization rates the radiolabeled compounds have to be scanned by an automatic TLC linear analyzer or to be scraped off and analyzed.

Sometimes acetonitrile/acetic acid extracts were used for the determination of HNE and HNA by an HPLC-method [292,293]. The disappearance of initial HNE levels, such as 100, 10 or 1 μM HNE, can be measured only by an HPLC-method (determination of remaining HNE during the whole experiment). The conditions for this "direct HNE and HNA analysis" are: C18 column, acetonitrile/water (4:6; v/v), 1 mL/min, 223 nm detection wavelength. Quantification of the acid was based on the comparison of the hydroxynonenoic acid peak with formation of NADH in experiments in which HNE reacted with aldehyde dehydrogenase. In those experiments, the rate of formation of NADH, which was observed spectrophotometrically at 340 nm, was equal to the consumption rate of HNE (measured by HPLC at 223 nm) and to the rate of formation of HNA. From the NADH formation rate (equal to the HNE consumption rate) and the peak integral of the HNA produced, one can calculate the extinction coefficient of the acid or the amount of HNA representing a defined peak area in the HPLC separation procedure [293]. HNE quantification was done by means of authentic HNE samples of known concentration, dissolved in methanol-water 80:20 (v/v) and comparison of the peak height of the standard chromatogram to sample chromatograms. A chromatogram of this direct HNE and HNA determination is shown in Figure 9.

The GSH-HNE-adduct was measured as an isoindol derivative after the reaction of the adduct existing in neutralized perchloric acid extracts with o-phthaldialdehyde in the presence of mercaptoethanol (Figure 10).

The reaction is started by addition of a sample aliquot to the o-phthaldiadehyde solution. The reaction mixture is transferred into a Hamilton syringe and the reaction is stopped by injection of the mixture into the HPLC system after 2 min. HPLC separation is carried out in an isocratic manner with 65% methanol-0.1 M sodium acetate buffer, pH 6.5 (v/v) in the presence of an ion-pair reagent (tetraethylammoniumhydroxide). Isoindol is detected by means of a fluorescence detector at wavelengths of 345 nm and 445 nm for excitation and emission, respectively. The formation of enantiomer products in the reaction of GSH-HNE adduct with o-phthaldialdehyde [347] led to the appearance of a double peak of the isoindol adduct (Figure 10). The reaction product of free reduced glutathione (GSH) with o-phthaldialdehyde eluted much earlier than the isoindol adduct within the isocratic HPLC separation. Precise measurement of the adduct concentration requires an extensive preparation of the neutralized perchloric acid extract of suspensions with the aim of removing disturbing influences of amino acid derivatives after HPLC separation. For the purification of the extracts, Bond Elut C18 columns and a Baker-10 Extraction system can be recommended [293]. The columns should be rinsed with methanol and equilibrated with water. After injection of the sample the column should be rinsed with 0.1 M sodium acetate buffer, pH 5, then rinsed with hexane. The sample is then eluted with methanol, evaporated to dryness, and redissolved in a defined volume of methanol [293]. The protein precipitate is washed with physiological saline solution, then dissolved in Protosol-Tissue and Gel Solubilizer and added to toluene scintillation fluid and measured by means of a β-counter if using ^{3}H-labeled HNE [292,297].

Different HPLC separations were applied in tracer kinetic experiments and in measurements of non-labelled HNE intermediates [293,312,319]. Urinary end products of HNE such as HNE mercapturic acid, 1,4-dihydroxynonene mercapturic acid, 4-hydroxynonenoic mercapturic acid, and the corresponding lactone were isolated and identified by mass spectrometry [319]. The HPLC-separation used by Srivastava *et al.*, 1998 was combined with different detection methods: electrospray ionization MS for the detection of HNE-GSH and DHN-GSH, GC-chemical ionization MS for HNA [312]. Inhibitors were applied to modify the HPLC peak pattern. Sorbinil, an inhibitor of aldose reductase, attenuated the DHN-GSH peak, and cyanamide, an inhibitor of aldehyde dehydrogenase (ALDH), attenuated HNA formation.

Various enzyme inhibitors were useful in studies on HNE metabolism. In investigations of HNE metabolic pathways of vascular smooth muscle cells it was shown that the formation of HNA was inhibited by cyanamide; indicating that the acid is derived from an ALDH-catalyzed pathway; while the overall rate of HNE metabolism was insensitive to inhibition of aldose reductase or aldehyde dehydrogenase. Nevetheless; inhibition of HNA formation by cyanamide led to a corresponding increase in the fraction of HNE metabolized by the GSH-linked pathway; indicating that ALDH-catalyzed oxidation competes with glutathione conjugation [311].

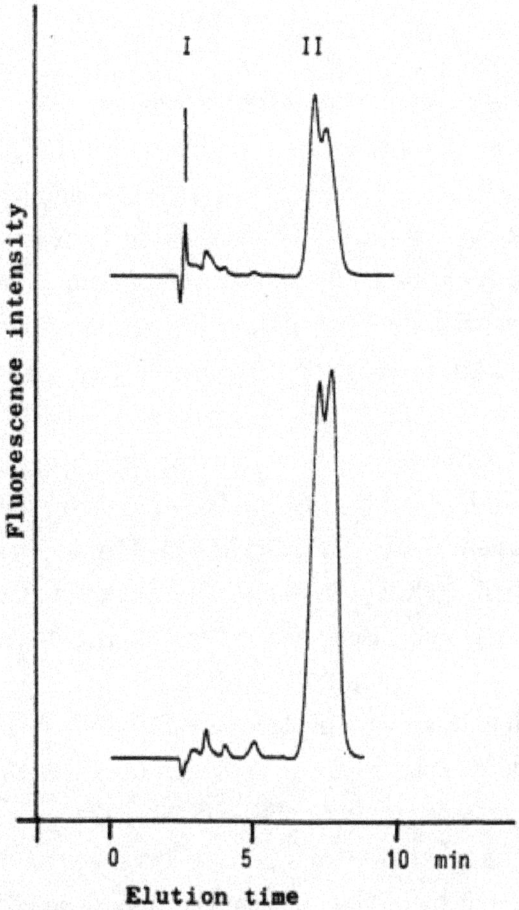

Figure 10. HPLC chromatogram of the isoindol derivative of the HNE-GSH conjugate (reaction product in presence of o-phthalaldehyde). Fluorimetric detection at 345 nm (excitation) and 445 nm (emission). Peaks: I GSH derivative; II HNE-GSH derivative; upper panel: Standard solution (45 µM); lower panel: extract from hepatocyte suspension incubated in presence of 100 µM 4-HNE [301].

7.6. HNE Metabolism as a Component of the Antioxidative Defense System

The catabolism of HNE may be regarded as a very important part of the antioxidative defense system of cells and organisms. This is due to the fact, that HNE like other aldehydic products of LPO is able to exert cytotoxic, mutagenic, signal, and carcinogenic effects as described above. The degradation of HNE to less toxic intermediates diminished the reactions between HNE and biomolecules and therefore effectively contributes to the antioxidative protection of cells and organisms. This protective effect is the more effective the more rapidly the metabolism of HNE leads to stable HNE products, which can be easily excreted. The primary importance of HNE-degrading pathways which function both at physiological and pathophysiological HNE levels, seems to be the protection of proteins from modification by aldehydic LPO products. This should especially be valid in regions with high HNE formation rate such as in postischemic or reperfused tissues or in synovial tissue of joints during rheumatoid arthritis and other inflammatory diseases or in lymphedematous tissue, *etc.*

7.7. HNE Intermediates as Potential Biomarkers of LPO

Two groups of HNE derivatives seem to be useful as possible biomarkers in clinical chemistry and laboratory medicine. The first are the mercapturic acid conjugates including the derivatives of HNE, HNA, and DHN. These conjugates are excreted with the urine and, therefore can be analyzed from urine. The second are HNE-protein adducts of blood serum or plasma. The HNE-modified plasma proteins are already used for evaluation of rheumatic and autoimmune diseases [348]. Both body fluids—urine and serum or plasma—are easily obtained from patients. Both groups of derivatives are relatively stable. While the mercapturic acid assays are time- and equipment-consuming, the determination of HNE-modified plasma proteins is quite easy to handle. Within the mercapturic acid conjugates Peiro *et al.*, 2005 favor the 1,4-dihydroxynonene mercapturic acid (DHN-MA) [349]. The authors compared three different urinary biomarkers of LPO during bromotrichloromethane induced oxidative stress in rats: MDA, the isoprostane 8-iso-PGF2α and DHN-MA as the major HNE urinary metabolite. The rats received a single dose of $BrCCl_3$ *per os* and LPO was estimated every day for a 4-day-period. All three parameters significantly increased in response to oxidative stress. DHN-MA showed the highest increase during the 24–48 h period after treatment.

The quantitation of the glutathione conjugate of HNE (HNE-GSH) in the human post-mortem brain was reported by Volkel *et al.*, 2006 applying the specific and very sensitive method of electrospray ionization triple quadrupole mass spectrometry (ESI-MS-MS) [350]. Levels of HNE-GSH conjugates were determined in hippocampus, entorhinal cortex, substantia innominata, frontal and temporal cortex, as well as cerebellum from patients with Alzheimer's disease and controls matched for age, gender, post mortem delay and storage time. The levels of HNE-GSH determined ranged between 1 and 500 pmol/g fresh weight in the brain areas examined. The brain specimen from patients with clinically and neuropathologically probable Alzheimer's disease diagnosed showed increased levels of HNE-GSH in the temporal and frontal cortex. Classification of disease severity revealed highest levels of HNE-GSH in the substantia innominata and the hippocampus, two brain regions known to be preferentially affected in Alzheimer's disease [4,9,350].

7.8. Further Medical Applications of HNE Metabolism

Several studies indicate that the glutathione conjugates of HNE metabolism are mediators of cell signalling and growth. A modification of the metabolic pattern, possibly by pharmacological interventions, could be of therapeutical importance [4]. Addition of HNE or the cell-permeable esters of HNE-GSH or DHN-GSH to cultures of rat aortic smooth muscle cells stimulated protein kinase C, NFκB, and AP-1, and increased cell growth. The mitogenic effects of HNE, but not HNE-GSH or DHN-GSH, were abolished by glutathione depletion. Inhibition of aldose reductase which catalyzes the reduction of HNE-GSH to DHN-GSH prevented protein kinase C, NFκB, and AP-1 stimulation and the increase in cell growth caused by HNE and HNE-GSH, but not DHN-GSH. Furthermore, the growth stimulating effect of DHN-GSH was enhanced in cells treated with antibodies directed against the glutathione conjugate transporters RLIP76 or the multidrug resistance protein-2 [351].

The demonstration that aldose reductase mediates endotoxin-induced inflammation and cardiomyopathy suggests that inhibition of this enzyme may be useful to attenuate maladaptive host responses and to treat acute cardiovascular dysfunction associated with endotoxic shock [335,352]. Srivastava *et al.*, 2006 also demonstrated a contribution of aldose reductase to diabetic hyperproliferation of vascular smooth muscle cells [353]. The authors showed that aldose reductase inhibitors such as tolrestat or sorbinil or antisense aldose reductase mRNA prevented hyperproliferation of cultured rat aortic smooth muscle cells induced by high glucose levels. The results suggest that inhibition of aldose reductase prevents glucose-induced stimulation of smooth muscle cell growth. It was therefore concluded that even though inhibition of aldose reductase increases vascular oxidative stress, this approach may be useful in preventing abnormal smooth muscle cell growth in vessels of diabetic patients.

Tammali *et al.* [354] showed that treatment of the human colon cancer cell line Caco-2 with HNE or HNE-GSH or the aldose reductase-catalyzed product of HNE-GSH, DHN-GSH, resulted in increased cyclooxygenase-2 (COX-2) expression and prostaglandin E2 production. Inhibition of aldose reductase prevented HNE- or HNE-GSH-induced but not DHN-GSH-induced up-regulation of COX-2 and PGE2. The finding that aldose reductase inhibition by siRNAs completely arrested tumor progression in nude mice bearing human colon adenocarcinoma cells suggests that aldose reductase is an obligatory mediator of growth factor-induced up-regulation of COX-2, PGE2, and growth of Caco-2 cells. These findings further suggest that inhibition of aldose reductase may be a novel therapeutic approach in preventing the progression of colon cancer [354].

Another oncological application of modulation of HNE metabolism was suggested in relation to the inhibition of cytosolic class 3 aldehyde dehydrogenase (ALDH3) by antisense oligonucleotides in hepatoma cells [59]. Several changes in aldehyde dehydrogenase isoenzyme expression take place in hepatoma cells. In particular cytosolic ALDH3, not expressed in normal hepatocytes, appears and increases with the degree of deviation. It has been demonstrated that cytosolic ALDH3 is important in determining the resistance of tumor cells to antitumor drugs, such as cyclophosphamide. Moreover, hepatoma-associated ALDH3 seems to be important in metabolizing aldehydes derived from LPO, and in particular the cytostatic aldehyde HNE. It was demonstrated previously that restoring endogenous LPO in hepatoma cells by enriching them with arachidonic acid causes a decrease of mRNA, protein and enzyme activity of ALDH3 and that this decrease reduces cell growth and/or causes cell death, depending on basal class 3 ALDH activity [59]. Applying antisense oligonucleotides the authors were able to demonstrate that ALDH3 mRNA and enzyme activity can be decreased by 90%. Concomitantly cell growth was reduced by about 70% in the JM2 hepatoma cell line with high ALDH3 activity [59].

The carbonyl scavenger *N*-acetylcysteine (NAC) is a pharmaceutical drug and nutritional supplement used primarily as a mucolytic agent and in the management of paracetamol (acetaminophen) overdose. It was used in various investigations to attenuate HNE-mediated organ damage, such as in alcoholic liver damage [355]. In these investigations NAC decreased enzyme release from damaged organs, decreased inflammatory reactivity and prevented significant loss of organ GSH content. NAC exerts also significant protective effects against HNE-induced neuronal death in cerebellar granule neurons. This neuroprotective effect is due, at least in part, to preservation of mitochondrial membrane potential and intracellular GSH levels [356]. An important medical application of modification of HNE metabolism is the development of therapeutic strategies for the prevention of a number of inflammatory diseases [33].

Further studies showed that caloric restriction in mammals and in flies (*Drosophila*) slows the age-related and the oxidative damage-related HNE accumulation [357]. Additionally suppression of apoptosis in aged kidney—where increased HNE and MDA levels were found—by caloric restriction was observed [358]. Additionally, a decrease of oxidative damage in skeletal muscle was demonstrated after caloric restriction of rhesus monkeys [359], and a decline of aldehyde dehydrogenase and glutathione S-transferase activities as pathways of mitochondrial HNE metabolism was prevented in dietary restricted animals [316].

8. Conclusions

This review tries to demonstrate that there is a multitude of possible targets of HNE within a cell. Several factors can be envisaged which determine the actual reactants within a cell: The concentration of HNE which can differentially modulate cell death, growth and differentiation [360], the concentration and nucleophilicity of the target, the polarity of the microenvironment, the accessibility within the tertiary structure, and the site of formation as well as the metabolic rate of HNE.

Abbreviations

ACC	acetyl CoA carboxylase
AcLDL	acetylated low-density lipoprotein
ACO2	mitochondrial aconitase
Acr-dG	adduct of deoxyguanosine and acrolein
AD	Alzheimer disease
ADH	alcohol dehydrogenase
ADP-iron	adenosine diphosphate-iron
AICAR	5-aminoimidazole-4-carboxyamide ribonucleoside
AIDS	acquired immuno-deficiency syndrome
AIE	approximate intracellular concentration (extinction)
AKR	aldo-keto reductase
ALD	alcoholic liver disease
ALDH	aldehyde dehydrogenase
Al-PEs	aldehyde-modified PEs
ALS	amyotrophic lateral sclerosis
AMD	age-related macular degeneration
AMPK	AMP-activated protein kinase
ANT	adenine nucleotide translocase
ApoB	apolipoprotein B-100
AR	aldose reductase
ARE	antioxidant response element
BAG3	Bcl-2- associated athanogene 3
BER	base excision repair

BSA	bovine serum albumin
BSO	L-buthionine-S,R-sulfoximine
CAR	carnosine
CCl_4	carbon tetrachloride
CERAD	consortium to establish a registry for Alzheimer disease
CH_2Cl_2	dichloromethane
CHO	Chinese-hamster ovary
CHOP	C/EBP homologous protein
Col II	collagen type II
COX-2	cyclooxygenase-2
CYP	cytochrome P450
ε-dA	1,*N*(6)-etheno-2'-deoxyadenosine
DHN	dihydroxynonene, non-2-ene-1,4-diol
DHN-MA	dihydroxynonene-mercapturic acid
DNA	desoxy-ribonucleic acid
DNA-PK	DNA-dependent protein kinase
DOX	doxorubicin
D3T	3H-1,2-dithiole-3-thione, inducing detoxification enzymes
eEF-2	Eukaryotic Elongation factor 2
EGFP	enhanced green fluorescent protein
EGFR	epidermal growth factor receptor
EHN	2,3-epoxy-4-hydroxy-nonanal
ERK1	extracellular signal-regulated kinase 1
ESI-MS-MS	electrospray ionization triple quadrupole mass spectrometry
FABP	fatty acid-binding protein
FAK	α1β1-integrin-focal adhesion kinase
FID	flame ionization detection
FITC	fluorescein isothiocyanate
GAPDH	glyceraldehyde-3-phosphate dehydrogenase
GC	gas chromatography
GC/MS	gas chromatography/mass spectrometry
GERD	gastroesophageal reflux disease
GRP78	glucose-regulated protein 78
GRX	glutaredoxin
GSH	glutathione
GSH-DHN	glutathione-dihydroxynonene conjugate
GSK3β	glycogen synthase kinase-3-β
GSSG	glutathione in the oxidized form, glutathione disulfide
GST	glutathione-S-transferase
HA-1	Chinese hamster fibroblast control cells
HAEC	human aortic endothelial cell
Hb	hemoglobin

HCC	hepatocellular carcinoma
HCV	hepatitis C virus
HMEC-1	immortalized human microvascular endothelial cells
HHE	4-hydroxy-hexenal
[^{3}H]HNE	tritium labeled hydroxynonenal
HNA	hydroxynonenoic acid
HNA-GSH	hydroxynonenoic acid-glutathione conjugate
HNE	4-hydroxy-nonenal
HNE-dG	1,*N*-2-propane adduct of guanine and HNE
HNE-H2A	HNE-modified histone-H2A
H2O2	hydrogen peroxide
HPLC	high performance liquid chromatography
HPNE	4-hydroperoxy-2-nonenal
HO-1	heme oxygenase-1
HSA	human serum albumin
HU	HU hydroxyurea
HUVEC	human umbilical vein endothelial cell
ICAM-1	intercellular adhesion molecule-1
IsoLGs	isolevuglandins
iTRAQ	isobaric Tags for Relative and Absolute Quantitation
LA	linolenic acid
LAD	late Alzheimer disease
LADH	lipoamide dehydrogenase
LCA	retinal dystrophy Leber congenital amaurosis
LC-MS/MS	liquid chromatography-tandem mass spectrometry
LDH	lactate dehydrogenase
LDL	low density lipoprotein
LKB1	liver kinase B1
LOX1	lectin-like oxidized low-density lipoprotein receptor-1
LPO	lipid peroxidation
MAL-6	2,2,6,6-tetramethyl-4-maleimidopiperidin-1-oxyl
MALDI-TOF	matrix-assisted laser desorption/ionization-time of flight
MAPK	mitogen-activated protein kinase
MBP-1	c-myc binding protein-1
MCI	mild cognitive impairment
MDA	malondialdehyde
MDM2	mouse double minute 2
MH1C1	rat hepatoma cells
MMP-13	matrix metalloproteinase 13
MRP-1	multidrug resistance protein-1
NAD+	nicotinamid dinucleotide, oxidized form
NADH	nicotinamid dinucleotide, reduced form

NADP	nicotinamid dinucleotide phosphate
NAFLD	non-alcoholic fatty liver disease
NEP	neprilysin
NER	nucleotide excision repair
Nrf2	NF-E2-related factor 2
OA	osteoarthritic
OC5 and OC14	Chinese hamster fibroblast H_2O_2-resistant cell lines
ONE	4-oxo-nonenal
oxLDL	oxidized low-density lipoprotein
ox-PAPC	oxidized PAPC
PAF	1-*O*-alkyl-2-acetyl-sn-glycero-3-phospho-choline
PAPC	1-palmitoyl-2-arachidonoyl-*sn*-glycero-3-phospho-choline
PCAD	preclinical Alzheimer disease
PDGF	platelet-derived growth factor
PDI	protein disulfide isomerase
PE	phosphatidyl-ethanolamine
PGE	prostaglandin E
Pin1	peptidyl-prolyl *cis/trans*-isomerase A1
PMF	peptide mass fingerprinting
POS	photoreceptor outer segment
PG(E2)	prostaglandin E2
PtdIns(3,4,5)P-3	phosphatidylinositol-3,4,5-trisphosphate
PTEN	phosphatase and tensin homolog deleted on chromosome 10
RAS	rennin/angiotensin system
RBC	erythrocytes
RDH	retinol dehydrogenase
RLIP76	Ral-binding GTPase activating protein
ROC	receiver operator curves
ROOH	lipid hydroperoxide
ROS	reactive oxygen species
RPE	retinal pigment epithelial
SAM	sterile-alpha motif
SCFAs	short chain fatty acids
SERCA1a	serco/endoplasmic reticulum Ca^{2+}-ATPase
SIRT3	sirtuin 3
SLE	systemic lupus erythematosus
SMC	smooth muscle cell
sXBP1	spliced X-box-binding protein-1
α-Syn	α-synuclein
TBA-RS	thiobarbituric acid-reactive substances
TBI	traumatic brain injury
TLC	thin layer chromatography

TLR	toll-like receptor
TRPV1	transient receptor potential vanilloid 1
TRX	thioredoxin
TrxR	thioredoxin reductase
UCP	uncoupling protein
UPR	unfolded protein response
VSL#3	a medical food delivering high amount of beneficial live bacteria
YAP1	Yeast Activator Protein 1
ZAK kinase	sterile alpha motif and leucine zipper containing kinase (AZK)

References

1. Schauenstein, E.; Esterbauer, H. Ueber die Reaktion von 9,12 Linolsaeureester in Wasser 6. Mitteilung: Chromatographische Untersuchungen ueber die Zusammensetzung des Praeparates LHPO und die Isolierung der einzelnen tumor-wirksamen Hydroperoxid-Komponenten (In Germany). *Chem. Mon.* **1963**, *94*, 11.
2. Halliwell, B.; Gutteridge, J. *Free Radicals in Biology and Medicine*, 4th ed.; Oxford University Press: Oxford, UK, 2007.
3. Catala, A. Lipid peroxidation of membrane phospholipids generates hydroxy-alkenals and oxidized phospholipids active in physiological and/or pathological conditions. *Chem. Phys. Lipids* **2009**, *157*, 1–11.
4. Poli, G.; Schaur, R.J.; Siems, W.G.; Leonarduzzi, G. 4-Hydroxynonenal: A membrane lipid oxidation product of medicinal interest. *Med. Res. Rev.* **2008**, *28*, 569–631.
5. Schaur, J.R. Basic aspects of the biochemical reactivity of 4-hydroxynonenal. *Mol. Asp. Med.* **2003**, *24*, 149–159.
6. Dubinina, E.E.; Dadali, V.A. Role of 4-hydroxy-*trans*-2-nonenal in cell functions. *Biochem. Moscow* **2010**, *75*, 1069–1087.
7. Fritz, K.S.; Petersen, D.R. An overview of the chemistry and biology of reactive aldehydes. *Free Rad. Biol. Med.* **2013**, *59*, 85–91.
8. Zarkovic, N.; Cipak, A.; Jaganjac, M.; Borovic, S.; Zarkovic, K. Pathophysiological relevance of aldehydic protein modifications. *J. Proteomics* **2013**, *92*, 239–247.
9. Perluigi, M.; Coccia, R.; Butterfield, D.A. 4-hydroxy-2-nonenal, a reactive product of lipid peroxidation, and neurodegenerative diseases: A toxic combination illuminated by redox proteomics studies. *Antioxid. Redox Signal.* **2012**, *17*, 1590–1609.
10. Riahi, Y.; Cohen, G.; Shamni, O.; Sasson, S. Signaling and cytotoxic functions of 4-hydroxyalkenals. *Am. J. Physiol. Endocrinol. Metab.* **2010**, *299*, E879–E886.
11. Forman, H.J. Reactive oxygen species and α,β-unsaturated aldehydes as second messengers in signal transduction. *Oxid. Nitrosative Stress Dis.* **2010**, *1203*, 35–44.
12. Higdon, A.; Diers, A.R.; Oh, J.Y.; Landar, A.; Darley-Usmar, V.M. Cell signalling by reactive lipid species: New concepts and molecular mechanisms. *Biochem. J.* **2012**, *442*, 453–464.
13. Dwivedi, S.; Sharma, A.; Patrick, B.; Sharma, R.; Awasthi, Y.C. Role of 4-hydroxynonenal and its metabolites in signaling. *Redox Rep.* **2007**, *12*, 4–10.

14. Ayala, A.; Munoz, M.F.; Argueelles, S. Lipid peroxidation: Production, metabolism, and signaling mechanisms of malondialdehyde and 4-hydroxy-2-nonenal. *Oxid. Med. Cell. Longev.* **2014**, doi:10.1155/2014/360438.
15. Spickett, C.M. The lipid peroxidation product 4-hydroxy-2-nonenal: Advances in chemistry and analysis. *Redox Biol.* **2013**, *1*, 145–152.
16. LoPachin, R.M.; Gavin, T.; Petersen, D.R.; Barber, D.S. Molecular mechanisms of 4-hydroxy-2-nonenal and acrolein toxicity: Nucleophilic targets and adduct formation. *Chem. Res. Toxicol.* **2009**, *22*, 1499–1508.
17. Mattson, M.P. Roles of the lipid peroxidation product 4-hydroxynonenal in obesity, the metabolic syndrome, and associated vascular and neurodegenerative disorders. *Exp. Gerontol.* **2009**, *44*, 625–633.
18. Guichardant, M.; Lagarde, M. Analysis of biomarkers from lipid peroxidation: A comparative study. *Eur. J. Lipid Sci. Technol.* **2009**, *111*, 75–82.
19. Tappel, A.L. Measurement and protection from *in vivo* lipid peroxidation. In *Free Radicals in Biology*; Pryor, W.A., Ed.; Academic Press: New York, NY, USA, 1980; pp. 1–47.
20. Aikens, J.; Dix, T.A. Perhydroxyl radical (HOO.) initiated lipid-peroxidation—The role of fatty-acid hydroperoxides. *J. Biol. Chem.* **1991**, *266*, 15091–15098.
21. Tsunada, S.; Iwakiri, R.; Noda, T.; Fujimoto, K.; Fuseler, J.; Rhoads, C.A.; Aw, T.Y. Chronic exposure to subtoxic levels of peroxidized lipids suppresses mucosal cell turnover in rat small intestine and reversal by glutathione. *Dig. Dis. Sci.* **2003**, *48*, 210–222.
22. Bochkov, V.N.; Kadl, A.; Huber, J.; Gruber, F.; Binder, B.R.; Leitinger, N. Protective role of phospholipid oxidation products in endotoxin-induced tissue damage. *Nature* **2002**, *419*, 77–81.
23. Yin, H.; Morrow, J.D.; Porter, N.A. Identification of a novel class of endoperoxides from arachidonate autoxidation. *J. Biol. Chem.* **2004**, *279*, 3766–3776.
24. Poli, G.; Schaur, R.J. 4-Hydroxynonenal in the pathomechanisms of oxidative stress. *Iubmb Life* **2000**, *50*, 315–321.
25. Pillon, N.J.; Soulere, L.; Vella, R.E.; Croze, M.; Care, B.R.; Soula, H.A.; Doutheau, A.; Lagarde, M.; Soulage, C.O. Quantitative structure-activity relationship for 4-hydroxy-2-alkenal induced cytotoxicity in L6 muscle cells. *Chem. Biol. Interact.* **2010**, *188*, 171–180.
26. Vazdar, M.; Jurkiewicz, P.; Hof, M.; Jungwirth, P.; Cwiklik, L. Behavior of 4-hydroxynonenal in phospholipid membranes. *J. Phys. Chem. B* **2012**, *116*, 6411–6415.
27. Dabrowski, M.J.; Zolnerciks, J.K.; Balogh, L.M.; Greene, R.J.; Kavanagh, T.J.; Atkins, W.M. Stereoselective Effects of 4-hydroxynonenal in cultured mouse hepatocytes. *Chem. Res. Toxicol.* **2010**, *23*, 1601–1607.
28. Hiratsuka, A.; Saito, H.; Watabe, T. (S)-preferential cytotoxicity of 4-hydroxy-2(E)-nonenal enantiomers in rat Clone 9 cells. *Toxicology* **2001**, *164*, 199.
29. Gueraud, F.; Crouzet, F.; Alary, J.; Rao, D.; Debrauwer, L.; Laurent, F.; Cravedi, J.P. Enantioselective metabolism of (R)- and (S)-4-hydroxy-2-nonenal in rat. *Biofactors* **2005**, *24*, 97–104.
30. Sadhukhan, S.; Han, Y.; Jin, Z.; Tochtrop, G.P.; Zhang, G.-F. Glutathionylated 4-hydroxy-2-(E)-alkenal enantiomers in rat organs and their contributions toward the disposal of 4-hydroxy-2-(E)-nonenal in rat liver. *Free Rad. Biol. Med.* **2014**, *70*, 78–85.

31. Wakita, C.; Maeshima, T.; Yamazaki, A.; Shibata, T.; Ito, S.; Akagawa, M.; Ojika, M.; Yodoi, J.; Uchida, K. Stereochemical configuration of 4-hydroxy-2-nonenal-cysteine adducts and their stereoselective formation in a redox-regulated protein. *J. Biol. Chem.* **2009**, *284*, 28810–28822.
32. Komisarski, M.; Kaczmarska, Z.; Kusmierek, J.T. Practical highly enantioselective synthesis of (R)- and (S)-(E)-4-hydroxynon-2-enal. *Acta Biochim. Polonica* **2009**, *56*, 189–193.
33. Yadav, U.C.S.; Ramana, K.V. Regulation of NF-kappaB-induced inflammatory signaling by lipid peroxidation-derived aldehydes. *Oxid. Med. Cell. Longevity* **2013**, doi:10.1155/2013/690545.
34. Singh, S.P.; Niemczyk, M.; Zimniak, L.; Zimniak, P. Fat accumulation in Caenorhabditis elegans triggered by the electrophilic lipid peroxidation product 4-Hydroxynonenal (4-HNE). *Aging* **2009**, *1*, 68–80.
35. Singh, S.P.; Niemczyk, M.; Saini, D.; Awasthi, Y.C.; Zimniak, L.; Zimniak, P. Role of the electrophilic lipid peroxidation product 4-hydroxynonenal in the development and maintenance of obesity in mice. *Biochemistry* **2008**, *47*, 3900–3911.
36. Schneider, C.; Brash, A.R. Monomeric and dimeric routes to formation of 4-hydroxynonenal during lipid peroxidation. *Abstr. Papers Am. Chem. Soc.* **2013**, *246*, Abstract 41-AGFD.
37. Long, E.K.; Picklo, M.J., Sr. *Trans*-4-hydroxy-2-hexenal, a product of *n*-3 fatty acid peroxidation: Make some room HNE. *Free Rad. Biol. Med.* **2010**, *49*, 1–8.
38. Wang, X.; Allen, T.D.; Yang, Y.; Moore, D.R.; Huycke, M.M. Cyclooxygenase-2 generates the endogenous mutagen *trans*-4-hydroxy-2-nonenal in *Enterococcus faecalis*-infected macrophages. *Cancer Prev. Res.* **2013**, *6*, 206–216.
39. Kang, S.C.; Kim, H.-W.; Kim, K.B.; Kwack, S.J.; Ahn, I.Y.; Bae, J.Y.; Lim, S.K.; Lee, B.M. Hepatotoxicity and nephrotoxicity produced by 4-hydroxy-2-nonenal (4-HNE) following 4-week oral administration to sprague-dawley rats. *J. Toxicol. Environ. Health Part A* **2011**, *74*, 779–789.
40. Zhang, Y.; Sano, M.; Shinmura, K.; Tamaki, K.; Katsumata, Y.; Matsuhashi, T.; Morizane, S.; Ito, H.; Hishiki, T.; Endo, J.; *et al.* 4-Hydroxy-2-nonenal protects against cardiac ischemia-reperfusion injury via the Nrf2-dependent pathway. *J. Mol. Cell. Cardiol.* **2010**, *49*, 576–586.
41. Mannervik, B. Five decades with glutathione and the GSTome. *J. Biol. Chem.* **2012,** *287*, 6072–6083.
42. Balogh, L.M.; Atkins, W.M. Interactions of glutathione transferases with 4-hydroxynonenal. *Drug Metab. Rev.* **2011**, *43*, 165–178.
43. Rudd, L.P.; Kabler, S.L.; Morrow, C.S.; Townsend, A.J. Enhanced glutathione depletion, protein adduct formation, and cytotoxicity following exposure to 4-hydroxy-2-nonenal (HNE) in cells expressing human multidrug resistance protein-1 (MRP1) together with human glutathione S-transferase-M1 (GSTM1). *Chem. Biol. Interact.* **2011**, *194*, 113–119.
44. Shireman, L.M.; Kripps, K.A.; Balogh, L.M.; Conner, K.P.; Whittington, D.; Atkins, W.M. Glutathione transferase A4-4 resists adduction by 4-hydroxynonenal. *Arch. Biochem. Biophys.* **2010**, *504*, 182–189.
45. Zhou, S.Y.; Decker, E.A. Ability of carnosine and other skeletal muscle components to quench unsaturated aldehydic lipid oxidation products. *J. Agric. Food Chem.* **1999**, *47*, 51–55.
46. Shearn, C.T.; Smathers, R.L.; Backos, D.S.; Reigan, P.; Orlicky, D.J.; Petersen, D.R. Increased carbonylation of the lipid phosphatase PTEN contributes to Akt2 activation in a murine model of early alcohol-induced steatosis. *Free Rad. Biol. Med.* **2013**, *65*, 680–692.

47. Zhu, Q.; Zheng, Z.-P.; Cheng, K.-W.; Wu, J.-J.; Zhang, S.; Tang, Y.S.; Sze, K.-H.; Chen, J.; Chen, F.; Wang, M. Natural polyphenols as direct trapping agents of lipid peroxidation-derived acrolein and 4-hydroxy-*trans*-2-nonenal. *Chem. Res. Toxicol.* **2009**, *22*, 1721–1727.
48. Stevens, J.F.; Sowell, J.D.; Frei, B. Ascorbic Acid Conjugates. US Patent US20090104705A1, 2009.
49. Uchida, K. Immunochemical detection of lipid peroxidation-specific epitopes. In *Biomarkers for Antioxidative Defense and Oxidative Damage*; Aldini, G., Yeum, K.J., Niki, E., Russell, R.M., Eds.; Blackwell Publishing: Hoboken, NJ, USA, 2010; pp. 157–171.
50. Wakita, C.; Honda, K.; Shibata, T.; Akagawa, M.; Uchida, K. A method for detection of 4-hydroxy-2-nonenal adducts in proteins. *Free Radic. Biol. Med.* **2011**, *51*, 1–4.
51. Zhang, B.; Shi, Z.; Duncan, D.T.; Prodduturi, N.; Marnett, L.J.; Liebler, D.C. Relating protein adduction to gene expression changes: A systems approach. *Mol. Biosyst.* **2011**, *7*, 2118–2127.
52. Dick, R.A.; Kwak, M.K.; Sutter, T.R.; Kensler, T.W. Antioxidative function and substrate specificity of NAD(P)H-dependent alkenal/one oxidoreductase—A new role for leukotriene B-4 12-hydroxydehydrogenase/15-oxoprostaglandin 13-reductase. *J. Biol. Chem.* **2001**, *276*, 40803–40810.
53. Chen, H.J.C.; Chung, F.L. Epoxidation of *trans*-4-hydroxy-2-nonenal by fatty acid hydroperoxides and hydrogen peroxide. *Chem. Res. Toxicol.* **1996**, *9*, 306–312.
54. Perry, E.A.; Castellani, R.J.; Moreira, P.I.; Nunomura, A.; Lui, Q.; Harris, P.L.R.; Sayre, L.M.; Szweda, P.A.; Szweda, L.I.; Zhu, X.; *et al.* Neurofilaments are the major neuronal target of hydroxynonenal-mediated protein cross-links. *Free Radic. Res.* **2013**, *47*, 507–510.
55. Singh, I.N.; Gilmer, L.K.; Miller, D.M.; Cebak, J.E.; Wang, J.A.; Hall, E.D. Phenelzine mitochondrial functional preservation and neuroprotection after traumatic brain injury related to scavenging of the lipid peroxidation-derived aldehyde 4-hydroxy-2-nonenal. *J. Cerebral Blood Flow Metab.* **2013**, *33*, 593–599.
56. Fournet, G.; Martin, G.; Quash, G. α,β-Acetylenic amino thiolester inhibitors of aldehyde dehydrogenases 1&3: Suppressors of apoptogenic aldehyde oxidation and activators of apoptosis. *Curr. Med. Chem.* **2013**, *20*, 527–533.
57. Guo, J.-M.; Liu, A.-J.; Zang, P.; Dong, W.-Z.; Ying, L.; Wang, W.; Xu, P.; Song, X.-R.; Cai, J.; Zhang, S.-Q.; *et al.* ALDH2 protects against stroke by clearing 4-HNE. *Cell Res.* **2013**, *23*, 915–930.
58. Ma, H.; Guo, R.; Yu, L.; Zhang, Y.; Ren, J. Aldehyde dehydrogenase 2 (ALDH2) rescues myocardial ischaemia/reperfusion injury: Role of autophagy paradox and toxic aldehyde. *Eur. Heart J.* **2011**, *32*, 1025–1038.
59. Muzio, G.; Maggiora, M.; Paiuzzi, E.; Oraldi, M.; Canuto, R.A. Aldehyde dehydrogenases and cell proliferation. *Free Radic. Biol. Med.* **2012**, *52*, 735–746.
60. Townsend, A.J.; Leone-Kabler, S.; Haynes, R.L.; Wu, Y.; Szweda, L.; Bunting, K.D. Selective protection by stably transfected human ALDH3A1 (but not human ALDH1A1) against toxicity of aliphatic aldehydes in V79 cells. *Chem. Biol. Interact.* **2001**, *130–132*, 261–273.
61. Black, W.; Chen, Y.; Matsumoto, A.; Thompson, D.C.; Lassen, N.; Pappa, A.; Vasiliou, V. Molecular mechanisms of ALDH3A1-mediated cellular protection against 4-hydroxy-2-nonenal. *Free Radic. Biol. Med.* **2012**, *52*, 1937–1944.

62. Yoval-Sanchez, B.; Rodriguez-Zavala, J.S. Differences in susceptibility to inactivation of human aldehyde dehydrogenases by lipid peroxidation byproducts. *Chem. Res. Toxicol.* **2012**, *25*, 722–729.
63. Murakami, Y.; Ohsawa, I.; Kasahara, T.; Ohta, S. Cytoprotective role of mitochondrial amyloid beta peptide-binding alcohol dehydrogenase against a cytotoxic aldehyde. *Neurobiol. Aging* **2009**, *30*, 325–329.
64. Srivastava, S.; Vladykovskaya, E.; Barski, O.A.; Kaiserova, M.S.K.; Petrash, J.M.; Chung, S.S.; Hunt, G.; Dawn, B.; Bhatnagar, A. Aldose reductase protects against early atherosclerotic lesion formation in apolipoprotein e-null mice. *Circ. Res.* **2009**, *105*, 793–802.
65. Keith, R.J.; Haberzettl, P.; Vladykovskaya, E.; Hill, B.G.; Kaiserova, K.; Srivastava, S.; Barski, O.; Bhatnagar, A. Aldose reductase decreases endoplasmic reticulum stress in ischemic hearts. *Chem. Biol. Interact.* **2009**, *178*, 242–249.
66. Li, D.; Ferrari, M.; Ellis, E.M. Human aldo-keto reductase AKR7A2 protects against the cytotoxicity and mutagenicity of reactive aldehydes and lowers intracellular reactive oxygen species in hamster V79-4 cells. *Chem. Biol. Interact.* **2012**, *195*, 25–34.
67. Jung, K.-A.; Kwak, M.-K. Enhanced 4-hydroxynonenal resistance in KEAP1 silenced human colon cancer cells. *Oxid. Med. Cell. Longevity* **2013**, doi:10.1155/2013/423965.
68. Lyon, R.C.; Li, D.; McGarvie, G.; Ellis, E.M. Aldo-keto reductases mediate constitutive and inducible protection against aldehyde toxicity in human neuroblastoma SH-SY5Y cells. *Neurochem. Int.* **2013**, *62*, 113–121.
69. Zhong, L.; Liu, Z.; Yan, R.; Johnson, S.; Zhao, Y.; Fang, X.; Cao, D. Aldo-keto reductase family 1 B10 protein detoxifies dietary and lipid-derived alpha, beta-unsaturated carbonyls at physiological levels. *Biochem. Biophys. Res. Commun.* **2009**, *387*, 245–250.
70. Martin, H.-J.; Maser, E. Role of human aldo-keto-reductase AKR1B10 in the protection against toxic aldehydes. *Chem. Biol. Interact.* **2009**, *178*, 145–150.
71. Endo, S.; Matsunaga, T.; Kuragano, T.; Ohno, S.; Kitade, Y.; Tajima, K.; el-Kabbani, O.; Hara, A. Properties and tissue distribution of a novel aldo-keto reductase encoding in a rat gene (Akr1b10). *Arch. Biochem. Biophys.* **2010**, *503*, 230–237.
72. Matsunaga, T.; Shinoda, Y.; Inoue, Y.; Endo, S.; el-Kabbani, O.; Hara, A. Protective effect of rat aldo-keto reductase (AKR1C15) on endothelial cell damage elicited by 4-hydroxy-2-nonenal. *Chem. Biol. Interact.* **2011**, *191*, 364–370.
73. Marchette, L.D.; Thompson, D.A.; Kravtsova, M.; Ngansop, T.N.; Mandal, M.N.A.; Kasus-Jacobi, A. Retinol dehydrogenase 12 detoxifies 4-hydroxynonenal in photoreceptor cells. *Free Radic. Biol. Med.* **2010**, *48*, 16–25.
74. Amunom, I.; Dieter, L.J.; Tamasi, V.; Cai, J.; Conklin, D.J.; Srivastava, S.; Martin, M.V.; Guengerich, F.P.; Prough, R.A. Cytochromes p450 catalyze the reduction of α,β-unsaturated aldehydes. *Chem. Res. Toxicol.* **2011**, *24*, 1223–1230.
75. Zhang, G.-F.; Kombu, R.S.; Kasumov, T.; Han, Y.; Sadhukhan, S.; Zhang, J.; Sayre, L.M.; Ray, D.; Gibson, K.M.; Anderson, V.A.; *et al.* Catabolism of 4-hydroxyacids and 4-hydroxynonenal via 4-hydroxy-4-phosphoacyl-CoAs. *J. Biol. Chem.* **2009**, *284*, 33521–33534.
76. Berthiaume, J.M.; Li, Q.; Sadhukhan, S.; Henry, F.; Tochtrop, G.P.; Brunengraber, H.; Zhang, G. Catabolism of 4-hydroxy-2(E)-nonenal (HNE) via omega oxidation in perfused rat livers. *FASEB J.* **2013**, *27*, Abstract No. 794.15.

77. Subramaniam, R.; Roediger, F.; Jordan, B.; Mattson, M.P.; Keller, J.N.; Waeg, G.; Butterfield, D.A. The lipid peroxidation product, 4-hydroxy-2-*trans*-nonenal, alters the conformation of cortical synaptosomal membrane proteins. *J. Neurochem.* **1997**, *69*, 1161–1169.
78. Chen, J.J.; Yu, B.P. Alterations in mitochondrial-membrane fluidity by lipid-peroxidation products. *Free Radic. Biol. Med.* **1994**, *17*, 411–418.
79. Fleuranceau-Morel, P.; Barrier, L.; Fauconneau, B.; Piriou, A.; Huguet, F. Origin of 4-hydroxynonenal incubation-induced inhibition of dopamine transporter and Na^+/K^+ adenosine triphosphate in rat striatal synaptosomes. *Neurosci. Lett.* **1999**, *277*, 91–94.
80. Liu, Q.; Ullery, J.; Zhu, J.; Liebler, D.C.; Marnett, L.J.; Zhang, B. RNA-seq data analysis at the gene and CDS levels provides a comprehensive view of transcriptome responses induced by 4-hydroxynonenal. *Mol. Biosyst.* **2013**, *9*, 3036–3046.
81. Schreier, S.M.; Muellner, M.K.; Steinkellner, H.; Hermann, M.; Esterbauer, H.; Exner, M.; Gmeiner, B.M.K.; Kapiotis, S.; Laggner, H. Hydrogen sulfide scavenges the cytotoxic lipid oxidation product 4-HNE. *Neurotox. Res.* **2010**, *17*, 249–256.
82. Siems, W.; Grune, T. Intracellular metabolism of 4-hydroxynonenal. *Mol. Asp. Med.* **2003**, *24*, 167–175.
83. Vila, A.; Tallman, K.A.; Jacobs, A.T.; Liebler, D.C.; Porter, N.A.; Marnett, L.J. Identification of protein targets of 4-hydroxynonenal using click chemistry for *ex vivo* biotinylation of azido and alkynyl derivatives. *Chem. Res. Toxicol.* **2008**, *21*, 432–444.
84. Tallman, K.A.; Vila, A.; Porter, N.A.; Marnett, L.J. Measuring electrophile stress. *Curr. Protoc. Toxicol.* **2009**, 11–17.
85. Chavez, J.; Chung, W.-G.; Miranda, C.L.; Singhal, M.; Stevens, J.F.; Maier, C.S. Site-specific protein adducts of 4-hydroxy-2(E)-nonenal in human THP-1 monocytic cells: Protein carbonylation is diminished by ascorbic acid. *Chem. Res. Toxicol.* **2010**, *23*, 37–47.
86. Codreanu, S.G.; Zhang, B.; Sobecki, S.M.; Billheimer, D.D.; Liebler, D.C. Global analysis of protein damage by the lipid electrophile 4-hydroxy-2-nonenal. *Mol. Cell. Proteomics* **2009**, *8*, 670–680.
87. Petersen, D.R.; Carbone, D.; Doorn, J. Hepatocellular targets of 4-hydroxy-2-nonenal modification. In Proceedings of the 3rd International Meeting of the HNE-Club, Genova, Italy, 16–18 June 2006; Abstract No. 6.
88. Hussain, S.N.A.; Matar, G.; Barreiro, E.; Florian, M.; Divangahi, M.; Vassilakopoulos, T. Modifications of proteins by 4-hydroxy-2-nonenal in the ventilatory muscles of rats. *Am. J. Physiol. Lung Cell. Mol. Physiol.* **2006**, *290*, L996–L1003.
89. Jung, T.; Engels, M.; Kaiser, B.; Grune, T. Distribution of oxidized and HNE-modified proteins in U87 cells. *Biofactors* **2005**, *24*, 165–170.
90. Carbone, D.L.; Doorn, J.A.; Kiebler, Z.; Ickes, B.R.; Petersen, D.R. Modification of heat shock protein 90 by 4-hydroxynonenal in a rat model of chronic alcoholic liver disease. *J. Pharmacol. Exp. Ther.* **2005**, *315*, 8–15.
91. Carbone, D.L.; Doorn, J.A.; Kiebler, Z.; Sampey, B.P.; Petersen, D.R. Inhibition of Hsp72-mediated protein refolding by 4-hydroxy-2-nonenal. *Chem. Res. Toxicol.* **2004**, *17*, 1459–1467.

92. Legards, J.F.; des Rosiers, C. Assay of 4-hydroxynonenal (HNE) adducts with various polyaminoacids (PAA) using gas chromatography-mass spectrometry (GCMS). In Proceedings of the 3rd International Meeting of the HNE-Club, Genova, Italy, 16–18 June 2006.
93. Doorn, J.A.; Petersen, D.R. Covalent adduction of nucleophilic amino acids by 4-hydroxynonenal and 4-oxononenal. *Chem. Biol. Interact.* **2003**, *143*, 93–100.
94. Isom, A.L.; Barnes, S.; Wilson, L.; Kirk, M.; Coward, L.; Darley-Usmar, V. Modification of cytochrome c by 4-hydroxy-2-nonenal: Evidence for histidine, lysine, and arginine-aldehyde adducts. *J. Am. Soc. Mass Spectrom.* **2004**, *15*, 1136–1147.
95. Liu, Z.F.; Minkler, P.E.; Sayre, L.A. Mass spectroscopic characterization of protein modification by 4-Hydroxy-2-(E)-nonenal and 4-Oxo-2-(E)-nonenal. *Chem. Res. Toxicol.* **2003**, *16*, 901–911.
96. Aldini, G.; Dalle-Donne, I.; Vistoli, G.; Facino, R.M.; Carini, M. Covalent modification of actin by 4-hydroxy-*trans*-2-nonenal (HNE): LC-ESI-MS/MS evidence for Cys374 Michael adduction. *J. Mass Spectrom.* **2005**, *40*, 946–954.
97. Bennaars-Eiden, A.; Higgins, L.A.; Hertzel, A.V.; Kapphahn, R.J.; Ferrington, D.; Bernlohr, D.A. Covalent modification of epithelial fatty acid binding protein by 4-hydroxynonenal *in vitro* and *in vivo*: Evidence for a role in antioxidant biology. *FASEB J.* **2003**, *17*, Abstract No. 845.847.
98. Carbone, D.L.; Doorn, J.A.; Kiebler, Z.; Petersen, D.R. Cysteine modification by lipid peroxidation products inhibits protein disulfide isomerase. *Chem. Res. Toxicol.* **2005**, *18*, 1324–1331.
99. Asselin, C.; Bouchard, B.; Tardif, J.C.; des Rosiers, C. Circulating 4-hydroxynonenal-protein thioether adducts assessed by gas chromatography-mass spectrometry are increased with disease progression and aging in spontaneously hypertensive rats. *Free Radic. Biol. Med.* **2006**, *41*, 97–105.
100. Moreau, R.; Nguyen, B.T.; Doneanu, C.E.; Hagen, T.M. Reversal by aminoguanidine of the age-related increase in glycoxidation and lipoxidation in the cardiovascular system of Fischer 344 rats. *Biochem. Pharmacol.* **2005**, *69*, 29–40.
101. Kim, C.H.; Zou, Y.N.; Kim, D.H.; Kim, N.D.; Yu, B.P.; Chung, H.Y. Proteomic analysis of nitrated and 4-hydroxy-2-nonenal-modified serum proteins during aging. *J. Gerontol. Ser. A Biol. Sci. Med. Sci.* **2006**, *61*, 332–338.
102. Petersen, D.R.; Doorn, J.A. Reactions of 4-hydroxynonenal with proteins and cellular targets. *Free Radic. Biol.Med.* **2004**, *37*, 937–945.
103. Renner, A.; Sagstetter, M.R.; Harms, H.; Lange, V.; Gotz, M.E.; Elert, O. Formation of 4-hydroxy-2-nonenal protein adducts in the ischemic rat heart after transplantation. *J. Heart Lung Transpl.* **2005**, *24*, 730–736.
104. Galligan, J.J.; Smathers, R.L.; Fritz, K.S.; Epperson, L.E.; Hunter, L.E.; Petersen, D.R. Protein carbonylation in a murine model for early alcoholic liver disease. *Chem. Res. Toxicol.* **2012**, *25*, 1012–1021.
105. Aoyama, K.; Matsubara, K.; Kobayashi, S. Aging and oxidative stress in progressive supranuclear palsy. *Eur. J. Neurol.* **2006**, *13*, 89–92.
106. Grune, T. HNE-modified proteins: Formation, distribution and fate. In Proceedings of the 3rd International Meeting of the HNE-Club, Genova, Italy, 2006; Abstract No. 17.
107. Carbone, D.L.; Doorn, J.A.; Petersen, D.R. 4-hydroxynonenal regulates 26S proteasomal degradation of alcohol dehydrogenase. *Free Radic. Biol. Med.* **2004**, *37*, 1430–1439.

108. Okada, K.; Wangpoengtrakul, C.; Osawa, T.; Toyokuni, S.; Tanaka, K.; Uchida, K. 4-hydroxy-2-nonenal-mediated impairment of intracellular proteolysis during oxidative stress—Identification of proteasomes as target molecules. *J. Biol. Chem.* **1999**, *274*, 23787–23793.
109. Hyun, D.H.; Lee, M.H.; Halliwell, B.; Jenner, P. Proteasomal dysfunction induced by 4-hydroxy-2,3-*trans*-nonenal, an end-product of lipid peroxidation: A mechanism contributing to neurodegeneration? *J. Neurochem.* **2002**, *83*, 360–370.
110. Bardag-Gorce, F.; Li, J.; French, B.A.; French, S.W. The effect of ethanol-induced CYP2E1 on proteasome activity: The role of 4-hydroxynonenal. *Exp. Mol. Pathol.* **2005**, *78*, 109–115.
111. Kessova, I.G.; Cederbaum, A.I. The effect of CYP2E1-dependent oxidant stress on activity of proteasomes in HepG2 cells. *J. Pharmacol. Exp. Ther.* **2005**, *315*, 304–312.
112. Ferrington, D.A.; Kapphahn, R.J. Catalytic site-specific inhibition of the 20S proteasome by 4-hydroxynonenal. *FEBS Lett.* **2004**, *578*, 217–223.
113. Tsuchiya, Y.; Yamaguchi, M.; Chikuma, T.; Hojo, H. Degradation of glyceraldehyde-3-phosphate dehydrogenase triggered by 4-hydroxy-2-nonenal and 4-hydroxy-2-hexenal. *Arch. Biochem. Biophys.* **2005**, *438*, 217–222.
114. Xu, G.Z.; Liu, Y.H.; Sayre, L.M. Independent synthesis, solution behavior, and studies on the mechanism of formation of a primary amine-derived fluorophore representing cross-linking of proteins by (E)-4-hydroxy-2-nonenal. *J. Org. Chem.* **1999**, *64*, 5732–5745.
115. Zarkovic, N. Protein-aldehydic adducts as biomarkers of oxidative stress, lipid peroxidation and oxidative homeostasis. *Free Radic. Res.* **2009**, *43*, 33.
116. Wang, J.-F.; Tan, H. Oxidative protein modification of soluble *N*-ethylmaleimide-sensitive factor attachment protein receptors. *FASEB J.* **2013**, *27*, 890–897.
117. Gallagher, E.P.; Gardner, J.L.; Barber, D.S. Several glutathione S-transferase isozymes that protect against oxidative injury are expressed in human liver mitochondria. *Biochem. Pharmacol.* **2006**, *71*, 1619–1628.
118. Alary, J.; Fernandez, Y.; Debrauwer, L.; Perdu, E.; Gueraud, F. Identification of intermediate pathways of 4-hydroxynonenal metabolism in the rat. *Chem. Res.Toxicol.* **2003**, *16*, 320–327.
119. Lucas, J.H.; Wheeler, D.G.; Guan, Z.; Suntres, Z.; Stokes, B.T. Effect of glutathione augmentation on lipid peroxidation after spinal cord injury. *J. Neurotrauma* **2002**, *19*, 763–775.
120. Yang, Y.S.; Sharma, R.; Sharma, A.; Awasthi, S.; Awasthi, Y.C. Lipid peroxidation and cell cycle signaling: 4-Hydroxynonenal, a key molecule in stress mediated signaling. *Acta Biochim. Polonica* **2003**, *50*, 319–336.
121. Patrick, B.; Li, J.; Jeyabal, P.V.S.; Reddy, P.; Yang, Y.S.; Sharma, R.; Sinha, M.; Luxon, B.; Zimniak, P.; Awasthi, S.; *et al.* Depletion of 4-hydroxynonenal in hGSTA4-transfected HLE B-3 cells results in profound changes in gene expression. *Biochem. Biophys. Res. Commun.* **2005**, *334*, 425–432.
122. Dwivedi, S.; Sharma, R.; Sharma, A.; Zimniak, P.; Ceci, J.D.; Awasthi, Y.C.; Boor, P.J. The course of CCl4 induced hepatotoxicity is altered in mGSTA4-4 null (−/−) mice. *Toxicology* **2006**, *218*, 58–66.
123. Engle, M.R.; Singh, S.P.; Czernik, P.J.; Gadd, D.; Montague, D.C.; Ceci, J.D.; Yang, Y.S.; Awasthi, S.; Awasthi, Y.C.; Zimniak, P. Physiological role of mGSTA4-4, a glutathione

S-transferase metabolizing 4-hydroxynonenal: Generation and analysis of mGSTA4 null mouse. *Toxicol. Appl. Pharmacol.* **2004**, *194*, 296–308.

124. Knoll, N.; Ruhe, C.; Veeriah, S.; Sauer, J.; Glei, M.; Gallagher, E.P.; Pool-Zobel, B.L. Genotoxicity of 4-hydroxy-2-nonenal in human colon tumor cells is associated with cellular levels of glutathione and the modulation of glutathione S-transferase A4 expression by butyrate. *Toxicol. Sci.* **2005**, *86*, 27–35.
125. Srivastava, S.; Ramana, K.V.; Bhatnagar, A.; Srivastava, S.K. Synthesis, quantification, characterization, and signaling properties of glutathionyl conjugates of enals. *Methods Enzymol.* **2010**, *474*, 297–313.
126. Frohnert, B.I.; Long, E.K.; Hahn, W.S.; Bernlohr, D.A. Glutathionylated lipid aldehydes are products of adipocyte oxidative stress and activators of macrophage inflammation. *Diabetes* **2014**, *63*, 89–100.
127. Spite, M.; Summers, L.; Porter, T.F.; Srivastava, S.; Bhatnagar, A.; Serhan, C.N. Resolvin D1 controls inflammation initiated by glutathione-lipid conjugates formed during oxidative stress. *Br. J. Pharmacol.* **2009**, *158*, 1062–1073.
128. Aldini, G.; Facino, R.M.; Beretta, G.; Carini, M. Carnosine and related dipeptides as quenchers of reactive carbonyl species: From structural studies to therapeutic perspectives. *Biofactors* **2005**, *24*, 77–87.
129. Aldini, G.; Carini, M.; Beretta, G.; Bradamante, S.; Facino, R.M. Carnosine is a quencher of 4-hydroxy-nonenal: Through what mechanism of reaction? *Biochem. Biophys. Res. Commun.* **2002**, *298*, 699–706.
130. Barski, O.A.; Xie, Z.; Baba, S.P.; Sithu, S.D.; Agarwal, A.; Cai, J.; Bhatnagar, A.; Srivastava, S. Dietary carnosine prevents early atherosclerotic lesion formation in apolipoprotein e-null mice. *Arterioscler. Thromb. Vasc. Biol.* **2013**, *33*, 1162–1170.
131. Wang, M.; Fang, H.; Han, X. Shotgun lipidomics analysis of 4-hydroxyalkenal species directly from lipid extracts after one-step *in situ* derivatization. *Anal. Chem.* **2012**, *84*, 4580–4586.
132. Guiotto, A.; Calderan, A.; Ruzza, P.; Osler, A.; Rubini, C.; Jo, D.G.; Mattson, M.P.; Borin, G. Synthesis and evaluation of neuroprotective alpha,beta-unsaturated aldehyde scavenger histidyl-containing analogues of carnosine. *J. Med. Chem.* **2005**, *48*, 6156–6161.
133. Guiotto, A.; Calderan, A.; Ruzza, P.; Borin, G. Carnosine and carnosine-related antioxidants: A review. *Curr. Med. Chem.* **2005**, *12*, 2293–2315.
134. Vistoli, G.; Orioli, M.; Pedretti, A.; Regazzoni, L.; Canevotti, R.; Negrisoli, G.; Carini, M.; Aldini, G. Design, synthesis, and evaluation of carnosine derivatives as selective and efficient sequestering agents of cytotoxic reactive carbonyl species. *Chemmedchem* **2009**, *4*, 967–975.
135. Bertinaria, M.; Rolando, B.; Giorgis, M.; Montanaro, G.; Guglielmo, S.; Buonsanti, M.F.; Carabelli, V.; Gavello, D.; Daniele, P.G.; Fruttero, R.; *et al.* Synthesis, physicochemical characterization, and biological activities of new carnosine derivatives stable in human serum as potential neuroprotective agents. *J. Med. Chem.* **2011**, *54*, 611–621.
136. Sugano, E.; Isago, H.; Murayama, N.; Tamai, M.; Tomita, H. Different anti-oxidant effects of thioredoxin 1 and thioredoxin 2 in retinal epithelial cells. *Cell Struct. Funct.* **2013**, *38*, 81–88.

137. Zhang, H.; Du, Y.; Zhang, X.; Lu, J.; Holmgren, A. Glutaredoxin 2 reduces both thioredoxin 2 and thioredoxin 1 and protects cells from apoptosis induced by auranofin and 4-hydroxynonenal. *Antioxid. Redox Signal.* **2014**, *21*, 669–681.
138. Tang, X.; Sayre, L.M.; Tochtrop, G.P. A mass spectrometric analysis of 4-hydroxy-2-(E)-nonenal modification of cytochrome c. *J. Mass Spectrom.* **2011**, *46*, 290–297.
139. Gardner, H.W.; Deighton, N. Effect of 4-hydroxy-2(E)-nonenal on soybean lipoxygenase-1. *Lipids* **2001**, *36*, 623–628.
140. Uchida, K.; Stadtman, E.R. Covalent attachment of 4-hydroxynonenal to glyceraldehyde-3-phosphate dehydrogenase—A possible involvement of intramolecular and intermolecular cross-linking reaction. *J. Biol. Chem.* **1993**, *268*, 6388–6393.
141. Ishii, T.; Tatsuda, E.; Kumazawa, S.; Nakayama, T.; Uchida, K. Molecular basis of enzyme inactivation by an endogenous electrophile 4-hydroxy-2-nonenal: Identification of modification sites in glyceraldehyde-3-phosphate dehydrogenase. *Biochemistry* **2003**, *42*, 3474–3480.
142. Van Iersel, M.L.P.S.; Ploemen, J.-P.H.T.M.; lo Bello, M.; Federici, G.; van Bladeren, P.J. Interactions of α,β-unsaturated aldehydes and ketones with human glutathione S-transferase P1-1. *Chem. Biol. Interact.* **1997**, *108*, 67–78.
143. Bosch-Morell, F.; Flohe, L.; Marin, N.; Romero, F.J. 4-Hydroxynonenal inhibits glutathione peroxidase: Protection by glutathione. *Free Radic. Biol. Med.* **1999**, *26*, 1383–1387.
144. Korotchkina, L.G.; Yang, H.S.; Tirosh, O.; Packer, L.; Patel, M.S. Protection by thiols of the mitochondrial complexes from 4-hydroxy-2-nonenal. *Free Radic. Biol. Med.* **2001**, *30*, 992–999.
145. Del Corso, A.; dal Monte, M.; Vilardo, P.G.; Cecconi, I.; Moschini, R.; Banditelli, S.; Cappiello, M.; Tsai, L.; Mura, U. Site-specific inactivation of aldose reductase by 4-hydroxynonenal. *Arch. Biochem. Biophys.* **1998**, *350*, 245–248.
146. Yu, M.K.; Moos, P.J.; Cassidy, P.; Wade, M.; Fitzpatrick, F.A. Conditional expression of 15-lipoxygenase-1 inhibits the selenoenzyme thioredoxin reductase—Modulation of selenoproteins by lipoxygenase enzymes. *J. Biol. Chem.* **2004**, *279*, 28028–28035.
147. Fang, J.G.; Holmgren, A. Inhibition of thioredoxin and thioredoxin reductase by 4-hydroxy-2-nonenal *in vitro* and *in vivo*. *J. Am. Chem. Soc.* **2006**, *128*, 1879–1885.
148. Chen, Z.H.; Saito, Y.; Yoshida, Y.; Sekine, A.; Noguchi, N.; Niki, E. 4-Hydroxynonenal induces adaptive response and enhances PC12 cell tolerance primarily through induction of thioredoxin reductase 1 via activation of Nrf2. *J. Biol. Chem.* **2005**, *280*, 41921–41927.
149. Chen, Z.H.; Yoshida, Y.; Saito, Y.; Noguchi, N.; Niki, E. Adaptive response induced by lipid peroxidation products in cell cultures. *FEBS Lett.* **2006**, *580*, 479–483.
150. Morquette, B.; Shi, Q.; Lavigne, P.; Ranger, P.; Fernandes, J.C.; Benderdour, M. Production of lipid peroxidation products in osteoarthritic tissues—New evidence linking 4-hydroxynonenal to cartilage degradation. *Arthritis Rheum.* **2006**, *54*, 271–281.
151. Ramanathan, R.; Mancini, R.A.; Suman, S.P.; Beach, C.M. Covalent binding of 4-hydroxy-2-nonenal to lactate dehydrogenase decreases NADH formation and metmyoglobin reducing activity. *J. Agric. Food Chem.* **2014**, *62*, 2112–2117.
152. Schlisser, A.E.; Yan, J.; Hales, B.F. Teratogen-induced oxidative stress targets glyceraldehyde-3-phosphate dehydrogenase in the organogenesis stage mouse embryo. *Toxicol. Sci.* **2010**, *118*, 686–695.

153. Markesbery, W.R.; Carney, J.M. Oxidative alterations in Alzheimer's disease. *Brain Pathol.* **1999**, *9*, 133–146.
154. Butterfield, D.A.; Lauderback, C.M. Lipid peroxidation and protein oxidation in Alzheimer's disease brain: Potential causes and consequences involving amyloid β-peptide-associated free radical oxidative stress1,2. *Free Radic. Biol. Med.* **2002**, *32*, 1050–1060.
155. Markesbery, W.R.; Lovell, M.A. Four-hydroxynonenal, a product of lipid peroxidation, is increased in the brain in Alzheimer's disease. *Neurobiol. Aging* **1998**, *19*, 33–36.
156. Sultana, R.; Butterfield, D.A. Oxidatively modified GST and MRP1 in Alzheimer's disease brain: Implications for accumulation of reactive lipid peroxidation products. *Neurochem. Res.* **2004**, *29*, 2215–2220.
157. Huang, Y.; Li, W.; Kong, A.-N.T. Anti-oxidative stress regulator NF-E2-related factor 2 mediates the adaptive induction of antioxidant and detoxifying enzymes by lipid peroxidation metabolite 4-hydroxynonenal. *Cell Biosci.* **2012**, *2*, 1–7.
158. Balogh, L.M.; le Trong, I.; Kripps, K.A.; Shireman, L.M.; Stenkamp, R.E.; Zhang, W.; Mannervik, B.; Atkins, W.M. Substrate specificity combined with stereopromiscuity in glutathione transferase A4-4-dependent metabolism of 4-hydroxynonenal. *Biochemistry* **2010**, *49*, 1541–1548.
159. Calamaras, T.D.; Lee, C.; Lan, F.; Ido, Y.; Siwik, D.A.; Colucci, W.S. Post-translational modification of serine/threonine kinase LKB1 via adduction of the reactive lipid species 4-hydroxy-*trans*-2-nonenal (HNE) at lysine residue 97 directly inhibits kinase activity. *J. Biol. Chem.* **2012**, *287*, 42400–42406.
160. Shearn, C.T.; Backos, D.S.; Orlicky, D.J.; Smathers-McCullough, R.L.; Petersen, D.R. Identification of 5' AMP-activated kinase as a target of reactive aldehydes during chronic ingestion of high concentrations of ethanol. *J. Biol. Chem.* **2014**, *289*, 15449–15462.
161. Wang, C.; Weerapana, E.; Blewett, M.M.; Cravatt, B.F. A chemoproteomic platform to quantitatively map targets of lipid-derived electrophiles. *Nat. Methods* **2014**, *11*, 79–85.
162. Shearn, C.T.; Fritz, K.S.; Reigan, P.; Petersen, D.R. Modification of Akt2 by 4-hydroxynonenal inhibits insulin-dependent Akt signaling in HepG2 cells. *Biochemistry* **2011**, *50*, 3984–3996.
163. MacDonald, M.J.; Langberg, E.C.; Tibell, A.; Sabat, G.; Kendrick, M.A.; Szweda, L.I.; Ostenson, C.G. Identification of ATP synthase as a lipid peroxide protein adduct in pancreatic islets from humans with and without type 2 Diabetes mellitus. *J. Clin. Endocrinol. Metab.* **2013**, *98*, E727–E731.
164. Guo, J.; Prokai-Tatrai, K.; Vien, N.; Rauniyar, N.; Ughy, B.; Prokai, L. Protein targets for carbonylation by 4-hydroxy-2-nonenal in rat liver mitochondria. *J. Proteomics* **2011**, *74*, 2370–2379.
165. Shearn, C.T.; Smathers, R.L.; Stewart, B.J.; Fritz, K.S.; Galligan, J.J.; Hail, N., Jr.; Petersen, D.R. Phosphatase and tensin homolog deleted on chromosome 10 (PTEN) inhibition by 4-hydroxynonenal leads to increased Akt activation in hepatocytes. *Mol. Pharmacol.* **2011**, *79*, 941–952.
166. Fang, X.; Fu, Y.; Long, M.J.C.; Haegele, J.A.; Ge, E.J.; Parvez, S.; Aye, Y. Temporally controlled targeting of 4-hydroxynonenal to specific proteins in living cells. *J. Am. Chem. Soc.* **2013**, *135*, 14496–14499.

167. Los, G.V.; Encell, L.P.; McDougall, M.G.; Hartzell, D.D.; Karassina, N.; Zimprich, C.; Wood, M.G.; Learish, R.; Ohane, R.F.; Urh, M.; *et al.* HatoTag: A novel protein labeling technology for cell imaging and protein analysis. *ACS Chem. Biol.* **2008**, *3*, 373–382.

168. Fritz, K.S.; Galligan, J.J.; Smathers, R.L.; Roede, J.R.; Shearn, C.T.; Reigan, P.; Petersen, D.R. 4-hydroxynonenal inhibits SIRT3 via thiol-specific modification. *Chem. Res. Toxicol.* **2011**, *24*, 651–662.

169. Krohne, T.U.; Kaemmerer, E.; Holz, F.G.; Kopitz, J. Lipid peroxidation products reduce lysosomal protease activities in human retinal pigment epithelial cells via two different mechanisms of action. *Exp. Eye Res.* **2010**, *90*, 261–266.

170. Wang, D.S.; Iwata, N.; Hama, E.; Saido, T.C.; Dickson, K. Oxidized neprilysin in aging and Alzheimer's disease brains. *Biochem. Biophys. Res. Commun.* **2003**, *310*, 236–241.

171. Wang, R.; Wang, S.; Malter, J.S.; Wang, D.-S. Effects of HNE-modification induced by A beta on neprilysin expression and activity in SH-SY5Y cells. *J. Neurochem.* **2009**, *108*, 1072–1082.

172. Wang, R.; Malter, J.S.; Wang, D.-S. *N*-Acetylcysteine prevents 4-hydroxynonenal—And amyloid-β-induced modification and inactivation of neprilysin in SH-SY5Y cells. *J. Alzheimers Dis.* **2010**, *19*, 179–189.

173. Liu, Q.; Simpson, D.C.; Gronert, S. Carbonylation of mitochondrial aconitase with 4-hydroxy-2-(E)-nonenal: Localization and relative reactivity of addition sites. *Biochim. Biophys. Acta Proteins Proteomics* **2013**, *1834*, 1144–1154.

174. Zhao, Y.; Miriyala, S.; Miao, L.; Mitov, M.; Schnell, D.; Dhar, S.K.; Cai, J.; Klein, J.B.; Sultana, R.; Butterfield, D.A.; *et al.* Redox proteomic identification of HNE-bound mitochondrial proteins in cardiac tissues reveals a systemic effect on energy metabolism after doxorubicin treatment. *Free Radic. Biol. Med.* **2014**, *72*, 55–65.

175. Gentile, F.; Pizzimenti, S.; Arcaro, A.; Pettazzoni, P.; Minelli, R.; D'Angelo, D.; Mamone, G.; Ferranti, P.; Toaldo, C.; Cetrangolo, G.; *et al.* Exposure of HL-60 human leukaemic cells to 4-hydroxynonenal promotes the formation of adduct(s) with alpha-enolase devoid of plasminogen binding activity. *Biochem. J.* **2009**, *422*, 285–294.

176. Reed, T.T.; Pierce, W.M.; Markesbery, W.R.; Butterfield, D.A. Proteomic identification of HNE-bound proteins in early Alzheimer disease: Insights into the role of lipid peroxidation in the progression of AD. *Brain Res.* **2009**, *1274*, 66–76.

177. Muller, C.; Bandemer, J.; Vindis, C.; Camare, C.; Mucher, E.; Gueraud, F.; Larroque-Cardoso, P.; Bernis, C.; Auge, N.; Salvayre, R.; *et al.* Protein disulfide isomerase modification and inhibition contribute to ER stress and apoptosis induced by oxidized low density lipoproteins. *Antioxid. Redox Signal.* **2013**, *18*, 731–742.

178. Aluise, C.D.; Rose, K.; Boiani, M.; Reyzer, M.L.; Manna, J.D.; Tallman, K.; Porter, N.A.; Marnett, L.J. Peptidyl-prolyl *cis/trans*-isomerase A1 (Pin1) is a target for modification by lipid electrophiles. *Chem. Res. Toxicol.* **2013**, *26*, 270–279.

179. Tanito, M.; Haniu, H.; Elliott, M.H.; Singh, A.K.; Matsumoto, H.; Anderson, R.E. Identification of 4-hydroxynonenal-modified retinal proteins induced by photooxidative stress prior to retinal degeneration. *Free Radic. Biol. Med.* **2006**, *41*, 1847–1859.

180. Toyokuni, S.; Yamada, S.; Kashima, M.; Ihara, Y.; Yamada, Y.; Tanaka, T.; Hiai, H.; Seino, Y.; Uchida, K. Serum 4-hydroxy-2-nonenal-modified albumin is elevated in patients with type 2 Diabetes Mellitus. *Antioxid. Redox Signal.* **2000**, *2*, 681–685.
181. Aldini, G.; Gamberoni, L.; Orioli, M.; Beretta, G.; Regazzoni, L.; Facino, R.M.; Carini, M. Mass spectrometric characterization of covalent modification of human serum albumin by 4-hydroxy-*trans*-2-nonenal. *J. Mass Spectrom.* **2006**, *41*, 1149–1161.
182. Liu, Q.; Simpson, D.C.; Gronert, S. The reactivity of human serum albumin towards *trans*-4-hydroxy-2-nonenal. *J. Mass Spectrom.* **2012**, *47*, 411–424.
183. Khatoon, F.; Alam, K.; Ali, A. Physicochemical and immunological studies on 4-hydroxynonenal modified HSA: Implications of protein damage by lipid peroxidation products in the etiopathogenesis of SLE. *Hum. Immunol.* **2012**, *73*, 1132–1139.
184. Xiang, W.; Weisbach, V.; Sticht, H.; Seebahn, A.; Bussmann, J.; Zimmermann, R.; Becker, C.-M. Oxidative stress-induced posttranslational modifications of human hemoglobin in erythrocytes. *Arch. Biochem. Biophys.* **2013**, *529*, 34–44.
185. Naveena, B.M.; Faustman, C.; Tatiyaborworntham, N.; Yin, S.; Ramanathan, R.; Mancini, R.A. Detection of 4-hydroxy-2-nonenal adducts of turkey and chicken myoglobins using mass spectrometry. *Food Chem.* **2010**, *122*, 836–840.
186. Smathers, R.L.; Fritz, K.S.; Galligan, J.J.; Shearn, C.T.; Reigan, P.; Marks, M.J.; Petersen, D.R. Characterization of 4-HNE modified L-FABP reveals alterations in structural and functional dynamics. *PLoS ONE* **2012**, *7*, e38459.
187. Hellberg, K.; Grimsrud, P.A.; Kruse, A.C.; Banaszak, L.J.; Ohlendorf, D.H.; Bernlohr, D.A. X-ray crystallographic analysis of adipocyte fatty acid binding protein (aP2) modified with 4-hydroxy-2-nonenal. *Protein Sci.* **2010**, *19*, 1480–1489.
188. Stewart, B.J.; Roede, J.R.; Doorn, J.A.; Petersen, D.R. Lipid aldehyde-mediated cross-linking of apolipoprotein B-100 inhibits secretion from HepG2 cells. *Biochim. Biophys. Acta Mol. Cell Biol. Lipids* **2009**, *1791*, 772–780.
189. Shao, B.; Pennathur, S.; Pagani, I.; Oda, M.N.; Witztum, J.L.; Oram, J.F.; Heinecke, J.W. Modifying apolipoprotein A–I by malondialdehyde, but not by an array of other reactive carbonyls, blocks cholesterol efflux by the ABCA1 pathway. *J. Biol. Chem.* **2010**, *285*, 18473–18484.
190. Zhu, X.; Tang, X.; Anderson, V.E.; Sayre, L.M. Mass spectrometric characterization of protein modification by the products of nonenzymatic oxidation of linoleic acid. *Chem. Res. Toxicol.* **2009**, *22*, 1386–1397.
191. Ou, J.J.; Zhang, Y.L.; Montine, T.J. *In vivo* assessment of lipid peroxidation products associated with age-related neurodegenerative diseases. *Exp. Neurol.* **2002**, *175*, 363–369.
192. Mattson, M.P.; Chan, S.L. Neuronal and glial calcium signaling in Alzheimer's disease. *Cell Calcium* **2003**, *34*, 385–397.
193. Pedersen, W.A.; Fu, W.; Keller, J.N.; Markesbery, W.R.; Appel, S.; Smith, R.G.; Kasarskis, E.; Mattson, M.P. Protein modification by the lipid peroxidation product 4-hydroxynonenal in the spinal cords of amyotrophic lateral sclerosis patients. *Ann. Neurol.* **1998**, *44*, 819–824.
194. Lovell, M.A.; Bradley, M.A.; Fister, S.X. 4-Hydroxyhexenal (HHE) impairs glutamate transport in astrocyte cultures. *J. Alzheimers Dis.* **2012**, *32*, 139–146.

195. Maroteaux, L.; Campanelli, J.T.; Scheller, R.H. Synuclein: A neuron-specific protein localized to the nucleus and presynaptic nerve terminal. *J. Neurosci.* **1988**, *8*, 2804–2815.
196. Cabin, D.E.; Shimazu, K.; Murphy, D.; Cole, N.B.; Gottschalk, W.; McIlwain, K.L.; Orrison, B.; Chen, A.; Ellis, C.E.; Paylor, R.; *et al.* Synaptic vesicle depletion correlates with attenuated synaptic responses to prolonged repetitive stimulation in mice lacking alpha-synuclein. *J. Neurosci.* **2002**, *22*, 8797–8807.
197. Murphy, D.D.; Rueter, S.M.; Trojanowski, J.Q.; Lee, V.M. Synucleins are developmentally expressed, and alpha-synuclein regulates the size of the presynaptic vesicular pool in primary hippocampal neurons. *J. Neurosci.* **2000**, *20*, 3214–3220.
198. Lotharius, J.; Brundin, P. Impaired dopamine storage resulting from α-synuclein mutations may contribute to the pathogenesis of Parkinson's disease. *Hum. Mol. Genet.* **2002**, *11*, 2395–2407.
199. Jomova, K.; Vondrakova, D.; Lawson, M.; Valko, M. Metals, oxidative stress and neurodegenerative disorders. *Mol. Cell. Biochem.* **2010**, *345*, 91–104.
200. Xiang, W.; Schlachetzki, J.C.M.; Helling, S.; Bussmann, J.C.; Berlinghof, M.; Schaeffer, T.E.; Marcus, K.; Winkler, J.; Klucken, J.; Becker, C.-M. Oxidative stress-induced posttranslational modifications of alpha-synuclein: Specific modification of alpha-synuclein by 4-hydroxy-2-nonenal increases dopaminergic toxicity. *Mol. Cell. Neurosci.* **2013**, *54*, 71–83.
201. Bae, E.-J.; Ho, D.-H.; Park, E.; Jung, J.W.; Cho, K.; Hong, J.H.; Lee, H.-J.; Kim, K.P.; Lee, S.-J. Lipid peroxidation product 4-hydroxy-2-nonenal promotes seeding-capable oligomer formation and cell-to-cell transfer of a-synuclein. *Antioxid. Redox Signal.* **2013**, *18*, 770–783.
202. Nasstrom, T.; Fagerqvist, T.; Barbu, M.; Karlsson, M.; Nikolajeff, F.; Kasrayan, A.; Ekberg, M.; Lannfelt, L.; Ingelsson, M.; Bergstrom, J. The lipid peroxidation products 4-oxo-2-nonenal and 4-hydroxy-2-nonenal promote the formation of α-synuclein oligomers with distinct biochemical, morphological, and functional properties. *Free Radic. Biol. Med.* **2011**, *50*, 428–437.
203. Shibata, N.; Inose, Y.; Toi, S.; Hiroi, A.; Yamamoto, T.; Kobayashi, M. Involvement of 4-hydroxy-2-nonenal accumulation in multiple system atrophy. *Acta Histochem. Cytochem.* **2010**, *43*, 69–75.
204. Hortigon-Vinagre, M.P.; Chardonnet, S.; Montigny, C.; Gutierrez-Martin, Y.; Champeil, P.; Henao, F. Inhibition by 4-hydroxynonenal (HNE) of Ca^{2+} transport by SERCA1a: Low concentrations of HNE open protein-mediated leaks in the membrane. *Free Radic. Biol. Med.* **2011**, *50*, 323–336.
205. Gunthorpe, M.J.; Benham, C.D.; Randall, A.; Davis, J.B. The diversity in the vanilloid (TRPV) receptor family of ion channels. *Trends Pharmacol. Sci.* **2002**, *23*, 183–191.
206. Kishimoto, E.; Naito, Y.; Handa, O.; Okada, H.; Mizushima, K.; Hirai, Y.; Nakabe, N.; Uchiyama, K.; Ishikawa, T.; Takagi, T.; *et al.* Oxidative stress-induced posttranslational modification of TRPV1 expressed in esophageal epithelial cells. *Am. J. Physiol. Gastrointest. Liver Physiol.* **2011**, *301*, G230–G238.
207. Vindis, C.; Escargueil-Blanc, I.; Elbaz, M.; Marcheix, B.; Grazide, M.H.; Uchida, K.; Salvayre, R.; Negre-Salvayre, A. Desensitization of platelet-derived growth factor receptor-beta by oxidized lipids in vascular cells and atherosclerotic lesions—Prevention by Aldehyde Scavengers. *Circ. Res.* **2006**, *98*, 785–792.

208. Kumano-Kuramochi, M.; Shimozu, Y.; Wakita, C.; Ohnishi-Kameyama, M.; Shibata, T.; Matsunaga, S.; Takano-Ishikawa, Y.; Watanabe, J.; Goto, M.; Xie, Q.; *et al.* Identification of 4-hydroxy-2-nonenal-histidine adducts that serve as ligands for human lectin-like oxidized LDL receptor-1. *Biochem. J.* **2012**, *442*, 171–180.
209. Kim, Y.S.; Park, Z.Y.; Kim, S.Y.; Jeong, E.; Lee, Y. Alteration of Toll-like receptor 4 activation by 4-hydroxy-2-nonenal mediated by the suppression of receptor homodimerization. *Chem. Biol. Interact.* **2009**, *182*, 59–66.
210. Liu, Q.; Smith, M.A.; Avila, J.; DeBernardis, J.; Kansal, M.; Takeda, A.; Zhu, X.W.; Nunomura, A.; Honda, K.; Moreira, P.I.; *et al.* Alzheimer-specific epitopes of tau represent lipid peroxidation-induced conformations. *Free Radic. Biol. Med.* **2005**, *38*, 746–754.
211. Santa-Maria, I.; Hernandez, F.; Martin, C.P.; Avila, J.; Moreno, F.J. Quinones facilitate the self-assembly of the phosphorylated tubulin binding region of tau into fibrillar polymers. *Biochemistry* **2004**, *43*, 2888–2897.
212. Mendez, D.; Hernaez, M.L.; Kamali, A.N.; Diez, A.; Puyet, A.; Bautista, J.M. Differential carbonylation of cytoskeletal proteins in blood group O erythrocytes: Potential role in protection against severe malaria. *Infect. Genet. Evol.* **2012**, *12*, 1780–1787.
213. Arashiki, N.; Otsuka, Y.; Ito, D.; Yang, M.; Komatsu, T.; Sato, K.; Inaba, M. The covalent modification of spectrin in red cell membranes by the lipid peroxidation product 4-hydroxy-2-nonenal. *Biochem. Biophys. Res. Commun.* **2010**, *391*, 1543–1547.
214. Mendez, D.; Hernaez, M.L.; Diez, A.; Puyet, A.; Bautista, J.M. Combined proteomic approaches for the identification of specific amino acid residues modified by 4-hydroxy-2-nonenal under physiological conditions. *J. Proteome Res.* **2010**, *9*, 5770–5781.
215. Yamashima, T. Hsp70.1 and related lysosomal factors for necrotic neuronal death. *J. Neurochem.* **2012**, *120*, 477–494.
216. Perluigi, M.; Poon, H.F.; Hensley, K.; Pierce, W.M.; Klein, J.B.; Calabrese, V.; de Marco, C.; Butterfield, D.A. Proteomic analysis of 4-hydroxy-2-nonenal-modified proteins in G93A-SOD1 transgenic mice—A model of familial amyotrophic lateral sclerosis. *Free Radic. Biol. Med.* **2005**, *38*, 960–968.
217. Sahara, S.; Yamashima, T. Calpain-mediated Hsp70.1 cleavage in hippocampal CA1 neuronal death. *Biochem. Biophys. Res. Commun.* **2010**, *393*, 806–811.
218. Connor, R.E.; Marnett, L.J.; Liebler, D.C. Protein-selective capture to analyze electrophile adduction of Hsp90 by 4-hydroxynonenal. *Chem. Res. Toxicol.* **2011**, *24*, 1275–1282.
219. Smathers, R.L.; Galligan, J.J.; Stewart, B.J.; Petersen, D.R. Overview of lipid peroxidation products and hepatic protein modification in alcoholic liver disease. *Chem. Biol. Interact.* **2011**, *192*, 107–112.
220. Jacobs, A.T.; Marnett, L.J. Systems analysis of protein modification and cellular responses induced by electrophile stress. *Acc. Chem. Res.* **2010**, *43*, 673–683.
221. Galligan, J.J.; Fritz, K.S.; Backos, D.S.; Shearn, C.T.; Smathers, R.L.; Jiang, H.; MacLean, K.N.; Reigan, P.R.; Petersen, D.R. Oxidative stress-mediated aldehyde adduction of GRP78 in a mouse model of alcoholic liver disease: Functional independence of ATPase activity and chaperone function. *Free Radic. Biol. Med.* **2014**, *73*, 411–420.

222. Brand, M.D.; Buckingham, J.A.; Esteves, T.C.; Green, K.; Lambert, A.J.; Miwa, S.; Murphy, M.P.; Pakay, J.L.; Talbot, D.A.; Echtay, K.S. Mitochondrial superoxide and aging: Uncoupling-protein activity and superoxide production. *Free Radic.* **2004**, 203–213.
223. Esteves, T.C.; Brand, M.D. The reactions catalysed by the mitochondrial uncoupling proteins UCP2 and UCP3. *Biochim. Biophys. Acta Bioenerget.* **2005**, *1709*, 35–44.
224. Smith, A.M.O.; Ratcliffe, R.G.; Sweetlove, L.J. Activation and function of mitochondrial uncoupling protein in plants. *J. Biol. Chem.* **2004**, *279*, 51944–51952.
225. Echtay, K.S.; Esteves, T.C.; Pakay, J.L.; Jekabsons, M.B.; Lambert, A.J.; Portero-Otin, M.; Pamplona, R.; Vidal-Puig, A.J.; Wang, S.; Roebuck, S.J.; *et al.* A signalling role for 4-hydroxy-2-nonenal in regulation of mitochondrial uncoupling. *EMBO J.* **2003**, *22*, 4103–4110.
226. Echtay, K.S.; Pakay, J.L.; Esteves, T.C.; Brand, M.D. Hydroxynonenal and uncoupling proteins: A model for protection against oxidative damage. *Biofactors* **2005**, *24*, 119–130.
227. Malingriaux, E.A.; Rupprecht, A.; Gille, L.; Jovanovic, O.; Jezek, P.; Jaburek, M.; Pohl, E.E. Fatty acids are key in 4-hydroxy-2-nonenal-mediated activation of uncoupling proteins 1 and 2. *PLoS ONE* **2013**, *8*, e77786.
228. Woyda-Ploszczyca, A.; Jarmuszkiewicz, W. Hydroxynonenal-stimulated activity of the uncoupling protein in *Acanthamoeba castellanii* mitochondria under phosphorylating conditions. *Biol. Chem.* **2013**, *394*, 649–658.
229. Aguirre, E.; Cadenas, S. GDP and carboxyatractylate inhibit 4-hydroxynonenal-activated proton conductance to differing degrees in mitochondria from skeletal muscle and heart. *Biochim. Biophys. Acta Bioenerget.* **2010**, *1797*, 1716–1726.
230. Jin, H.; Daly, D.S.; Marks, J.R.; Zangar, R.C. Oxidatively modified proteins as plasma biomarkers in breast cancer. *Cancer Biomarkers* **2013**, *13*, 193–200.
231. Pillon, N.J.; Vella, R.E.; Soulere, L.; Becchi, M.; Lagarde, M.; Soulage, C.O. Structural and functional changes in human insulin induced by the lipid peroxidation byproducts 4-hydroxy-2-nonenal and 4-hydroxy-2-hexenal. *Chem. Res.Toxicol.* **2011**, *24*, 752–762.
232. Lee, S.H.; Takahashi, R.; Goto, T.; Oe, T. Mass spectrometric characterization of modifications to angiotensin II by lipid peroxidation products, 4-oxo-2(E)-nonenal and 4-hydroxy-2(E)-nonenal. *Chem. Res. Toxicol.* **2010**, *23*, 1771–1785.
233. El-Bikai, R.; Welman, M.; Margaron, Y.; Cote, J.-F.; Macqueen, L.; Buschmann, M.D.; Fahmi, H.; Shi, Q.; Maghni, K.; Fernandes, J.C.; *et al.* Perturbation of adhesion molecule-mediated chondrocyte-matrix interactions by 4-hydroxynonenal binding: Implication in osteoarthritis pathogenesis. *Arthritis Res. Ther.* **2010**, *12*, doi:10.1186/ar3173.
234. Alzolibani, A.A.; al Robaee, A.A.; al-Shobaili, H.A.; Rasheed, Z. 4-Hydroxy-2-nonenal modified histone-H2A: A possible antigenic stimulus for systemic lupus erythematosus autoantibodies. *Cell. Immunol.* **2013**, *284*, 154–162.
235. Guichardant, M.; Taibi-Tronche, P.; Fay, L.B.; Lagarde, M. Covalent modifications of aminophospholipids by 4-hydroxynonenal. *Free Radic. Biol. Med.* **1998**, *25*, 1049–1056.
236. Guo, L.; Chen, Z.; Amarnath, V.; Davies, S.S. Identification of novel bioactive aldehyde-modified phosphatidylethanolamines formed by lipid peroxidation. *Free Radic. Biol. Med.* **2012**, *53*, 1226–1238.

237. Sowell, J.; Frei, B.; Stevens, J.F. Vitamin C conjugates of genotoxic lipid peroxidation products: Structural characterization and detection in human plasma. (Retracted article, see *Proc. Natl. Acad. Sci. USA* **2007**, *104*, 14543–14543) *Proc. Natl. Acad. Sci. USA* **2004**, *101*, 17964–17969.
238. Sowell, J.; Frei, B.; Stevens, J.F. Vitamin C conjugates of genotoxic lipid peroxidation products: Structural characterization and detection in human plasma. *Proc. Natl. Acad. Sci. USA* **2007**, *104*, 14543.
239. Miranda, C.L.; Reed, R.L.; Kuiper, H.C.; Alber, S.; Stevens, J.F. Ascorbic acid promotes detoxification and elimination of 4-hydroxy-2(E)-nonenal in human monocytic THP-1 cells. *Chem. Res. Toxicol.* **2009**, *22*, 863–874.
240. Amarnath, V.; Amarnath, K.; Davies, S.; Roberts, L.J. Pyridoxamine: An extremely potent scavenger of 1,4-dicarbonyls. *Chem. Res. Toxicol.* **2004**, *17*, 410–415.
241. Hardas, S.S.; Sultana, R.; Clark, A.M.; Beckett, T.L.; Szweda, L.I.; Murphy, M.P.; Butterfield, D.A. Oxidative modification of lipoic acid by HNE in Alzheimer disease brain. *Redox Biol.* **2013**, *1*, 80–85.
242. Voulgaridou, G.-P.; Anestopoulos, I.; Franco, R.; Panayiotidis, M.I.; Pappa, A. DNA damage induced by endogenous aldehydes: Current state of knowledge. *Mutat. Res. Fund. Mol. Mech. Mutagen.* **2011**, *711*, 13–27.
243. Minko, I.G.; Kozekov, I.D.; Harris, T.A.; Rizzo, C.J.; Lloyd, R.S.; Stone, M.P. Chemistry and biology of DNA containing 1,N^2-deoxyguanosine adducts of the α,β-unsaturated aldehydes acrolein, crotonaldehyde, and 4-hydroxynonenal. *Chem. Res. Toxicol.* **2009**, *22*, 759–778.
244. Eckl, P.M.; Ortner, A.; Esterbauer, H. Genotoxic properties of 4-hydroxyalkenals and analogous aldehydes. *Mutat. Res. Fund. Mol. Mech. Mutagen.* **1993**, *290*, 183–192.
245. Karlhuber, G.M.; Bauer, H.C.; Eckl, P.M. Cytotoxic and genotoxic effects of 4-hydroxynonenal in cerebral endothelial cells. *Mutat. Res. Fund. Mol. Mech. Mutagen.* **1997**, *381*, 209–216.
246. Wang, H.; Kozekov, I.D.; Harris, T.M.; Rizzo, C.J. Site-specific synthesis and reactivity of oligonucleotides containing stereochemically defined 1,N^2-deoxyguanosine adducts of the lipid peroxidation product *trans*-4-hydroxynonenal. *J. Am. Chem. Soc.* **2003**, *125*, 5687–5700.
247. Wacker, M.; Wanek, P.; Eder, E. Detection of 1,N^2-propanodeoxyguanosine adducts of *trans*-4-hydroxy-2-nonenal after gavage of *trans*-4-hydroxy-2-nonenal or induction of lipid peroxidation with carbon tetrachloride in F344 rats. *Chem. Biol. Interact.* **2001**, *137*, 269–283.
248. Kozekov, I.D.; Turesky, R.J.; Alas, G.R.; Harris, C.M.; Harris, T.M.; Rizzo, C.J. Formation of deoxyguanosine cross-links from calf thymus DNA treated with acrolein and 4-hydroxy-2-nonenal. *Chem. Res. Toxicol.* **2010**, *23*, 1701–1713.
249. Huang, H.; Kozekov, I.D.; Kozekova, A.; Wang, H.; Lloyd, R.S.; Rizzo, C.J.; Stone, M.P. DNA Cross-link induced by *trans*-4-Hydroxynonenal. *Environ. Mol. Mutagen.* **2010**, *51*, 625–634.
250. Huang, H.; Wang, H.; Kozekova, A.; Rizzo, C.J.; Stone, M.P. Formation of a N^2-dG:N^2-dG carbinolamine DNA cross-link by the *trans*-4-Hydroxynonenal-derived (6S,8R,11S) 1,N^2-dG adduct. *J. Am. Chem. Soc.* **2011**, *133*, 16101–16110.
251. Choudhury, S.; Dyba, M.; Pan, J.; Roy, R.; Chung, F.-L. Repair kinetics of acrolein- and (E)-4-hydroxy-2-nonenal-derived DNA adducts in human colon cell extracts. *Mutat. Res. Fund. Mol. Mech. Mutagen.* **2013**, *751*, 15–23.

252. Choudhury, S.; Pan, J.; Amin, S.; Chung, F.L.; Roy, R. Repair kinetics of *trans*-4-hydroxynonenal-induced cyclic 1,*N*²-propanodeoxyguanine DNA adducts by human cell nuclear extracts. *Biochemistry* **2004**, *43*, 7514–7521.
253. Winczura, A.; Zdzalik, D.; Tudek, B. Damage of DNA and proteins by major lipid peroxidation products in genome stability. *Free Radic. Res.* **2012**, *46*, 442–459.
254. Janowska, B.; Komisarski, M.; Prorok, P.; Sokolowska, B.; Kusmierek, J.; Janion, C.; Tudek, B. Nucleotide excision repair and recombination are engaged in repair of *trans*-4-hydroxy-2-nonenal adducts to DNA bases in *Escherichia coli*. *Int. J. Biol. Sci.* **2009**, *5*, 611–620.
255. Banerjee, S.; Christov, P.P.; Kozekova, A.; Rizzo, C.J.; Egli, M.; Stone, M.P. Replication bypass of the *trans*-4-Hydroxynonenal-derived (6S,8R,11S)-1,*N*²-deoxyguanosine DNA adduct by the *Sulfolobus solfataricus* DNA polymerase IV. *Chem. Res. Toxicol.* **2012**, *25*, 422–435.
256. Janowska, B.; Kurpios-Piec, D.; Prorok, P.; Szparecki, G.; Komisarski, M.; Kowalczyk, P.; Janion, C.; Tudek, B. Role of damage-specific DNA polymerases in M13 phage mutagenesis induced by a major lipid peroxidation product *trans*-4-hydroxy-2-nonenal. *Mutat. Res. Fund. Mol. Mech. Mutagen.* **2012**, *729*, 41–51.
257. Chen, H.J.C.; Gonzalez, F.J.; Shou, M.G.; Chung, F.L. 2,3-Epoxy-4-hydroxynonanal, a potential lipid peroxidation product for etheno adduct formation, is not a substrate of human epoxide hydrolase. *Carcinogenesis* **1998**, *19*, 939–943.
258. Lee, S.H.; Arora, J.A.; Oe, T.; Blair, I.A. 4-Hydroperoxy-2-nonenal-induced formation of 1,*N*²-etheno-2'-deoxyguanosine adducts. *Chem. Res. Toxicol.* **2005**, *18*, 780–786.
259. Wang, Y.; Millonig, G.; Nair, J.; Patsenker, E.; Stickel, F.; Mueller, S.; Bartsch, H.; Seitz, H.K. Ethanol-induced cytochrome P4502E1 causes carcinogenic etheno-DNA lesions in alcoholic liver disease. *Hepatology* **2009**, *50*, 453–461.
260. Nair, J.; Srivatanakul, P.; Haas, C.; Jedpiyawongse, A.; Khuhaprema, T.; Seitz, H.K.; Bartsch, H. High urinary excretion of lipid peroxidation-derived DNA damage in patients with cancer-prone liver diseases. *Mutat. Res. Fund. Mol. Mech. Mutagen.* **2010**, *683*, 23–28.
261. Grune, T.; Muller, K.; Zollner, S.; Haseloff, R.; Blasig, I.E.; David, H.; Siems, W. Evaluation of purine nucleotide loss, lipid peroxidation and ultrastructural alterations in post-hypoxic hepatocytes. *J. Physiol. Lond.* **1997**, *498*, 511–522.
262. Grune, T.; Sommerburg, O.; Petras, T.; Siems, W.G. Postanoxic formation of aldehydic lipid-peroxidation products in human renal tubular cells. *Free Radic. Biol. Med.* **1995**, *18*, 21–27.
263. Blasig, I.E.; Grune, T.; Schonheit, K.; Rohde, E.; Jakstadt, M.; Haseloff, R.F.; Siems, W.G. 4-hydroxynonenal, a novel indicator of lipid-peroxidation for reperfusion injury of the myocardium. *Am. J. Physiol. Heart Circ. Physiol.* **1995**, *269*, H14–H22.
264. Grune, T.; Siems, W.G.; Zollner, H.; Esterbauer, H. Metabolism of 4-hydroxynonenal, a cytotoxic lipid-peroxidation product, in ehrlich mouse ascites-cells at different proliferation stages. *Cancer Res.* **1994**, *54*, 5231–5235.
265. Blasig, I.E.; Schoenheit, K.; Siems, W.G. Formation of 4-hydroxyalkenals by the reperfusion-injured rat-heart. *Cell. Biochem. Mol. Asp. Reperfus. Injury* **1994**, *723*, 462–465.
266. Siems, W.G.; Grune, T.; Esterbauer, H. 4-hydroxynonenal formation during ischemia and reperfusion of rat small-intestine. *Life Sci.* **1995**, *57*, 785–789.

267. Poli, G.; Dianzani, M.U.; Cheeseman, K.H.; Slater, T.F.; Lang, J.; Esterbauer, H. Separation and characterization of the aldehydic products of lipid-peroxidation stimulated by carbon-tetrachloride or adp iron in isolated rat hepatocytes and rat-liver microsomal suspensions. *Biochem. J.* **1985**, *227*, 629–638.
268. Shao, C.X.; Roberts, K.N.; Markesbery, W.R.; Scheff, S.W.; Lovell, M.A. Oxidative stress in head trauma in aging. *Free Radic. Biol. Med.* **2006**, *41*, 77–85.
269. Williams, T.I.; Lynn, B.C.; Markesbery, W.R.; Lovell, M.A. Increased levels of 4-hydroxynonenal and acrolein, neurotoxic markers of lipid peroxidation, in the brain in Mild Cognitive Impairment and early Alzheimer's disease. *Neurobiol. Aging* **2006**, *27*, 1094–1099.
270. Negre-Salvayre, A.; Auge, N.; Ayala, V.; Basaga, H.; Boada, J.; Brenke, R.; Chapple, S.; Cohen, G.; Feher, J.; Grune, T.; *et al.* Pathological aspects of lipid peroxidation. *Free Radic. Res.* **2010**, *44*, 1125–1171.
271. Grune, T.; Siems, W.G.; Schneider, W. Accumulation of aldehydic lipid peroxidation products during postanoxic reoxygenation of isolated rat hepatocytes. *Free Radic. Biol. Med.* **1993,** *15*, 125–132.
272. Schwarzer, E.; Muller, O.; Arese, P.; Siems, W.G.; Grune, T. Increased levels of 4-hydroxynonenal in human monocytes fed with malarial pigment hemozoin—A possible clue for hemozoin toxicity. *FEBS Lett.* **1996**, *388*, 119–122.
273. Gil, L.; Siems, W.; Mazurek, B.; Gross, J.; Schroeder, P.; Voss, P.; Grune, T. Age-associated analysis of oxidative stress parameters in human plasma and erythrocytes. *Free Radic. Res.* **2006**, *40*, 495–505.
274. Selley, M.L.; Bartlett, M.R.; McGuiness, J.A.; Hapel, A.J.; Ardlie, N.G.; Lacey, M.J. Determination of the lipid-peroxidation product *trans*-4-hydroxy-2-nonenal in biological samples by high-performance liquid-chromatography and combined capillary column gas chromatography negative-ion chemical ionization mass-spectrometry. *J. Chromatogr. Biomed. Appl.* **1989**, *488*, 329–340.
275. Schmidt, H.; Grune, T.; Muller, R.; Siems, W.G.; Wauer, R.R. Increased levels of lipid peroxidation products malondialdehyde and 4-hydroxynonenal after perinatal hypoxia. *Pediatr. Res.* **1996**, *40*, 15–20.
276. Michel, P.; Eggert, W.; AlbrechtNebe, H.; Grune, T. Increased lipid peroxidation in children with autoimmune diseases. *Acta Paediatr.* **1997**, *86*, 609–612.
277. Uchida, K.; Itakura, K.; Kawakishi, S.; Hiai, H.; Toyokuni, S.; Stadtman, E.R. Characterization of epitopes recognized by 4-hydroxy-2-nonenal specific antibodies. *Arch. Biochem. Biophys.* **1995**, *324*, 241–248.
278. Yoritaka, A.; Hattori, N.; Uchida, K.; Tanaka, M.; Stadtman, E.R.; Mizuno, Y. Immunohistochemical detection of 4-hydroxynonenal protein adducts in Parkinson disease. *Proc. Natl. Acad. Sci. USA* **1996**, *93*, 2696–2701.
279. Okamoto, K.; Toyokuni, S.; Uchida, K.; Ogawa, O.; Takenewa, J.; Kakehi, Y.; Kinoshita, H.; Hattorinakakuki, Y.; Hiai, H.; Yoshida, O. Formation of 8-hydroxy-2'-deoxyguanosine and 4-hydroxy-2-nonenal-modified proteins in human renal-cell carcinoma. *Int. J. Cancer* **1994**, *58*, 825–829.

280. Ando, Y.; Nyhlin, N.; Suhr, O.; Holmgren, G.; Uchida, K.; ElSahly, M.; Yamashita, T.; Terasaki, H.; Nakamura, M.; Uchino, M.; *et al.* Oxidative stress is found in amyloid deposits in systemic amyloidosis. *Biochem. Biophys. Res. Commun.* **1997**, *232*, 497–502.
281. Butterfield, D.A.; Reed, T.; Perluigi, M.; de Marco, C.; Coccia, R.; Cini, C.; Sultana, R. Elevated protein-bound levels of the lipid peroxidation product, 4-hydroxy-2-nonenal, in brain from persons with mild cognitive impairment. *Neurosci. Lett.* **2006**, *397*, 170–173.
282. Morikawa, S.; Kurauchi, O.; Tanaka, M.; Yoneda, M.; Uchida, K.; Itakura, A.; Furugori, K.; Mizutani, S.; Tomoda, Y. Increased mitochondrial damage by lipid peroxidation in trophoblast cells of preeclamptic placentas. *Biochem. Mol. Biol. Int.* **1997**, *41*, 767–775.
283. Hamilton, R.F.; Li, L.; Eschenbacher, W.L.; Szweda, L.; Holian, A. Potential involvement of 4-hydroxynonenal in the response of human lung cells to ozone. *Am. J. Physiol. Lung Cell. Mol. Physiol.* **1998**, *274*, L8–L16.
284. Liu, X.L.; Lovell, M.A.; Lynn, B.C. Detection and quantification of endogenous cyclic DNA adducts derived from *trans*-4-hydroxy-2-nonenal in human brain tissue by isotope dilution capillary liquid chromatography nanoelectrospray tandem mass spectrometry. *Chem. Res. Toxicol.* **2006**, *19*, 710–718.
285. Surh, J.; Kwon, H. Estimation of daily exposure to 4-hydroxy-2-alkenals in Korean foods containing n-3 and n-6 polyunsaturated fatty acids. *Food Addit. Contam.* **2005**, *22*, 701–708.
286. Ito, Y.; Kosuge, Y.; Sakikubo, T.; Horie, K.; Ishikawa, N.; Obokata, N.; Yokoyama, E.; Yamashina, K.; Yamamoto, M.; Saito, H.; *et al.* Protective effect of S-allyl-L-cysteine, a garlic compound, on amyloid beta-protein-induced cell death in nerve growth factor-differentiated PC12 cells. *Neurosci. Res.* **2003**, *46*, 119–125.
287. Glei, M.; Hofmann, T.; Kuster, K.; Hollmann, J.; Lindhauer, M.G.; Pool-Zobel, B.L. Both wheat (*Triticum aestivum*) bran arabinoxylans and gut flora-mediated fermentation products protect human colon cells from genotoxic activities of 4-hydroxynonenal and hydrogen peroxide. *J. Agric. Food Chem.* **2006**, *54*, 2088–2095.
288. Loguercio, C.; Federico, A.; Tuccillo, C.; Terracciano, F.; D'Auria, M.V.; de Simone, C.; Blanco, C.D. Beneficial effects of a probiotic VSL#3 on parameters of liver dysfunction in chronic liver diseases. *J. Clin. Gastroenterol.* **2005**, *39*, 540–543.
289. Chowdhury, P.; Soulsby, M. Lipid peroxidation in rat brain is increased by simulated weightlessness and decreased by a soy-protein diet. *Ann. Clin. Lab. Sci.* **2002**, *32*, 188–192.
290. Skrzydlewska, E.; Ostrowska, J.; Farbiszewski, R.; Michalak, K. Protective effect of green tea against lipid peroxidation in the rat liver, blood serum and the brain. *Phytomedicine* **2002**, *9*, 232–238.
291. Leiphon, L.J.; Picklo, M.J., Sr. Inhibition of aldehyde detoxification in CNS mitochondria by fungicides. *Neurotoxicology* **2007**, *28*, 143–149.
292. Ullrich, O.; Grune, T.; Henke, W.; Esterbauer, H.; Siems, W.G. Identification of metabolic pathways of the lipid-peroxidation product 4-hydroxynonenal by mitochondria isolated from rat-kidney cortex. *FEBS Lett.* **1994**, *352*, 84–86.
293. Siems, W.G.; Zollner, H.; Grune, T.; Esterbauer, H. Metabolic fate of 4-hydroxynonenal in hepatocytes: 1,4-dihydroxynonene is not the main product. *J. Lipid Res.* **1997**, *38*, 612–622.

294. Siems, W.G.; Grune, T. High capacity of secondary antioxidative pathways in thymocytes: Rapid detoxification of 4-hydroxynonenal. *Int. J. Thymol.* **1998**, *6*, 518–529.
295. Ullrich, O.; Huser, H.; Ehrlich, W.; Grune, T. Intracellular metabolism of 4-hydroxynonenal in primary cultures of rabbit synovial fibroblasts. *Free Radic. Biol. Med.* **1997**, *22*, 1153–1157.
296. Grune, T.; Siems, W.G.; Petras, T. Identification of metabolic pathways of the lipid peroxidation product 4-hydroxynonenal in *in situ* perfused rat kidney. *J. Lipid Res.* **1997**, *38*, 1660–1665.
297. Grune, T.; Siems, W.; Kowalewski, J.; Zollner, H.; Esterbauer, H. Identification of metabolic pathways of the lipid-peroxidation product 4-hydroxynonenal by enterocytes of rat small-intestine. *Biochem. Int.* **1991**, *25*, 963–971.
298. Petras, T.; Siems, W.G.; Grune, T. 4-hydroxynonenal is degraded to mercapturic acid conjugate in rat-kidney. *Free Radic. Biol. Med.* **1995**, *19*, 685–688.
299. Kinter, M.; Roberts, R.J. Glutathione consumption and glutathione peroxidase inactivation in fibroblast cell lines by 4-hydroxy-2-nonenal. *Free Radic. Biol. Med.* **1996**, *21*, 457–462.
300. Esterbauer, H.; Zollner, H.; Lang, J. Metabolism of the lipid-peroxidation product 4-hydroxynonenal by isolated hepatocytes and by liver cytosolic fractions. *Biochem. J.* **1985**, *228*, 363–373.
301. Fauler, G. Investigations on the Metabolism of 4-Hydroxy-Alkenals. Ph.D. Thesis, University of Graz, Graz, Austria, 1987.
302. Siems, W.G.; Zollner, H.; Esterbauer, H. Metabolic pathways of lipid peroxidation product 4-hydroxynonenal in hepatocytes quantitative assessment of an antioxidative defense system. *Free Radic. Biol. Med.* **1990**, *9*, 110.
303. Schaur, R.J.; Zollner, H.; Esterbauer, H. Biological effects of aldehydes with particular attention to hydroxynonenal and malondialdehyde. In *Membrane Lipid Peroxidation*; Vigo-Pelfrey, C., Ed.; CRC Press: Boca Raton, FL, USA, 1991; Volume 3, pp. 141–163.
304. Esterbauer, H.; Schaur, R.J.; Zollner, H. Chemistry and biochemistry of 4-hydroxynonenal, malonaldehyde and related aldehydes. *Free Radic. Biol. Med.* **1991**, *11*, 81–128.
305. Siems, W.; Grune, T.; Zollner, H.; Esterbauer, H. Formation and metabolism of the lipid peroxidation product 4-hydroxynonenal in liver and small intestine. In *Free Radicals: From Basic Science to Medicine*; Poli, G., Albano, E., Dianzani, M.U., Eds.; Birkhäuser Verlag: Basel, Switzerland, 1993; pp. 89–101.
306. Siems, W.G.; Zollner, H.; Grune, T.; Esterbauer, H. Qualitative and quantitative-determination of metabolites of the lipid-peroxidation product 4-hydroxynonenal from hepatocytes, enterocytes and tumor-cells. *Fresenius J. Anal. Chem.* **1992**, *343*, 75–76.
307. Canuto, R.A.; Ferro, M.; Muzio, G.; Bassi, A.M.; Leonarduzzi, G.; Maggiora, M.; Adamo, D.; Poli, G.; Lindahl, R. Effects of aldehyde products of lipid peroxidation on the activity of aldehyde metabolizing enzymes in hepatomas. *Adv. Exp. Med. Biol.* **1993**, *328*, 17–25.
308. Canuto, R.A.; Muzio, G.; Maggiora, M.; Poli, G.; Biasi, F.; Dianzani, M.U.; Ferro, M.; Bassi, A.M.; Penco, S.; Marinari, U.M. Ability of different hepatoma-cells to metabolize 4-hydroxynonenal. *Cell Biochem. Funct.* **1993**, *11*, 79–86.
309. Ferro, M.; Marinari, U.M.; Poli, G.; Dianzani, M.U.; Fauler, G.; Zollner, H.; Esterbauer, H. Metabolism of 4-hydroxynonenal by the rat hepatoma-cell line MH1C1. *Cell Biochem. Funct.* **1988**, *6*, 245–250.

310. Canuto, R.A.; Muzio, G.; Bassi, A.M.; Biocca, M.E.; Poli, G.; Esterbauer, H.; Ferro, M. Metabolism of 4 hydroxynonenal in hepatoma cell lines. In *Advances in Experimental Medicine and Biology, Vol. 284. Enzymology and Molecular Biology of Carbonyl Metabolism 3*, Proceedings of the Fifth International Workshop, West Lafayette, IN, USA, 12–15 June 1990; Weiner, H., Wermuth, M.B., Crabb, D.W., Eds.; Plenum Press: New York, NY, USA, 1991, pp. 75–84.
311. Srivastava, S.; Conklin, D.J.; Liu, S.Q.; Prakash, N.; Boor, P.J.; Srivastava, S.K.; Bhatnagar, A. Identification of biochemical pathways for the metabolism of oxidized low-density lipoprotein derived aldehyde-4-hydroxy *trans*-2-nonenal in vascular smooth muscle cells. *Atherosclerosis* **2001**, *158*, 339–350.
312. Srivastava, S.; Chandra, A.; Wang, L.F.; Seifert, W.E.; DaGue, B.B.; Ansari, N.H.; Srivastava, S.K.; Bhatnagar, A. Metabolism of the lipid peroxidation product, 4-hydroxy-*trans*-2-nonenal, in isolated perfused rat heart. *J. Biol. Chem.* **1998**, *273*, 10893–10900.
313. Spitz, D.R.; Malcolm, R.R.; Roberts, R.J. Cytotoxicity and metabolism of 4-hydroxy-2-nonenal and 2-nonenal in H_2O_2-resistant cell-lines—Do aldehydic by-products of lipid-peroxidation contribute to oxidative stress. *Biochem. J.* **1990**, *267*, 453–459.
314. Meyer, M.J.; Mosely, D.E.; Amarnath, V.; Picklo, M.J. Metabohom of 4-hydroxy-*trans*-2-nonenal by central nervous system mitochondria is dependent on age and NAD(+) availability. *Chem. Res. Toxicol.* **2004**, *17*, 1272–1279.
315. Canuto, R.A.; Muzio, G.; Biocca, M.E.; Dianzani, M.U. Oxidative-metabolism of 4-hydroxy-2, 3-nonenal during diethyl-nitrosamine-induced carcinogenesis in rat-liver. *Cancer Lett.* **1989**, *46*, 7–13.
316. Chen, J.J.; Yu, B.P. Detoxification of reactive aldehydes in mitochondria: Effects of age and dietary restriction. *Aging Clin. Exp. Res.* **1996**, *8*, 334–340.
317. Honzatko, A.; Brichac, J.; Murphy, T.C.; Reberg, A.; Kubatova, A.; Smoliakova, I.P.; Picklo, M.J. Enantioselective metabolism of *trans*-4-hydroxy-2-nonenal by brain mitochondria. *Free Radic. Biol. Med.* **2005**, *39*, 913–924.
318. Alary, J.; Gueraud, F.; Cravedi, J.-P. Fate of 4-hydroxynonenal *in vivo*: Disposition and metabolic pathways. *Mol. Asp. Med.* **2003**, *24*, 177–187.
319. Alary, J.; Bravais, F.; Cravedi, J.P.; Debrauwer, L.; Rao, D.; Bories, G. Mercapturic acid conjugates as urinary end metabolites of the lipid-peroxidation product 4-hydroxy-2-nonenal in the rat. *Chem. Res. Toxicol.* **1995**, *8*, 34–39.
320. Winter, C.K.; Segall, H.J.; Jones, A.D. Distribution of *trans*-4-hydroxy-2-hexenal and tandem mass-spectrometric detection of its urinary mercapturic acid in the rat. *Drug Metab. Dispos.* **1987**, *15*, 608–612.
321. Alin, P.; Danielson, U.H.; Mannervik, B. 4-hydroxyalk-2-enals are substrates for glutathione transferase. *FEBS Lett.* **1985**, *179*, 267–270.
322. Danielson, U.H.; Esterbauer, H.; Mannervik, B. Structure-activity-relationships of 4-hydroxyalkenals in the conjugation catalyzed by mammalian glutathione transferases. *Biochem. J.* **1987**, *247*, 707–713.
323. Mitchell, D.Y.; Petersen, D.R. The oxidation of α,β-unsaturated aldehydic products of lipid-peroxidation by rat-liver aldehyde dehydrogenases. *Toxicol. Appl. Pharmacol.* **1987**, *87*, 403–410.

324. Mitchell, D.Y.; Petersen, D.R. Inhibition of rat hepatic mitochondrial aldehyde dehydrogenase mediated acetaldehyde oxidation by *trans*-4-hydroxy-2-nonenal. *Hepatology* **1991**, *13*, 728–734.
325. Hartley, D.P.; Ruth, J.A.; Petersen, D.R. The hepatocellular metabolism of 4-hydroxynonenal by alcohol-dehydrogenase, aldehyde dehydrogenase, and glutathione-S-transferase. *Arch. Biochem. Biophys.* **1995**, *316*, 197–205.
326. Singhal, S.S.; Zimniak, P.; Awasthi, S.; Piper, J.T.; He, N.G.; Teng, J.I.; Petersen, D.R.; Awasthi, Y.C. Several closely-related glutathione-S-transferase isozymes catalyzing conjugation of 4-hydroxynonenal are differentially expressed in human tissues. *Arch. Biochem. Biophys.* **1994**, *311*, 242–250.
327. Berhane, K.; Widersten, M.; Engstrom, A.; Kozarich, J.W.; Mannervik, B. Detoxication of base propenals and other α,β-unsaturated aldehyde products of radical reactions and lipid-peroxidation by human glutathione transferases. *Proc. Natl. Acad. Sci. USA* **1994**, *91*, 1480–1484.
328. Fukuda, A.; Nakamura, Y.; Ohigashi, H.; Osawa, T.; Uchida, K. Cellular response to the redox active lipid peroxidation products: Induction of glutathione S-transferase P by 4-hydroxy-2-nonenal. *Biochem. Biophys. Res. Commun.* **1997**, *236*, 505–509.
329. Yang, Y.Z.; Yang, Y.S.; Trent, M.B.; He, N.G.; Lick, S.D.; Zimniake, P.; Awasthi, Y.C.; Boor, P.J. Glutathione-S-transferase A4-4 modulates oxidative stress in endothelium: Possible role in human atherosclerosis. *Atherosclerosis* **2004**, *173*, 211–221.
330. Cao, Z.; Hardej, D.; Trombetta, L.D.; Li, Y. The role of chemically induced glutathione and glutathione S-transferase in protecting against 4-hydroxy-2-nonenal-mediated cytotoxicity in vascular smooth muscle cells. *Cardiovas. Toxicol.* **2003**, *3*, 165–177.
331. Kaiserova, K.; Srivastava, S.; Hoetker, J.D.; Awe, S.O.; Tang, X.L.; Cai, J.; Bhatnagar, A. Redox activation of aldose reductase in the ischemic heart. *J. Biol. Chem.* **2006**, *281*, 15110–15120.
332. Bander Jagt, D.L.; Kolb, N.S.; bander Jagt, T.J.; Chino, J.; Martinez, F.J.; Hunsaker, L.A.; Royer, R.E. Substrate specificity of human aldose reductase: Identification of 4-hydroxynonenal as an endogenous substrate. *Biochim. Biophys. Acta* **1995**, *1249*, 117–126.
333. Srivastava, S.K.; Ansari, N.H.; Hair, G.A.; Das, B. Aldose and aldehyde reductases in human-tissues. *Biochim. Biophys. Acta* **1984**, *800*, 220–227.
334. He, Q.; Khanna, P.; Srivastava, S.; van Kuijk, F.; Ansari, N.H. Reduction of 4-hydroxynonenal and 4-hydroxyhexenal by retinal aldose reductase. *Biochem. Biophys. Res. Commun.* **1998**, *247*, 719–722.
335. Srivastava, S.K.; Yadav, U.C.S.; Reddy, A.B.M.; Saxena, A.; Tammali, R.; Shoeb, M.; Ansari, N.H.; Bhatnagar, A.; Petrash, M.J.; Srivastava, S.; *et al.* Aldose reductase inhibition suppresses oxidative stress-induced inflammatory disorders. *Chem. Biol. Interact.* **2011**, *191*, 330–338.
336. Koh, Y.H.; Park, Y.S.; Takahashi, M.; Suzuki, K.; Taniguchi, N. Aldehyde reductase gene expression by lipid peroxidation end products, MDA and HNE. *Free Radic. Res.* **2000**, *33*, 739–746.
337. Hoehn, A.; Koenig, J.; Grune, T. Protein oxidation in aging and the removal of oxidized proteins. *J. Proteomics* **2013**, *92*, 132–159.
338. Grune, T.; Davies, K.J.A. Breakdown of oxidized proteins as a part of secondary antioxidant defenses in mammalian cells. *Biofactors* **1997**, *6*, 165–172.
339. Grune, T.; Reinheckel, T.; Davies, K.J.A. Degradation of oxidized proteins in mammalian cells. *FASEB J.* **1997**, *11*, 526–534.

340. Laurent, A.; Perdu-Durand, E.; Alary, J.; Debrauwer, L.; Cravedi, J.P. Metabolism of 4-hydroxynonenal, a cytotoxic product of lipid peroxidation, in rat precision-cut liver slices. *Toxicol. Lett.* **2000**, *114*, 203–214.
341. Alary, J.; Debrauwer, L.; Fernandez, Y.; Paris, A.; Cravedi, J.P.; Dolo, L.; Rao, D.; Bories, G. Identification of novel urinary metabolites of the lipid peroxidation product 4-hydroxy-2-nonenal in rats. *Chem. Res. Toxicol.* **1998**, *11*, 1368–1376.
342. Li, Q.; Sadhukhan, S.; Berthiaume, J.M.; Ibarra, R.A.; Tang, H.; Deng, S.; Hamilton, E.; Nagy, L.E.; Tochtrop, G.P.; Zhang, G.-F. 4-Hydroxy-2(E)-nonenal (HNE) catabolism and formation of HNE adducts are modulated by beta oxidation of fatty acids in the isolated rat heart. *Free Rad. Biol. Med.* **2013**, *58*, 35–44.
343. Kubatova, A.; Murphy, T.C.; Combs, C.; Picklo, M.J. Astrocytic biotransformation of *trans*-4-hydroxy-2-nonenal is dose-dependent. *Chem. Res. Toxicol.* **2006**, *19*, 844–851.
344. Alary, J.; Debrauwer, L.; Fernandez, Y.; Cravedi, J.P.; Rao, D.; Bories, G. 1,4-dihydroxynonene mercapturic acid, the major end metabolite of exogenous 4-hydroxy-2-nonenal, is a physiological component of rat and human urine. *Chem. Res. Toxicol.* **1998**, *11*, 130–135.
345. Laurent, A.; Alary, J.; Debrauwer, L.; Cravedi, J.P. Analysis in the rat of 4-hydroxynonenal metabolites excreted in bile: Evidence of enterohepatic circulation of these byproducts of lipid peroxidation. *Chem. Res. Toxicol.* **1999**, *12*, 887–894.
346. Ji, B.; Ito, K.; Suzuki, H.; Sugiyama, Y.; Horie, T. Multidrug resistance-associated protein2 (MRP2) plays an important role in the biliary excretion of glutathione conjugates of 4-hydroxynonenal. *Free Rad. Biol. Med.* **2002**, *33*, 370–378.
347. Esterbauer, H.; Ertl, A.; Scholz, N. Reaction of cysteine with α,β-unsaturated aldehydes. *Tetrahedron* **1976**, *32*, 285–289.
348. Grune, T.; Michel, P.; Sitte, N.; Eggert, W.; AlbrechtNebe, H.; Esterbauer, H.; Siems, W.G. Increased levels of 4-hydroxynonenal modified proteins in plasma of children with autoimmune diseases. *Free Rad. Biol. Med.* **1997**, *23*, 357–360.
349. Peiro, G.; Alary, J.; Cravedi, J.P.; Rathahao, E.; Steghens, J.P.; Gueraud, F. Dihydroxynonene mercapturic acid, a urinary metabolite of 4-hydroxynonenal, as a biomarker of lipid peroxidation. *Biofactors* **2005**, *24*, 89–96.
350. Volkel, W.; Sicilia, T.; Pahler, A.; Gsell, W.; Tatschner, T.; Jellinger, K.; Leblhuber, F.; Riederer, P.; Lutz, W.K.; Gotz, M.E. Increased brain levels of 4-hydroxy-2-nonenal glutathione conjugates in severe Alzheimer's disease. *Neurochem. Int.* **2006**, *48*, 679–686.
351. Ramana, K.V.; Bhatnagar, A.; Srivastava, S.; Yadav, U.C.; Awasthi, S.; Awasthi, Y.C.; Srivastava, S.K. Mitogenic responses of vascular smooth muscle cells to lipid peroxidation-derived aldehyde 4-hydroxy-*trans*-2-nonenal (HNE)—Role of aldose reductase-catalyzed reduction of the HNE-glutathione conjugates in regulating cell growth. *J. Biol. Chem.* **2006**, *281*, 17652–17660.
352. Ramana, K.V.; Willis, M.S.; White, M.D.; Horton, J.W.; DiMaio, J.M.; Srivastava, D.; Bhatnagar, A.; Srivastava, S.K. Endotoxin-induced cardiomyopathy and systemic inflammation in mice is prevented by aldose reductase inhibition. *Circulation* **2006**, *114*, 1838–1846.
353. Srivastava, S.; Ramana, K.V.; Tammali, R.; Srivastava, S.K.; Bhatnagar, A. Contribution of aldose reductase to diabetic hyperproliferation of vascular smooth muscle cells. *Diabetes* **2006**, *55*, 901–910.

354. Tammali, R.; Ramana, K.V.; Singhal, S.S.; Awasthi, S.; Srivastava, S.K. Aldose reductase regulates growth factor-induced cyclooxygenase-2 expression and prostaglandin E2 production in human colon cancer cells. *Cancer Res.* **2006**, *66*, 9705–9713.
355. Ronis, M.J.J.; Butura, A.; Sampey, B.P.; Shankar, K.; Prior, R.L.; Korourian, S.; Albano, E.; Ingelman-Sundberg, M.; Petersen, D.R.; Badger, T.M. Effects of *N*-acetylcysteine on ethanol-induced hepatotoxicity in rats fed via total enteral nutrition. *Free Rad. Biol. Med.* **2005**, *39*, 619–630.
356. Arakawa, M.; Ushimaru, N.; Osada, N.; Oda, T.; Ishige, K.; Ito, Y. *N*-acetylcysteine selectively protects cerebellar granule cells from 4-hydroxynonenal-induced cell death. *Neurosci. Res.* **2006**, *55*, 255–263.
357. Zheng, J.Y.; Mutcherson, R.; Helfand, S.L. Calorie restriction delays lipid oxidative damage in *Drosophila melanogaster*. *Aging Cell* **2005**, *4*, 209–216.
358. Lee, J.H.; Jung, K.J.; Kim, J.W.; Kim, H.J.; Yu, B.P.; Chung, H.Y. Suppression of apoptosis by calorie restriction in aged kidney. *Exp. Gerontol.* **2004**, *39*, 1361–1368.
359. Zainal, T.A.; Oberley, T.D.; Allison, D.B.; Szweda, L.I.; Weindruch, R. Caloric restriction of rhesus monkeys lowers oxidative damage in skeletal muscle. *FASEB J.* **2000**, *14*, 1825–1836.
360. Shoeb, M.; Ansari, N.H.; Srivastava, S.K.; Ramana, K.V. 4-Hydroxynonenal in the pathogenesis and progression of human diseases. *Curr. Med. Chem.* **2014**, *21*, 230–237.

4

Oxidative Stress Responses in the Human Fungal Pathogen, *Candida albicans*

Alessandra da Silva Dantas [1,†], Alison Day [2,†], Mélanie Ikeh [2,†], Iaroslava Kos [2,†], Beatrice Achan [2,†] and Janet Quinn [2,*]

[1] Departamento de Biologia Celular e Genética, Instituto de Biologia Roberto Alcantara Gomes, Universidade do Estado do Rio de Janeiro (UERJ), Rio de Janeiro 20550-013, Brazil; E-Mail: alesdantas@gmail.com

[2] Institute for Cell and Molecular Biosciences, Faculty of Medical Sciences, Newcastle University, Newcastle upon Tyne NE2 4HH, UK; E-Mails: a.m.day@ncl.ac.uk (A.D.); m.ikeh@ncl.ac.uk (M.I.); i.kos@ncl.ac.uk (I.K.); b.achan@ncl.ac.uk (B.A.)

† These authors contributed equally to this work.

* Author to whom correspondence should be addressed; E-Mail: janet.quinn@ncl.ac.uk;

Academic Editors: Michael Breitenbach and Peter Eckl

Abstract: *Candida albicans* is a major fungal pathogen of humans, causing approximately 400,000 life-threatening systemic infections world-wide each year in severely immunocompromised patients. An important fungicidal mechanism employed by innate immune cells involves the generation of toxic reactive oxygen species (ROS), such as superoxide and hydrogen peroxide. Consequently, there is much interest in the strategies employed by *C. albicans* to evade the oxidative killing by macrophages and neutrophils. Our understanding of how *C. albicans* senses and responds to ROS has significantly increased in recent years. Key findings include the observations that hydrogen peroxide triggers the filamentation of this polymorphic fungus and that a superoxide dismutase enzyme with a novel mode of action is expressed at the cell surface of *C. albicans*. Furthermore, recent studies have indicated that combinations of the chemical stresses generated by phagocytes can actively prevent *C. albicans* oxidative stress responses through a mechanism termed the stress pathway interference. In this review, we present an up-date of our current understanding of the role and regulation of oxidative stress responses in this important human fungal pathogen.

Keywords: fungal pathogenesis; *Candida albicans*; oxidative stress; stress signaling

1. *Candida albicans* Is a Major Fungal Pathogen of Humans

The polymorphic fungus, *Candida albicans*, is a constituent of the normal human microbiome. This fungus, together with other *Candida* family members, is present on the skin and in the oral cavity and gastrointestinal and urogenital tracts of most healthy individuals [1,2]. In the healthy host, *C. albicans* normally exists as a benign commensal organism. However, as an opportunistic pathogen, this fungus can also cause superficial infections, such as oral or vaginal candidiasis, or life-threatening systemic infections [2]. Perturbation of the microbiome through antibiotic usage or mild to severe defects in immune defences, such as in patients with HIV, can result in superficial oral and vaginal infections (thrush), termed oral (OC) and vulvovaginal (VVC) candidiasis, respectively. OC occurs in about 90% of HIV-infected persons as an AIDS-defining illness [3]. Defective immunity in premature infants and the elderly can also result in OC [4,5]. Significantly, 75% of women of childbearing age suffer from VVC, 45% of whom go on to have a least one recurrent infection [6]. Superficial candidiasis can also manifest as chronic infections of the skin and nails, resulting in mucocutaneous candidiasis (CMC) [7]. Although superficial infections are remarkably commonplace, they are non-life threatening and can be easily treated.

In contrast, systemic candidiasis is associated with unacceptably high crude and attributable mortality rates of 42 and 27%, respectively, despite the availability of antifungal drugs, such as the polyenes, azoles and echinocandins. These mortality rates exceed those attributed to sepsis caused by the most aggressive bacterial and viral pathogens [8] and are attributed to difficulties in diagnosing fungal systemic infections and the consequential delays in treatment [9]. Patients who are severely immunocompromised, such as those on immunosuppressive treatments for cancer or transplant surgery, are at risk of systemic candidiasis [10]. In such patients, the innate defence mechanisms, which are vital to prevent invasive disease, are significantly compromised [11]. Consequently, the fungus can survive in the bloodstream and subsequently colonise a number of internal organs [2]. Other risk factors include invasive clinical procedures or trauma, which disrupt the protective anatomical barrier of the mucosa, and the use of venous catheters, which can allow access of the fungus to the bloodstream [10]. Indeed, overall, *Candida* spp. are the fourth most common nosocomial (hospital acquired) systemic infection in the United States [8]. Clearly, *C. albicans* poses a significant medical problem, and thus, it is important that we understand what makes this fungus such a successful pathogen.

2. Reactive Oxygen Species Are a Core Component of the Immune Cell Armoury

In healthy hosts, the first line of defence against *C. albicans* is through phagocytosis by innate immune cells, including macrophages and neutrophils. A major antimicrobial defence mechanism mounted by these phagocytes is the production of reactive oxygen species (ROS) through a process known as the respiratory burst. Following stimulation by cytokines, phagocytic cells activate the assembly of the NADPH oxidase complex, which results in the generation of superoxide ($O_2^{-}\bullet$). Given the potency of the ROS produced by NADPH oxidase [12], activation of this multi-subunit enzyme is tightly regulated. The NADPH oxidase complex consists of the Nox2 (gp91^{phox}) catalytic subunit, the $p22^{phox}$ transmembrane protein and three cytosolic subunits, $p47^{phox}$, $p67^{phox}$ and $p40^{phox}$. Nox2 and $p22^{phox}$ make up the membrane-associated cytochrome b_{558} heterodimer. Activation of Nox2 is dependent on the interaction with the cytosolic components, in particular $p67^{phox}$, which translocate to the membrane following phagocytosis [13]. This interaction is dependent on the binding of the small GTPase Rac to $p67^{phox}$, which induces a conformation change in this subunit, thus promoting its interaction with Nox2 [14]. Activation of Nox2 drives the production of superoxide via the NADPH-driven reduction of molecular oxygen. This is generated at an extremely high rate of 5 to 10 nmol per s within the neutrophil phagosome [15], and it has been estimated that approximately 4 mol L^{-1} of O_2^{-}• is produced per bacterium engulfed in the phagocytic vacuole [16]. The superoxide is then dismutated to hydrogen peroxide (H_2O_2) by superoxide dismutase or to hydroxyl anions (OH^-) and hydroxyl radicals (OH) via the Haber-Weiss reaction. The importance of the NADPH oxidase-mediated respiratory burst as an antimicrobial mechanism is manifested in patients with chronic granulomatous disease (CGD). CGD is a human genetic disorder characterized by a deficiency in the NAPDH oxidase complex and is associated with recurrent and life-threatening bacterial and fungal infections [17]. Significantly, patients with CGD have an increased susceptibility to *Candida* infections [18]. Interestingly, in addition to the fungicidal roles of ROS, recent work has revealed that the ROS produced by NADPH oxidase also functions to recruit phagocytes to *C. albicans* infection foci. This NADPH oxidase-regulated recruitment of phagocytes is important for efficient phagocytosis, containment of the fungus within the phagocyte and survival of the host [19].

Other toxic chemicals are subsequently derived from the ROS in the phagosome [20]. For example, H_2O_2 can react with chloride ions (Cl^-) to form hypochlorous acid (HOCl) in a reaction catalysed by myeloperoxidase (MPO). In addition, the nitric oxide radical generated by the action of the inducible nitric oxide synthase (iNOS) interacts with superoxide to produce the highly toxic peroxynitrite (ONOO) [21]. Recently, work has also revealed that the combination of reactive oxygen species together with the cationic stress generated during phagocyte maturation underlies the potency of phagocytes in *C. albicans* killing [22]. Thus, phagocytic cells synthesize an array of toxic chemicals that work in combination to promote fungal killing. It is also noteworthy that, in addition to ROS production within the phagosome, phagocytes secrete ROS into the external milieu [23]. Consistent with this, *C. albicans* cells have been shown to mount an oxidative stress response prior to phagocytosis [24]. Furthermore, *C. albicans* will also come in contact with ROS produced by H_2O_2-producing bacteria in the mouth and gut. Several commensal bacteria, for example *Enterococcus faecalis* [25] and *Lactobacillus* species [26], secrete ROS into their surroundings, and this may have an inhibitory effect on the growth of *C. albicans* in host niches, other than the phagosomal environment. Consistent with this, using a *Caenorhabditis elegans* model of polymicrobial infection, *E. faecalis* was shown to reduce the virulence of *C. albicans* [27].

The ROS generated within the phagosome creates a toxic environment that induces oxidative stress in *C. albicans*. Indeed, exogenous ROS can induce programmed cell death in this fungal pathogen [28]. ROS interact with proteins, lipids and nucleic acids [29], causing irreversible damage to the pathogen. DNA damage caused by ROS can result in chemical base changes, structural alterations, single- and double-strand breaks and cross-linkage. Lipid peroxidation occurs by a free radical chain reaction, which culminates in peroxidation events at many fatty acid side chains, leading to the damage of the cell membrane. ROS reactions with proteins can lead to the formation of protein-protein cross-links, oxidation of the peptide backbone and reversible or irreversible oxidation of amino acid side chains. Although this can be deleterious to protein function, as discussed below, several oxidative stress-sensing proteins are activated by the reversible oxidation of cysteine residues.

3. Response of *Candida albicans* to ROS

3.1. Transcriptional Responses to ROS

A well-characterized response of eukaryotic microbes to ROS is the rapid induction of mRNAs that encode oxidative stress detoxification and repair proteins. Interestingly, *C. albicans* is considerably more resistant to oxidative stress than the benign model yeasts, *Schizosaccharomyces pombe* and *Saccharomyces cerevisiae* [30,31]. However, the basis for this resistance does not appear to be due to differences in transcriptional responses to oxidative stress, as all three fungi appear to induce a similar set of core antioxidant genes following exposure to H_2O_2 [32,33]. These include catalase (*CAT1*), glutathione peroxidase (*GPX*) and superoxide dismutase (*SOD*) antioxidant-encoding genes, in addition to genes encoding components of the glutathione/glutaredoxin (*GSH1*, *TTR1*) and thioredoxin (*TSA1*, *TRX1*, *TRR1*) systems, which play critical roles in repairing oxidatively-damaged proteins, protein folding and sulphur metabolism. Such oxidative stress-responsive genes are also induced in *C. albicans* following exposure to macrophages or neutrophils [34–38], illustrating that this pathogen induces the respiratory burst in these phagocytes. The analyses of GFP-reporter fusions, under the control of oxidative stress-responsive promoters, have also revealed that *C. albicans* is exposed to significant levels of ROS prior to phagocytosis [24]. In contrast, however, oxidative stress responses do not appear to be induced once *C. albicans* cells have established systemic kidney infections [34,39,40]. Thus, inducible oxidative stress responses appear vital for *C. albicans* to survive phagocytosis by innate immune cells, but are seemingly less important for the fungus to develop systemic infections. Indeed, whilst a number of genes encoding key antioxidants (such as *CAT1*, *TRX1*, *GRX2*, *SOD1*, *SOD5*) are important for virulence in systemic models of infection [41–45], others (including *TSA1*, *GPX*s) are dispensable [46,47].

3.2. Transcriptional Responses to ROS Are Inhibited in the Presence of Cationic Stress

In healthy individuals, *C. albicans* cannot evade the oxidative-killing mechanisms mounted by innate immune cells. Such cells prevent infection by employing a battery of toxic chemicals in addition to ROS. For example, phagocytes expose *C. albicans* to cationic fluxes (K^+) and acidification, as well as to superoxide anions [16,21,48]. However, as *C. albicans* is resistant to each of these individual stresses *in vitro* [30,31], a key question, therefore, is: what accounts for the potency of innate immune defences? Although host microenvironments are complex and dynamic, our understanding of *C. albicans* stress responses is based on studies of individual stresses. Significantly, however, recent work has revealed that *C. albicans* is exquisitely sensitive to combinations of oxidative and cationic stresses [49], which are encountered following phagocytosis. Cationic stress can be imposed *in vitro* by exposure of the fungus to either NaCl or KCl and in the phagocyte is caused by increased flux of K^+ into the phagosome [16]. Strikingly, exposure to cationic stress results in the inhibition of *C. albicans* oxidative stress responses. This phenomenon has been termed "stress pathway interference" [22]. The combinatorial stress-mediated synergistic killing of *C. albicans* contrasts starkly with the stress cross-protection described in model yeasts, whereby exposure to one stress protects against subsequent exposure to a different stresses [50]. The existence of stress pathway interference was revealed through gene expression analysis in *C. albicans*. Transcript profiling showed that H_2O_2-induced gene expression is severely attenuated, and intracellular ROS levels increase dramatically, following combinatorial oxidative and cationic stress. For example, key antioxidant genes, such as *CAT1* encoding catalase and *TRR1* encoding thioredoxin reductase, which are significantly induced following H_2O_2 stress, fail to be induced following exposure of cells to H_2O_2 in the presence of cationic stress [22]. This cationic stress-mediated inhibition of oxidative stress responses appears to be of physiological relevance, as the high fungicidal activity of human neutrophils is dependent on the combinatorial effects of the oxidative burst and cationic fluxes [22]. However, as discussed above, oxidative stress-responsive genes are induced following co-culture of *C. albicans* with phagocytes [34,35], so how is this reconciled with the combinatorial stress-mediated inactivation of such genes? It has been suggested [22] that the activation of *C. albicans* antioxidant genes during interaction with phagocytes may be due to exposure to extracellular ROS prior to engulfment [24]. Furthermore, as exposure of cells to combinatorial oxidative and cationic stresses prevents the normal activation of oxidative stress-responsive genes, this may explain why *C. albicans* oxidative stress genes are not expressed in certain host niches, such as during systemic infections of the kidney, despite the presence of neutrophil infiltrates [34].

3.3. Extracellular Antioxidant Enzymes as a Pathogen-Specific Adaptation Mechanism

As *C. albicans* appears to mount standard transcription responses to oxidative stress, the high level of resistance of this pathogen to ROS could, instead, be related to the evolutionary expansion of the SOD family and the fact that this pathogen expresses SODs and other antioxidant enzymes on the cell surface. *C. albicans* contains six SOD enzymes distributed between different cellular compartments. Sod1–3 are intracellular enzymes, while Sod4–6 are glycosylphosphatidylinositol (GPI)-anchored cell wall-associated enzymes. The Cu-/Zn-containing Sod1 is induced following phagocytosis and is required for *C. albicans* to resist macrophage-mediated killing [43]. The extracellular Sods also have vital roles in the detoxification of superoxide radicals generated by phagocytes; co-culture of macrophages with *C. albicans* cells lacking Sod4 and Sod5 leads to massive extracellular ROS accumulation *in vitro* [23]. Consequently, inactivation of Sod4 and Sod5 results in *C. albicans* cells that are exquisitely more susceptible to phagocyte-mediated killing [23,35]. Interestingly, the expression of Sod4 and Sod5 is dependent on the morphology of *C. albicans*, as Sod4 is expressed in yeast cells, whereas Sod5 is a hyphal-induced gene [44,51]. Sod5 is also induced following phagocytosis by neutrophils independently of hyphae formation [35]. Recently, structural analysis of Sod5 revealed that it represents a novel class of superoxide dismutases that only depends on Cu for activity. Furthermore, it is secreted in its apo-form and can readily capture extracellular copper without the aid of a Cu chaperone, which rapidly induces activity [52]. It is suggested that this novel mode of activation is uniquely adapted to the host environment, as macrophages release copper in an attempt to kill invading microbes through copper toxicity [53]. In addition to specific Sods, two key peroxidase enzymes have also been found at the cell surface of *C. albicans*; the thiol-specific peroxidase Tsa1 [47,54] and the peroxide detoxifying enzyme catalase [54]. Tsa1 and Cat1 were identified as major plasminogen-binding proteins in isolated cell wall protein preparations [54], and the cell wall localization of Tsa1 has also been illustrated using fluorescence microscopy [47]. These extracellular mechanisms for protection against ROS likely reflect an adaptation of this pathogenic fungus to prevent the intracellular accumulation of toxic levels of ROS.

3.4. Morphogenesis as an Oxidative Stress Response

Following phagocytosis, *C. albicans* can evade oxidative-killing by macrophages and neutrophils by switching from budding to filamentous cells, which can pierce the phagosomal membrane [37]. Not only does this allow the pathogen to escape, but this also results in the *C. albicans*-mediated killing of the phagocyte [55]. It has recently been demonstrated that the ability of *C. albicans* to mount robust oxidative stress responses is vital for this polymorphic pathogen to filament in the phagosome [46,56]. *C. albicans* mutants that are sensitive to ROS *in vitro* fail to filament once phagocytosed and, thus, are trapped within the macrophage and unable to evade phagocyte-mediated killing. Consistent with the requirement of fungal oxidative stress defences to allow filamentation and macrophage escape, the phagocyte NADPH oxidase is important in inhibiting filamentation *in vivo* [57]. The ROS produced by the NADPH oxidase also function to recruit phagocytes, thereby increasing phagocytosis and inhibiting filamentation [19]. Thus, the outcome of the battle between *C. albicans* and innate immune cells appears dependent on the NADPH oxidase-regulated functions of the phagocyte and the robustness of the fungal oxidative stress responses.

The mechanisms underlying *C. albicans* filamentation following phagocytosis remain to be fully explored. A recent study reported that the ROS-induced induction of arginine biosynthesis genes is important for hyphal formation following phagocytosis of *C. albicans* [58]. Moreover, exposure of *C. albicans* to the ROS H_2O_2 triggers the filamentation of this polymorphic fungus *in vitro* [42,59]. A close examination of the morphology of these cells revealed that H_2O_2-induced filaments are hyperpolarized buds, which are morphologically distinct from hyphae and pseudohyphae filamentous forms [42]. The hyperpolarised bud is a relatively recently characterized filamentous form of *C. albicans* and is normally associated with either mutations or chemicals that perturb cell cycle progression [60,61]. The observation that H_2O_2 stimulates hyperpolarised bud formation provided the first example of a physiologically relevant condition that induces this filamentous form of growth in *C. albicans.* Does exposure of ROS following phagocytosis trigger the formation of hyperpolarised buds allowing this pathogen to pierce the phagocyte membrane and escape? Evidence so far indicates that ROS-stimulated hyperpolarized bud formation may not contribute to *C. albicans* filamentation within the macrophage. For example, ROS-sensitive *C. albicans* mutants that cannot filament within the macrophage [56] can form H_2O_2-induced hyperpolarized buds *in vitro* [42]. Nonetheless, ROS-stimulated filamentation of *C. albicans* may be beneficial for survival in other host niches, such as the gut, where *C. albicans* co-exists with ROS-producing bacteria.

4. Signalling Pathways that Mediate *C. albicans* Responses to ROS

To date, three signalling pathways have been demonstrated to be directly activated in response to ROS in *C. albicans*. These include the Cap1 transcription factor, the Hog1 stress-activated protein kinase and the Rad53 DNA damage checkpoint kinase. Here, we discuss the role and regulation of these pathways in oxidative stress responses in *C. albicans*. Other signalling pathways not known to be activated by ROS, but which contribute to oxidative stress tolerance in *C. albicans* are also briefly summarized.

4.1. The Cap1 ROS-Responsive Transcription Factor

In *C. albicans*, the Cap1 transcription factor is the major regulator of the oxidative stress-induced transcriptome and proteome, both *in vitro* [62,63] and *ex vivo*, following exposure to neutrophils [35]. Cap1 is a bZip transcription factor of the AP-1 family and is closely related to the *S. cerevisiae* Yap1 and *S. pombe* Pap1 proteins, which have well-characterized roles in oxidative stress and multi-drug resistance [64,65]. Similarly, *C. albicans cap1Δ* cells are sensitive to several reactive oxygen species and drugs [33,66,67]. Chromatin immunoprecipitation (CHiP) analysis to determine direct targets of Cap1 identified many key antioxidant genes, including *CTA1* and *TRX1*, and those involved in the response to drugs, such as *MDR1* [68]. Cap1 plays a role in recruiting the Ada2 component of the SAGA/ADA histone acetylase co-activator complex to the promoters of oxidative stress and drug-responsive target genes [69,70]. Cells lacking Ada2 are highly sensitive to ROS, and the oxidative stress-induced transcription of key Cap1 target genes is significantly impaired; therefore, Cap1 recruitment of the SAGA complex appears to be a vital component of the oxidative stress response in *C. albicans*.

4.1.1. Regulation of Cap1

Similar to that reported for *S. cerevisiae* Yap1, *C. albicans* Cap1 rapidly accumulates in the nucleus in response to H_2O_2 [56,67]. Under non-stressed conditions, Yap1 shuttles between the cytoplasm and the nucleus due to the interaction of a nuclear export sequence (NES), located at the C-terminus of these transcription factors, with the Crm1 nuclear export factor [71]. However, following exposure to H_2O_2, Yap1 is activated by oxidation of specific cysteine residues, resulting in disulphide bond formation between two cysteine-rich domains (n-CRD and c-CRD). This triggers a conformational change within Yap1 that masks the NES, thereby preventing its interaction with Crm1. The inability to be recognized by Crm1 leads to the nuclear accumulation of Yap1, the nuclear-dependent phosphorylation of this transcription factor and the induction of Yap1-dependent genes [72]. Conversely, activation of Yap1 is counteracted by the thioredoxins Trx1 and Trx2, which function to reduce oxidised Yap1 [72]. This basic mechanism of regulation is conserved in *C. albicans* (Figure 1). Mutation of the c-CRD affects Cap1 regulation [67], and Cap1 is rapidly oxidised following exposure to H_2O_2 [42]. In addition, following the nuclear accumulation of Cap1, this transcription factor becomes phosphorylated, and the induction of Cap1-dependent genes is observed. Furthermore, as seen in *S. cerevisiae*, thioredoxin functions to reverse the H_2O_2-induced oxidation and activation of Cap1 [42].

Fungal AP-1-like transcription factors are not directly oxidised by H_2O_2, but instead, specific peroxidase enzymes sense and transduce the H_2O_2 signal to these transcription factors (Figure 1). Similar to that observed in *S. cerevisiae* [72], Cap1 oxidation requires Gpx3, a glutathione peroxidase (Gpx)-like enzyme [56]. Studies with Yap1 showed that this transcription factor undergoes multiple oxidation events, with Gpx3 initiating Yap1 oxidation [73–75]. Similarly, multiple oxidized forms of Cap1 are also observed [56]. Gpx3-mediated oxidation of Yap1 and Cap1 also requires a second protein, Ybp1, which binds to and forms a complex with the AP-1-like factors [56,76]. A recent study has provided insight into an additional function of Ybp1 in both *C. albicans* and *S. cerevisiae*, as Cap1 and Yap1 are highly unstable in *ybp1Δ* cells [56]. Ubiquitin-mediated degradation of oxidised AP-1-like factors has recently been shown to be an important regulatory mechanism [77,78]; therefore, Ybp1 binding to the reduced cytoplasmic pools of Yap1 or Cap1 possibly functions to prevent this proteasome-mediated degradation [56].

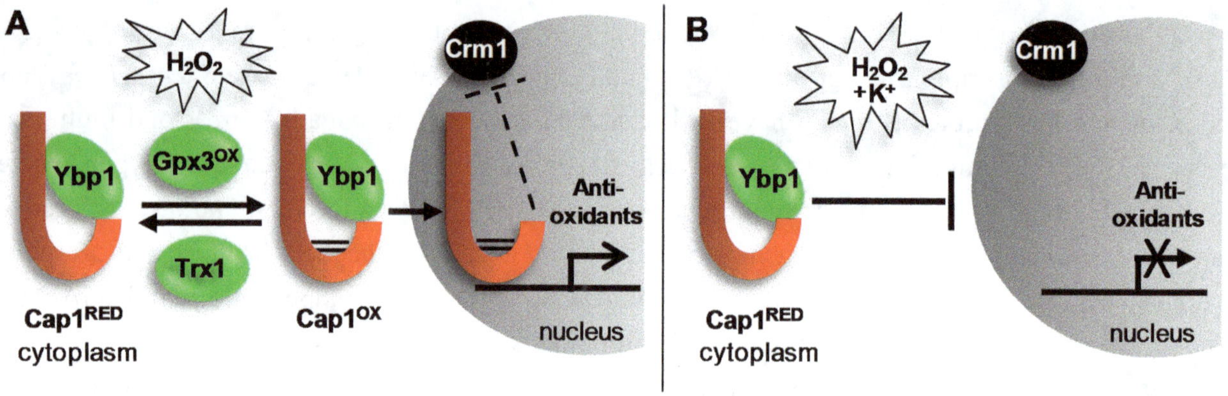

Figure 1. H_2O_2-induced activation of Cap1 is inhibited in the presence of cations. (**A**) Exposure of *C. albicans* to H_2O_2 promotes the Gpx3/Ybp1-mediated oxidation and activation the Cap1 transcription factor (Cap1^{ox}). Cap1^{ox} can no longer interact with the Crm1 nuclear export factor resulting in its nuclear accumulation, and the subsequent Cap1-dependent induction of genes with antioxidant functions necessary for cell survival. Following cellular adaptation, Cap1^{ox} is returned to the inactive reduced form (Cap1^{RED}) by thioredoxin (Trx1); (**B**) Remarkably, when *C. albicans* cells are exposed to H_2O_2 in the presence of cations, Cap1 fails to accumulate in the nucleus and therefore antioxidant gene expression is not induced leading to cell death. This is important as, following phagocytosis, *C. albicans* is exposed simultaneously to ROS and cationic fluxes. See text for details.

In *S. cerevisiae*, Yap1 functions alongside the Skn7 response regulator transcription factor, to regulate antioxidant gene expression [79]. An orthologue of Skn7 has been identified in *C. albicans* [80]. The overall domain architecture is conserved and comprised of a DNA-binding domain, a coiled-coil domain and a receiver domain (analogous to those in response regulator proteins of two-component signal transduction pathways). It is not known whether Skn7 acts alongside Cap1 in *C. albicans*. However, *C. albicans* cells lacking Skn7 display increased sensitivity to ROS, including H_2O_2, consistent with this transcription factor regulating oxidative stress-induced gene expression [80].

Interestingly, Cap1 fails to be activated following exposure to combinatorial oxidative and cationic stress (Figure 1), which underlies the lack of antioxidant gene expression following this combinatorial stress treatment (Section 3.2). In contrast with that seen following oxidative stress, following combinatorial cationic plus oxidative stress treatments, Cap1 fails to accumulate in the nucleus [22]. Consequently, Cap1 is not phosphorylated, and Cap1-dependent oxidative stress genes are not induced [22]. The impact of stress pathway interference upon Cap1 signalling underlies the potency of combinatorial cationic plus oxidative stress, as ectopic expression of the Cap1-dependent catalase gene, *CAT1*, rescues the hypersensitivity to the combinatorial stress [22]. However, the mechanism underlying combinatorial stress-mediated inactivation of Cap1 is not known. Cations inhibit catalase function, which results in high levels of intracellular ROS [22]. Whether high levels of ROS result in Cap1 inactivation or whether cations inhibit Cap1 activation in other ways remains to be determined.

4.1.2. Role of Cap1 in Virulence

Loss of Cap1 or its regulators Gpx3 and Ybp1 attenuates virulence in some, but not all infection models. For example, cells lacking Cap1, Gpx3 or Ybp1 are unable to kill macrophages, due to the inability of these mutant strains to filament following phagocytosis [56]. Consequently, cells lacking Cap1 or its regulators are sensitive to macrophage- and neutrophil-mediated killing [35,56,81]. Cap1, Gpx3 and Ybp1 are also vital for *C. albicans* virulence in a *Galleria mellonella* model of infection [56], and Cap1 is important for virulence in a *Caenorhabditis elegans* infection model in nematode hosts that have a functional NADPH oxidase [81]. In contrast, Cap1, Gpx3 and Ybp1 are dispensable for *C. albicans* virulence in murine systemic infection models [56,81]. Similar findings were reported for Skn7 [80]. The observation that Cap1 is dispensable for virulence in murine systemic models of infection was unexpected, as certain genes that are induced by Cap1 in response to H_2O_2, such as *CTA1* and *TRX1*, are important for *C. albicans* survival in such models [42,45]. This indicates that Cap1-independent basal levels of such genes may be important for virulence in such models and that Cap1-mediated gene expression is not vital for the establishment of systemic infections.

4.2. The Hog1 SAPK

Stress-activated MAPKs are conserved signalling molecules that promote the ability of cells to adapt to environmental change [82]. They are components of a three-tiered core signalling module that comprises the SAPK itself, a MAP kinase kinase (MAPKK) and a MAPKK kinase (MAPKKK). Activation of the MAPKKK results in the phosphorylation and activation of the MAPKK, which, in turn, culminates in the phosphorylation of the SAPK on conserved threonine and tyrosine residues located within the TGY motif in the phosphorylation lip of the catalytic domain. This induces the activation and nuclear accumulation of the kinase [83] and the proline-directed phosphorylation of Ser/Thr residues on diverse substrates, including transcription factors, kinases, cell cycle regulators and membrane proteins, thus eliciting appropriate cellular responses. In *C. albicans*, Hog1 is robustly phosphorylated and rapidly accumulates in the nucleus following exposure of cells to H_2O_2 [33]. In addition, cells lacking Hog1 display increased sensitivity to a range of ROS, indicating that Hog1 activation is a critical component of the oxidative stress response in *C. albicans* [84,85]. Interestingly, Hog1 is only activated following exposure of *C. albicans* cells to relatively high levels of H_2O_2 compared to the analogous Sty1 SAPK in the model yeast, *S. pombe*. This may reflect an adaption of this pathogenic fungus to restrict Hog1 activation to ROS-rich environments during infection [85]. Despite the increased H_2O_2 sensitivity exhibited by *hog1Δ* cells and significant phosphorylation of Hog1 in response to H_2O_2, transcript profiling experiments revealed that Hog1 is largely dispensable for H_2O_2-induced gene expression [33]. Although a small subset of H_2O_2-responsive genes were identified that showed Hog1-dependent induction, subsequent analysis failed to identify any genes coding for proteins with known antioxidant functions [33]. This is in contrast with *S. pombe*, where Sty1 is required for the activation of the core stress genes in response to H_2O_2, including genes encoding important antioxidants, such as catalase and glutathione peroxidase [86]. What, therefore, is the role of Hog1 in the *C. albicans* oxidative stress response if it is not required for the induction of antioxidant gene expression? One possibility is that Hog1 contributes to the oxidative stress response at a post-transcriptional level in *C. albicans*. Indeed, the *S. pombe* Sty1 SAPK has been shown to interact with translation factors [87]. However, Hog1 does not play a major role in regulating the oxidative stress-induced proteome, although proteomic experiments did indicate that Hog1 might be required to ensure the prolonged expression of some proteins during recovery from H_2O_2 stress [88]. Loss of Hog1 has been shown to affect respiratory function [89], although it is unclear whether this underlies the sensitivity of *hog1Δ* cells to ROS. One downstream target of Hog1 regulated by H_2O_2 stress is the Mkc1 cell integrity MAPK. Mkc1 is rapidly phosphorylated in response to H_2O_2 stress in a Hog1-dependent mechanism, although Mkc1 is not required for cell survival in response to H_2O_2 stress [90]. In addition, the Sko1 transcription factor is a target of the Hog1 SAPK in *C. albicans*, as this becomes phosphorylated following stress in a Hog1-dependent manner [91]. However, consistent with Hog1 not playing a major role in regulating oxidative stress-induced gene expression, the H_2O_2-induced transcriptome is not dependent on Sko1 [92]. Thus, in *C. albicans*, Hog1 regulation of the oxidative stress response must involve targets in addition to Mkc1 and Sko1 (Figure 2).

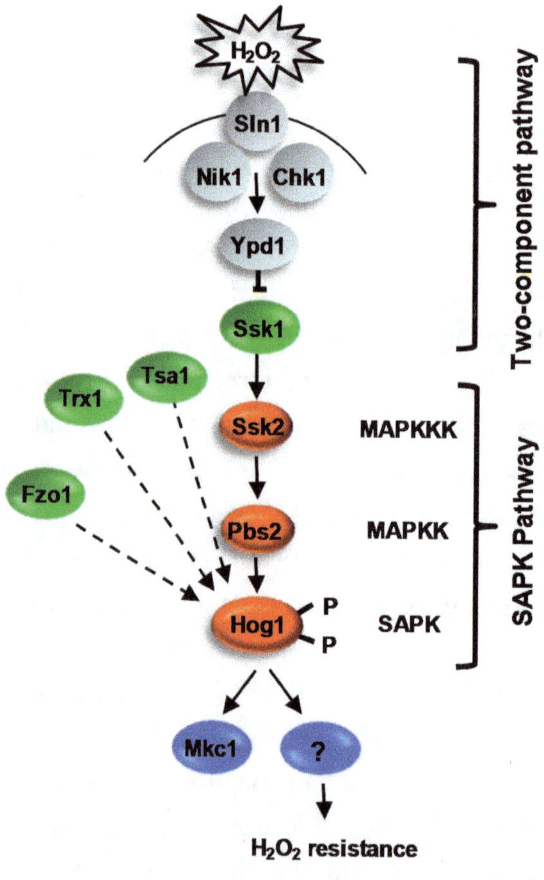

Figure 2. H_2O_2-induced activation of the Hog1 SAPK. In response to H_2O_2, Hog1 becomes rapidly phosphorylated and accumulates in the nucleus, and *C. albicans* cells lacking Hog1 are sensitive to oxidative stress. Proteins required for H_2O_2-induced activation of Hog1 are shown in green. These include the response regulator Ssk1 (but no other two-component protein), the redox sensitive antioxidants Tsa1 and Trx1, and the mitochondria biogenesis factor Fzo1. Following H_2O_2-induced activation, Hog1 phosphorylates the Mkc1 MAPK. However, cells lacking Mkc1 are not sensitive to oxidative stress, suggesting that an, as yet, unknown Hog1 substrate(s), mediates oxidative stress resistance.

4.2.1. Regulation of Hog1 in Response to ROS

Whilst little is known regarding the cellular targets of Hog and the cellular role(s) of this kinase in promoting oxidative stress tolerance, more progress has been made in delineating how H_2O_2 signals are relayed to Hog1 (Figure 2). Oxidative stress-induced activation of Hog1 is entirely dependent on the Pbs2 MAPKK [93], which, in turn, is regulated by a single MAPKKK Ssk2 [94]. Furthermore, a recent study reported that deletion of a mitochondrial biogenesis factor, Fzo1, significantly impairs the H_2O_2-induced activation of Hog1 [95]. Thus, intriguingly, functional mitochondria may play an important role in the regulation of the Hog1 pathway in response to oxidative stress. In addition, both two-component related proteins and redox-sensitive antioxidants are necessary for the activation of the Hog1 SAPK in response to oxidative stress, and these will be described in turn.

4.2.1.1. Two-Component Mediated Regulation of Hog1

In the model yeasts, two-component signalling pathways have been shown to play an important role in the sensing and transmission of stress signals to their respective SAPK pathways. Such pathways are comprised of a histidine kinase, an intermediary phosphorelay protein and a response regulator protein. In *S. cerevisiae*, the Sln1 histidine kinase is inactivated in response to osmotic stress. This halts phosphorelay through the Ypd1 phosphorelay protein, leading to the rapid dephosphorylation of the Ssk1 response regulator. Dephosphorylated Ssk1 is a potent activator of the Ssk2/Ssk22 MAPKKKs in *S. cerevisiae*, which regulate Hog1 activation [96,97]. In *C. albicans*, deletion of the analogous *SSK1* gene prevents Hog1 activation in response to oxidative stress, and consistent with this, *ssk1Δ* cells are sensitive to oxidative stress [98]. Although Ssk1 is involved in the transmission of oxidative stress signals to Hog1, the identity of the histidine kinase(s) responsible for sensing and signalling oxidative stress signals to Ssk1 in *C. albicans* remains elusive [99,100]. Of the three histidine kinases present in *C. albicans*, Chk1 would appear to be the most likely candidate for a potential peroxide-sensing histidine kinase, as this shows significant similarity to the *S. pombe* peroxide-sensing histidine kinases, Mak2 and Mak3 [101,102]. However, deletion of *CHK1* alone or in combination with either of the genes encoding the two remaining histidine kinases, *SLN1* or *NIK1*, does not impair H_2O_2-induced activation of Hog1 [100,103]. Hence, it is currently unclear as to which histidine kinase(s) senses oxidative stress and regulates phosphorelay to Ssk1. Moreover, observations that Hog1 activation is seen in cells expressing a non-phosphorylatable Ssk1 mutant [103] or in cells lacking the Ypd1 phosphorelay protein in which Ssk1 is predicted to be unphosphorylated [104] indicate that Ssk1 may relay H_2O_2 signals to Hog1 in a mechanism independent of two-component signalling. It is also noteworthy that a novel response regulator, named Crr1/Srr1, has been recently identified that is only present in fungi belonging to the *Candida* clade [105,106]. Cells lacking Crr1 or expressing a mutant lacking the predicted aspartate phosphorylation site are sensitive to H_2O_2 [105]. However, in contrast with Ssk1, Crr1 is not required for the H_2O_2-induced activation of Hog1 [105]. Thus, whilst this novel response regulator mediates the response of *C. albicans* to H_2O_2, it does so in a Hog1-independent manner. Finally, in *S. cerevisiae*, the transmembrane protein, Sho1, relays osmotic stress signals to the Hog1 SAPK in parallel with the Sln1-mediated two-component signalling pathway [107]. In *C. albicans*, the analogous Sho1 protein appears to have been reassigned to oxidative stress signalling [100]. However, it is not clear how this is mediated, as Sho1 is not required for ROS-stimulated activation of the Hog1 pathway [100].

4.2.1.2. Redox-Sensitive Antioxidant Proteins as Regulators of Hog1

It is now well recognized that redox-sensitive antioxidant proteins have important sensing and signalling roles in the cellular response to oxidative stress [108]. In *C. albicans*, the redox-sensitive thioredoxin peroxidase enzyme, Tsa1, is specifically required for H_2O_2-induced activation of Hog1 [42]. This is similar to that previously reported in *S. pombe*, as H_2O_2-induced activation of the Sty1 SAPK also requires the analogous thioredoxin peroxidase enzyme, Tpx1 [109]. In *S. pombe*, intermolecular disulphide bonds are formed between conserved cysteine residues in Sty1 and Tpx1 following H_2O_2 stress, which suggests that Tpx1 regulates Sty1 function directly. However, the mechanism of Tsa1 regulation of Hog1 in *C. albicans* may be different, as the conserved peroxidatic cysteine residue of Tsa1, which is essential for Tpx1 regulation of Sty1, is dispensable for Tsa1 regulation of Hog1 [42]. Furthermore, the thioredoxin enzyme, Trx1, which regulates the redox status of Tsa1, is also essential for the relay of oxidative stress signals to the Hog1 SAPK module [42]. eletion of *TRX1* or mutation of the catalytic cysteine residues of Trx1 drastically impairs Hog1 phosphorylation in response to H_2O_2. However, it would appear that Trx1 regulates Hog1 independently of Tsa1, as the catalytic cysteine residues of Tsa1, which are reduced by Trx1, are dispensable for Hog1 activation [42]. In mammalian systems, thioredoxin functions as a repressor of the Hog1-related JNK and p38 SAPK signalling cascades [110]. The upstream Ask1 MAPKKK in the mammalian SAPK pathways is activated via cysteine oxidation, and Trx1 negatively regulates this pathway by reducing the oxidized cysteines of Ask1 [111,112]. As Trx1 is a positive regulator of the Hog1 SAPK in *C. albicans*, it seems unlikely that a similar mechanism is in place. It is also interesting to note that protein tyrosine phosphatases, which are negative regulators of SAPKs, are susceptible to inactivation by oxidation of their catalytic cysteine residue [113]. Whether thioredoxin regulates such phosphatases that dephosphorylate Hog1 in *C. albicans*, however, remains to be established.

4.2.2. Role of the Hog1 SAPK in Virulence

The stress-activated MAPK Hog1 in *C. albicans* is phosphorylated and accumulates in the nucleus, in response to a range of stresses likely to be encountered in the host, including ROS, osmotic stress and anti-microbial peptides [114]. Cells lacking Hog1 display impaired virulence in a wide range of infection models, including murine systemic and commensal models [114–116], and are more susceptible to killing by macrophages or neutrophils [117]. As Hog1 regulates a number of distinct stress responses, it is difficult to dissect whether it is the role of Hog1 in oxidative stress responses or a different aspect of Hog1 signalling that is important for virulence in these models. Importantly, however, although Hog1 signalling has also been implicated in morphogenetic regulation, mutational analysis has inferred that the importance of Hog1 in virulence is due to its role in stress protection, rather than its role in repressing the yeast to hyphal transition [115].

4.3. The Rad53 DNA Damage Checkpoint Kinase

Following exposure to H_2O_2, *C. albicans* forms hyperpolarised buds, which are morphologically distinct from hyphae and pseudohyphae filamentous forms (Section 3.4). Consistent with this, H_2O_2-induced hyperpolarized bud formation occurs independently of the key hyphal regulators, Efg1 and Cph1, and, instead, depends on the activation of the Rad53 DNA damage checkpoint pathway [42,61] (Figure 3). A wide range of genotoxic stresses, including UV, methyl methanesulfonate (MMS) and the ribonucleotide reductase inhibitor hydroxyurea have been shown to activate the Rad53 kinase in *C. albicans* [61], and loss of Rad53 or upstream regulators of Rad53, prevents hyperpolarised bud formation [61,118]. ROS are also genotoxic agents due to the induction of DNA damage [119], which, in turn, triggers the activation of the Rad53 DNA checkpoint pathway [120]. Indeed, treatment of *C. albicans* cells with H_2O_2 elicits the phosphorylation of Rad53, and cells lacking *RAD53* fail to form hyperpolarised buds in response to H_2O_2 [42].

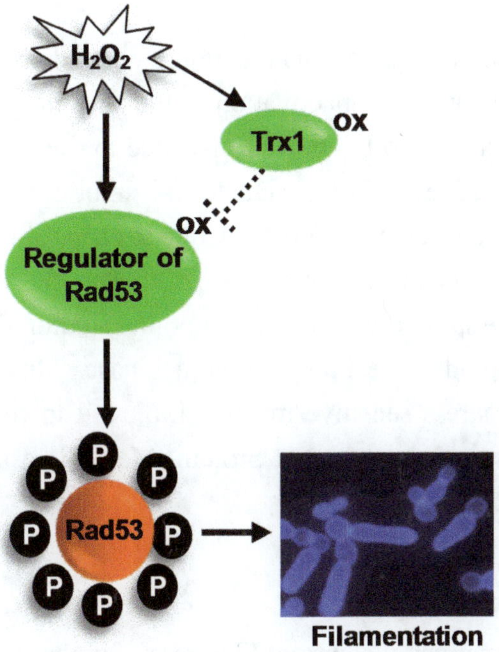

Figure 3. H_2O_2-induced activation of Rad53 triggers filamentation in *C. albicans*. The redox sensitive oxidoreductase Trx1 inhibits H_2O_2-induced activation of the DNA damage checkpoint kinase Rad53. This suggests that a regulator of Rad53 is activated by oxidation, and this active oxidised form is reduced by Trx1. Activation of the DNA damage checkpoint triggers the formation of hyperpolarised buds. See text for details.

Regulation of Rad53

Regarding the H_2O_2-mediated activation of Rad53, a recent study illustrated that H_2O_2-induced oxidation, and inactivation, of the thioredoxin protein Trx1 is important for the activation of Rad53 and polarized cell growth [42]. Rad53 is constitutively phosphorylated in cells lacking Trx1, which display a hyperpolarized bud morphology. Conversely, ectopic expression of the thioredoxin reductase gene, *TRR1*, which reduces oxidized Trx1, inhibited H_2O_2-induced filamentation [42]. Taken together, these results illustrate that oxidation of Trx1 following H_2O_2 exposure is key in the activation of Rad53 that drives hyperpolarised bud formation. The finding that Trx1 inhibits Rad53 activation under non-stressed conditions may be conserved in higher eukaryotes, as ectopic expression of thioredoxin inhibits the phosphorylation of the analogous DNA damage checkpoint kinase, Chk2, in mammalian cells [121]. However, the mechanism of Trx1 regulation of Rad53 is unclear. As Trx1 functions to reduce oxidised proteins, an attractive hypothesis is that Rad53, or a regulator of this kinase, is activated by oxidation (Figure 3). In this regard, it is interesting that the human homologue (ATM) of the fungal Tel1 DNA-damage sensing kinase, which regulates Rad53, has recently been shown to be activated by oxidation [122]. Further studies are needed to determine if Tel1 is similarly regulated to mediate H_2O_2-induced filamentation in *C. albicans*.

4.4. Other Signaling Pathways that Contribute to Oxidative Stress Resistance

The cAMP/PKA signalling pathway has a negative impact on oxidative stress responses in *C. albicans*. For example, induction of the pathway by inactivation of the phosphodiesterase, Pde2, which degrades cAMP, results in increased sensitivity to H_2O_2 [123]. Related to this, farnesol treatment of *C. albicans* cells results in increased resistance to H_2O_2, due to the inhibition of the cAMP/PKA signalling pathway [124]. Such changes in resistance are possibly due to changes in the levels of anti-oxidant gene expression [124]; however, the mechanism linking cAMP/PKA to their regulation is unknown.

There is also evidence that the spindle assembly checkpoint is required for *C. albicans* oxidative stress resistance. Cells lacking the spindle checkpoint protein kinase, Mps1, are sensitive to H_2O_2 [125] and, similar to other oxidative stress sensitive mutants [56], fail to filament following phagocytosis. Related to this, the spindle assembly checkpoint protein, Mad2, is essential for *C. albicans* survival in macrophages [126].

5. Conclusions and Future Perspectives

In this review, we have summarized the current literature of oxidative stress responses and how they are regulated in the human fungal pathogen *C. albicans*. This is an important area of research, as oxidative stress adaptation is emerging as an important virulence trait in this, and other, fungal pathogens. Table 1 summarizes studies that have documented the impact of the loss of oxidative stress regulatory proteins or antioxidant enzymes on *C. albicans* virulence in either a murine systemic infection model or a macrophage/neutrophil phagocyte-survival infection model. From this summary, a number of observations can be made. First of all, when examined, mutants that display an impaired tolerance to oxidative stress show an impaired ability to survive phagocyte killing. This is consistent with previous observations that *C. albicans* mounts a robust transcriptional response to oxidative stress following phagocytosis [34–38]. Secondly, not all oxidative stress-sensitive *C. albicans* mutants display attenuated virulence in a murine systemic infection model. Perhaps most striking is the observation that the major regulator of anti-oxidant gene expression, Cap1, is dispensable for virulence in such an infection model. This is particularly intriguing, as certain genes, which are dependent on Cap1 for induction following oxidative stress, are important for virulence in systemic models. Thirdly, the role of many other oxidative stress-responsive proteins in mediating *C. albicans* virulence, such as the Rad53-mediated DNA damage checkpoint pathway, have yet to be explored.

Table 1. Summary of the role of oxidative stress-responsive signalling proteins and antioxidant enzymes in *C. albicans* virulence. The importance of proteins needed for resistance to oxidative stress in mediating *C. albicans* virulence in either a systemic infection model (SIM) or phagocyte infection model (PIM) is indicated; +, important for virulence; −, dispensable for virulence; nd, not determined. For further explanation, see the text.

Protein	Function	SIM	PIM	References
Signalling Proteins				
Hog1	Stress-activated protein kinase	+	+	[114,115,117]
Ssk1	Response regulator	+	+	[127,128]
Cap1	Transcription factor	−	+	[56,81]
Ybp1	Cap1 regulator	−	+	[56]
Gpx3	Cap1 regulator	−	+	[56]
Skn7	Transcription factor	−	nd	[80]
Signalling Proteins				
Pde2	Phosphodiesterase	+	nd	[123]
Mps1	Spindle checkpoint	nd	+	[125]
Mad2	Spindle checkpoint	+	+	[126]
Antioxidant Enzymes				
Cat1	Catalase	+	+	[45]
Trx1	Thioredoxin	+	nd	[42]
Tsa1	Thioredoxin peroxidase	−	nd	[47]
Sod1	Superoxide dismutase	+	+	[43]
Sod5	Superoxide dismutase	+	+	[23,44]
Grx2	Glutaredoxin	+	nd	[41]
Gpx31-33	Glutathione peroxidases	nd	+	[46]

In addition to gaps in our knowledge regarding the relative importance of specific oxidative stress responses in mediating *C. albicans* virulence, there are also additional key questions that remain to be addressed. For example, what is the role of the Hog1 SAPK in mediating oxidative stress resistance in *C. albicans*, and how is this regulated? This is important, as, although Hog1 is an essential virulence determinant in *C. albicans*, the conservation with highly related SAPKs in human cells suggests that Hog1 itself may be unsuitable as a specific antifungal target. Thus, there is much interest in identifying fungal-specific SAPK regulators or substrates, as such proteins hold greater promise for future therapeutic strategies. In addition, the recent findings that *C. albicans* is exquisitely sensitive to combinations of stress that are encountered following phagocytosis represent a new unchartered area in the field of stress signalling. A key question for the future is how do combinations of stress imposed by the phagosome inhibit oxidative stress adaptation and survival of *C. albicans*? Addressing this question is critical to further our understanding of *Candida*-host interactions during disease progression.

Acknowledgments

We thank our friends and colleagues, in particular Al Brown, Lars Erwig, Ken Haynes, Brian Morgan and Elizabeth Veal, for stimulating discussions on oxidative stress signalling mechanisms and fungal pathogenesis. Research in the Janet Quinn lab is funded by the BBSRC BB/K016393/1, an MRC-DTP studentship, the NIHR Newcastle Biomedical Research Centre and the Wellcome Trust Strategic Award for Medical Mycology and Fungal Immunology 097377/Z/11/Z.

Author Contributions

All authors contributed to the initial draft of this review, which was edited by Janet Quinn.

References and Notes

1. Odds, F.C. *Candida and Candidosis*, 2nd ed.; Bailliere-Tindall: London, UK, 1988.
2. Calderone, R.A.; Clancy, C.J. *Candida and Candidiasis*; ASM Press: Washington, DC, USA, 2012.
3. Anaissie, E.J.; McGinnis, M.R.; Pfaller, M.A. *Clinical Mycology*; Churchill Livingstone: London, UK, 2009.
4. Flevari, A.; Theodorakopoulou, M.; Velegraki, A.; Armaganidis, A.; Dimopoulos, G. Treatment of invasive candidiasis in the elderly: A review. *Clin. Interv. Aging* **2013**, *8*, 1199–1208.
5. Healy, C.M.; Campbell, J.R.; Zaccaria, E.; Baker, C.J. Fluconazole prophylaxis in extremely low birth weight neonates reduces invasive candidiasis mortality rates without emergence of fluconazole-resistant candida species. *Pediatrics* **2008**, *121*, 703–710.
6. Sobel, J.D. Vulvovaginal candidosis. *Lancet* **2007**, *369*, 1961–1971.
7. Lilic, D. Unravelling fungal immunity through primary immune deficiencies. *Curr. Opin. Microbiol.* **2012**, *15*, 420–426.
8. Morgan, J. Global trends in candidemia: Review of reports from 1995–2005. *Curr. Infect. Dis. Rep.* **2005**, *7*, 429–439.
9. Brown, G.D.; Denning, D.W.; Gow, N.A.; Levitz, S.M.; Netea, M.G.; White, T.C. Hidden killers: Human fungal infections. *Sci. Transl. Med.* **2012**, doi:10.1126/scitranslmed.3004404.
10. Koh, A.Y.; Kohler, J.R.; Coggshall, K.T.; van Rooijen, N.; Pier, G.B. Mucosal damage and neutropenia are required for *Candida albicans* dissemination. *PLOS Pathog.* **2008**, *4*, e35.
11. Seider, K.; Heyken, A.; Luttich, A.; Miramon, P.; Hube, B. Interaction of pathogenic yeasts with phagocytes: Survival, persistence and escape. *Curr. Opin. Microbiol.* **2010**, *13*, 392–400.
12. Babior, B.M. NADPH oxidase. *Curr. Opin. Immunol.* **2004**, *16*, 42–47.
13. Bedard, K.; Krause, K.H. The NOX family of ROS-generating NADPH oxidases: Physiology and pathophysiology. *Physiol. Rev.* **2007**, *87*, 245–313.
14. Pick, E. Role of the Rho GTPase Rac in the activation of the phagocyte NADPH oxidase: Outsourcing a key task. *Small GTPases* **2014**, *5*, e27952.
15. Hampton, M.B.; Kettle, A.J.; Winterbourn, C.C. Involvement of superoxide and myeloperoxidase in oxygen-dependent killing of *Staphylococcus aureus* by neutrophils. *Infect. Immun.* **1996**, *64*, 3512–3517.
16. Reeves, E.P.; Lu, H.; Jacobs, H.L.; Messina, C.G.; Bolsover, S.; Gabella, G.; Potma, E.O.; Warley, A.; Roes, J.; Segal, A.W. Killing activity of neutrophils is mediated through activation of proteases by K^+ flux. *Nature* **2002**, *416*, 291–297.

17. Holland, S.M. Chronic granulomatous disease. *Clin. Rev. Allerg. Immunol.* **2010**, *38*, 3–10.
18. Cohen, M.S.; Isturiz, R.E.; Malech, H.L.; Root, R.K.; Wilfert, C.M.; Gutman, L.; Buckley, R.H. Fungal infection in chronic granulomatous disease. The importance of the phagocyte in defense against fungi. *Am. J. Med.* **1981**, *71*, 59–66.
19. Brothers, K.M.; Gratacap, R.L.; Barker, S.E.; Newman, Z.R.; Norum, A.; Wheeler, R.T. Nadph oxidase-driven phagocyte recruitment controls *Candida albicans* filamentous growth and prevents mortality. *PLOS Pathog.* **2013**, *9*, e1003634.
20. Brown, A.J.; Haynes, K.; Quinn, J. Nitrosative and oxidative stress responses in fungal pathogenicity. *Curr. Opin. Microbiol.* **2009**, *12*, 384–391.
21. Fang, F.C. Antimicrobial reactive oxygen and nitrogen species: Concepts and controversies. *Nat. Rev. Microbiol.* **2004**, *2*, 820–832.
22. Kaloriti, D.; Jacobsen, M.; Yin, Z.; Patterson, M.; Tillmann, A.; Smith, D.A.; Cook, E.; You, T.; Grimm, M.J.; Bohovych, I.; *et al.* Mechanisms underlying the exquisite sensitivity of *Candida albicans* to combinatorial cationic and oxidative stress that enhances the potent fungicidal activity of phagocytes. *MBio* **2014**, *5*, e01334–e01314.
23. Frohner, I.E.; Bourgeois, C.; Yatsyk, K.; Majer, O.; Kuchler, K. *Candida albicans* cell surface superoxide dismutases degrade host-derived reactive oxygen species to escape innate immune surveillance. *Mol. Microbiol.* **2009**, *71*, 240–252.
24. Miramon, P.; Dunker, C.; Windecker, H.; Bohovych, I.M.; Brown, A.J.; Kurzai, O.; Hube, B. Cellular responses of *Candida albicans* to phagocytosis and the extracellular activities of neutrophils are critical to counteract carbohydrate starvation, oxidative and nitrosative stress. *PLOS ONE* **2012**, *7*, e52850.
25. Huycke, M.M.; Abrams, V.; Moore, D.R. Enterococcus faecalis produces extracellular superoxide and hydrogen peroxide that damages colonic epithelial cell DNA. *Carcinogenesis* **2002**, *23*, 529–536.
26. Fitzsimmons, N.; Berry, D.R. Inhibition of *Candida albicans* by lactobacillus acidophilus: Evidence for the involvement of a peroxidase system. *Microbios* **1994**, *80*, 125–133.
27. Cruz, M.R.; Graham, C.E.; Gagliano, B.C.; Lorenz, M.C.; Garsin, D.A. Enterococcus faecalis inhibits hyphal morphogenesis and virulence of *Candida albicans*. *Infect. Immun.* **2013**, *81*, 189–200.
28. Phillips, A.J.; Sudbery, I.; Ramsdale, M. Apoptosis induced by environmental stresses and amphotericin B in *Candida albicans*. *Proc. Natl. Acad. Sci. USA* **2003**, *100*, 14327–14332.
29. Halliwell, B. Oxidative stress and cancer: Have we moved forward? *Biochem. J.* **2007**, *401*, 1–11.
30. Jamieson, D.J.; Stephen, D.W.; Terriere, E.C. Analysis of the adaptive oxidative stress response of *Candida albicans*. *FEMS Microbiol. Lett.* **1996**, *138*, 83–88.
31. Nikolaou, E.; Agrafioti, I.; Stumpf, M.; Quinn, J.; Stansfield, I.; Brown, A.J. Phylogenetic diversity of stress signalling pathways in fungi. *BMC Evol. Biol.* **2009**, doi:10.1186/1471-2148-9-44.
32. Enjalbert, B.; Nantel, A.; Whiteway, M. Stress-induced gene expression in *Candida albicans*: Absence of a general stress response. *Mol. Biol. Cell* **2003**, *14*, 1460–1467.
33. Enjalbert, B.; Smith, D.A.; Cornell, M.J.; Alam, I.; Nicholls, S.; Brown, A.J.; Quinn, J. Role of the Hog1 stress-activated protein kinase in the global transcriptional response to stress in the fungal pathogen *Candida albicans*. *Mol. Biol. Cell* **2006**, *17*, 1018–1032.
34. Enjalbert, B.; MacCallum, D.M.; Odds, F.C.; Brown, A.J. Niche-specific activation of the oxidative

stress response by the pathogenic fungus *Candida albicans*. *Infect. Immun.* **2007**, *75*, 2143–2151.

35. Fradin, C.; de Groot, P.; MacCallum, D.; Schaller, M.; Klis, F.; Odds, F.C.; Hube, B. Granulocytes govern the transcriptional response, morphology and proliferation of *Candida albicans* in human blood. *Mol. Microbiol.* **2005**, *56*, 397–415.
36. Fradin, C.; Kretschmar, M.; Nichterlein, T.; Gaillardin, C.; d'Enfert, C.; Hube, B. Stage-specific gene expression of *Candida albicans* in human blood. *Mol. Microbiol.* **2003**, *47*, 1523–1543.
37. Lorenz, M.C.; Bender, J.A.; Fink, G.R. Transcriptional response of *Candida albicans* upon internalization by macrophages. *Eukaryot. Cell* **2004**, *3*, 1076–1087.
38. Rubin-Bejerano, I.; Fraser, I.; Grisafi, P.; Fink, G.R. Phagocytosis by neutrophils induces an amino acid deprivation response in *Saccharomyces cerevisiae* and *Candida albicans*. *Proc. Natl. Acad. Sci. USA* **2003**, *100*, 11007–11012.
39. Thewes, S.; Kretschmar, M.; Park, H.; Schaller, M.; Filler, S.G.; Hube, B. *In vivo* and *ex vivo* comparative transcriptional profiling of invasive and non-invasive *Candida albicans* isolates identifies genes associated with tissue invasion. *Mol. Microbiol.* **2007**, *63*, 1606–1628.
40. Walker, L.A.; Maccallum, D.M.; Bertram, G.; Gow, N.A.; Odds, F.C.; Brown, A.J. Genome-wide analysis of *Candida albicans* gene expression patterns during infection of the mammalian kidney. *Fungal Genet. Biol.* **2009**, *46*, 210–219.
41. Chaves, G.M.; Bates, S.; Maccallum, D.M.; Odds, F.C. *Candida albicans GRX2*, encoding a putative glutaredoxin, is required for virulence in a murine model. *Genet. Mol. Res.* **2007**, *6*, 1051–1063.
42. Da Silva Dantas, A.; Patterson, M.J.; Smith, D.A.; Maccallum, D.M.; Erwig, L.P.; Morgan, B.A.; Quinn, J. Thioredoxin regulates multiple hydrogen peroxide-induced signaling pathways in *Candida albicans*. *Mol. Cell. Biol.* **2010**, *30*, 4550–4563.
43. Hwang, C.S.; Rhie, G.E.; Oh, J.H.; Huh, W.K.; Yim, H.S.; Kang, S.O. Copper- and zinc-containing superoxide dismutase (Cu/ZnSOD) is required for the protection of *Candida albicans* against oxidative stresses and the expression of its full virulence. *Microbiology* **2002**, *148*, 3705–3713.
44. Martchenko, M.; Alarco, A.M.; Harcus, D.; Whiteway, M. Superoxide dismutases in *Candida albicans*: Transcriptional regulation and functional characterization of the hyphal-induced *SOD5* gene. *Mol. Biol. Cell* **2004**, *15*, 456–467.
45. Wysong, D.R.; Christin, L.; Sugar, A.M.; Robbins, P.W.; Diamond, R.D. Cloning and sequencing of a *Candida albicans* catalase gene and effects of disruption of this gene. *Infect. Immun.* **1998**, *66*, 1953–1961.
46. Miramon, P.; Dunker, C.; Kasper, L.; Jacobsen, I.D.; Barz, D.; Kurzai, O.; Hube, B. A family of glutathione peroxidases contributes to oxidative stress resistance in *Candida albicans*. *Med. Mycol.* **2014**, *52*, 223–239.
47. Urban, C.; Xiong, X.; Sohn, K.; Schroppel, K.; Brunner, H.; Rupp, S. The moonlighting protein tsa1p is implicated in oxidative stress response and in cell wall biogenesis in *Candida albicans*. *Mol. Microbiol.* **2005**, *57*, 1318–1341.
48. Steinberg, B.E.; Huynh, K.K.; Grinstein, S. Phagosomal acidification: Measurement, manipulation and functional consequences. *Biochem. Soc. Trans.* **2007**, *35*, 1083–1087.
49. Kaloriti, D.; Tillmann, A.; Cook, E.; Jacobsen, M.; You, T.; Lenardon, M.; Ames, L.; Barahona, M.; Chandrasekaran, K.; Coghill, G.; *et al.* Combinatorial stresses kill pathogenic *Candida* species. *Med. Mycol.* **2012**, *50*, 699–709.

50. Schuller, C.; Brewster, J.L.; Alexander, M.R.; Gustin, M.C.; Ruis, H. The hog pathway controls osmotic regulation of transcription via the stress response element (STRE) of the *Saccharomyces cerevisiae* CTT1 gene. *EMBO J.* **1994**, *13*, 4382–4389.
51. Heilmann, C.J.; Sorgo, A.G.; Siliakus, A.R.; Dekker, H.L.; Brul, S.; de Koster, C.G.; de Koning, L.J.; Klis, F.M. Hyphal induction in the human fungal pathogen *Candida albicans* reveals a characteristic wall protein profile. *Microbiology* **2011**, *157*, 2297–2307.
52. Gleason, J.E.; Galaleldeen, A.; Peterson, R.L.; Taylor, A.B.; Holloway, S.P.; Waninger-Saroni, J.; Cormack, B.P.; Cabelli, D.E.; Hart, P.J.; Culotta, V.C. *Candida albicans* SOD5 represents the prototype of an unprecedented class of Cu-only superoxide dismutases required for pathogen defense. *Proc. Natl. Acad. Sci. USA* **2014**, *111*, 5866–5871.
53. Festa, R.A.; Thiele, D.J. Copper at the front line of the host-pathogen battle. *PLOS Pathog.* **2012**, *8*, e1002887.
54. Crowe, J.D.; Sievwright, I.K.; Auld, G.C.; Moore, N.R.; Gow, N.A.; Booth, N.A. *Candida albicans* binds human plasminogen: Identification of eight plasminogen-binding proteins. *Mol. Microbiol.* **2003**, *47*, 1637–1651.
55. Lewis, L.E.; Bain, J.M.; Okai, B.; Gow, N.A.; Erwig, L.P. Live-cell video microscopy of fungal pathogen phagocytosis. *J. Vis. Exp.* **2013**, doi:10.3791/50196.
56. Patterson, M.J.; McKenzie, C.G.; Smith, D.A.; da Silva Dantas, A.; Sherston, S.; Veal, E.A.; Morgan, B.A.; MacCallum, D.M.; Erwig, L.P.; Quinn, J. Ybp1 and Gpx3 signaling in *Candida albicans* govern hydrogen peroxide-induced oxidation of the Cap1 transcription factor and macrophage escape. *Antioxid. Redox Signal.* **2013**, *19*, 2244–2260.
57. Brothers, K.M.; Newman, Z.R.; Wheeler, R.T. Live imaging of disseminated candidiasis in zebrafish reveals role of phagocyte oxidase in limiting filamentous growth. *Eukaryot. Cell* **2011**, *10*, 932–944.
58. Jimenez-Lopez, C.; Collette, J.R.; Brothers, K.M.; Shepardson, K.M.; Cramer, R.A.; Wheeler, R.T.; Lorenz, M.C. *Candida albicans* induces arginine biosynthetic genes in response to host-derived reactive oxygen species. *Eukaryot. Cell* **2013**, *12*, 91–100.
59. Nasution, O.; Srinivasa, K.; Kim, M.; Kim, Y.J.; Kim, W.; Jeong, W.; Choi, W. Hydrogen peroxide induces hyphal differentiation in *Candida albicans*. *Eukaryot. Cell* **2008**, *7*, 2008–2011.
60. Bachewich, C.; Nantel, A.; Whiteway, M. Cell cycle arrest during S or M phase generates polarized growth via distinct signals in *Candida albicans*. *Mol. Microbiol.* **2005**, *57*, 942–959.
61. Shi, Q.M.; Wang, Y.M.; Zheng, X.D.; Lee, R.T.; Wang, Y. Critical role of DNA checkpoints in mediating genotoxic-stress-induced filamentous growth in *Candida albicans*. *Mol. Biol. Cell* **2007**, *18*, 815–826.
62. Kusch, H.; Engelmann, S.; Albrecht, D.; Morschhauser, J.; Hecker, M. Proteomic analysis of the oxidative stress response in *Candida albicans*. *Proteomics* **2007**, *7*, 686–697.
63. Wang, Y.; Cao, Y.Y.; Jia, X.M.; Cao, Y.B.; Gao, P.H.; Fu, X.P.; Ying, K.; Chen, W.S.; Jiang, Y.Y. Cap1p is involved in multiple pathways of oxidative stress response in *Candida albicans*. *Free Radic. Biol. Med.* **2006**, *40*, 1201–1209.
64. Moye-Rowley, W.S. Regulation of the transcriptional response to oxidative stress in fungi: Similarities and differences. *Eukaryot. Cell* **2003**, *2*, 381–389.

65. Toone, W.M.; Morgan, B.A.; Jones, N. Redox control of AP-1-like factors in yeast and beyond. *Oncogene* **2001**, *20*, 2336–2346.
66. Alarco, A.M.; Raymond, M. The bzip transcription factor Cap1p is involved in multidrug resistance and oxidative stress response in *Candida albicans*. *J. Bacteriol.* **1999**, *181*, 700–708.
67. Zhang, X.; de Micheli, M.; Coleman, S.T.; Sanglard, D.; Moye-Rowley, W.S. Analysis of the oxidative stress regulation of the *Candida albicans* transcription factor, Cap1p. *Mol. Microbiol.* **2000**, *36*, 618–629.
68. Znaidi, S.; Barker, K.S.; Weber, S.; Alarco, A.M.; Liu, T.T.; Boucher, G.; Rogers, P.D.; Raymond, M. Identification of the *Candida albicans* Cap1p regulon. *Eukaryot. Cell* **2009**, *8*, 806–820.
69. Ramirez-Zavala, B.; Mogavero, S.; Scholler, E.; Sasse, C.; Rogers, P.D.; Morschhauser, J. SAGA/ADA complex subunit Ada2 is required for Cap1- but not Mrr1-mediated upregulation of the *Candida albicans* multidrug efflux pump Mdr1. *Antimicrob. Agents Chemother.* **2014**, *58*, 5102–5110.
70. Sellam, A.; Askew, C.; Epp, E.; Lavoie, H.; Whiteway, M.; Nantel, A. Genome-wide mapping of the coactivator Ada2p yields insight into the functional roles of SAGA/ADA complex in *Candida albicans*. *Mol. Biol. Cell* **2009**, *20*, 2389–2400.
71. Delaunay, A.; Isnard, A.D.; Toledano, M.B. H_2O_2 sensing through oxidation of the Yap1 transcription factor. *EMBO J.* **2000**, *19*, 5157–5166.
72. Delaunay, A.; Pflieger, D.; Barrault, M.B.; Vinh, J.; Toledano, M.B. A thiol peroxidase is an H_2O_2 receptor and redox-transducer in gene activation. *Cell* **2002**, *111*, 471–481.
73. Okazaki, S.; Tachibana, T.; Naganuma, A.; Mano, N.; Kuge, S. Multistep disulfide bond formation in Yap1 is required for sensing and transduction of H_2O_2 stress signal. *Mol. Cell.* **2007**, *27*, 675–688.
74. Okazaki, S.; Naganuma, A.; Kuge, S. Peroxiredoxin-mediated redox regulation of the nuclear localization of Yap1, a transcription factor in budding yeast. *Antioxid. Redox Signal.* **2005**, *7*, 327–334.
75. Tachibana, T.; Okazaki, S.; Murayama, A.; Naganuma, A.; Nomoto, A.; Kuge, S. A major peroxiredoxin-induced activation of Yap1 transcription factor is mediated by reduction-sensitive disulfide bonds and reveals a low level of transcriptional activation. *J. Biol. Chem.* **2009**, *284*, 4464–4472.
76. Veal, E.A.; Ross, S.J.; Malakasi, P.; Peacock, E.; Morgan, B.A. Ybp1 is required for the hydrogen peroxide-induced oxidation of the Yap1 transcription factor. *J. Biol. Chem.* **2003**, *278*, 30896–30904.
77. Gulshan, K.; Thommandru, B.; Moye-Rowley, W.S. Proteolytic degradation of the Yap1 transcription factor is regulated by subcellular localization and the E3 ubiquitin ligase Not4. *J. Biol. Chem.* **2012**, *287*, 36796–36805.
78. Kitamura, K.; Taki, M.; Tanaka, N.; Yamashita, I. Fission yeast Ubr1 ubiquitin ligase influences the oxidative stress response via degradation of active Pap1 bZIP transcription factor in the nucleus. *Mol. Microbiol.* **2011**, *80*, 739–755.
79. Morgan, B.A.; Banks, G.R.; Toone, W.M.; Raitt, D.; Kuge, S.; Johnston, L.H. The Skn7 response regulator controls gene expression in the oxidative stress response of the budding yeast *Saccharomyces cerevisiae*. *EMBO J.* **1997**, *16*, 1035–1044.
80. Singh, P.; Chauhan, N.; Ghosh, A.; Dixon, F.; Calderone, R. Skn7 of *Candida albicans*: Mutant construction and phenotype analysis. *Infect. Immun.* **2004** *72*, 2390–2394.

81. Jain, C.; Pastor, K.; Gonzalez, A.Y.; Lorenz, M.C.; Rao, R.P. The role of *Candida albicans* AP-1 protein against host derived ROS in *in vivo* models of infection. *Virulence* **2013**, *4*, 67–76.
82. Smith, D.A.; Morgan, B.A.; Quinn, J. Stress signalling to fungal stress-activated protein kinase pathways. *FEMS Microbiol. Lett.* **2010**, *306*, 1–8.
83. Bellon, S.; Fitzgibbon, M.J.; Fox, T.; Hsiao, H.M.; Wilson, K.P. The structure of phosphorylated p38gamma is monomeric and reveals a conserved activation-loop conformation. *Structure* **1999**, *7*, 1057–1065.
84. Alonso-Monge, R.; Navarro-Garcia, F.; Roman, E.; Negredo, A.I.; Eisman, B.; Nombela, C.; Pla, J. The Hog1 mitogen-activated protein kinase is essential in the oxidative stress response and chlamydospore formation in *Candida albicans*. *Eukaryot. Cell* **2003**, *2*, 351–361.
85. Smith, D.A.; Nicholls, S.; Morgan, B.A.; Brown, A.J.; Quinn, J. A conserved stress-activated protein kinase regulates a core stress response in the human pathogen *Candida albicans*. *Mol. Biol. Cell* **2004**, *15*, 4179–4190.
86. Chen, D.; Toone, W.M.; Mata, J.; Lyne, R.; Burns, G.; Kivinen, K.; Brazma, A.; Jones, N.; Bahler, J. Global transcriptional responses of fission yeast to environmental stress. *Mol. Biol. Cell* **2003**, *14*, 214–229.
87. Asp, E.; Nilsson, D.; Sunnerhagen, P. Fission yeast mitogen-activated protein kinase Sty1 interacts with translation factors. *Eukaryot. Cell* **2008**, *7*, 328–338.
88. Yin, Z.; Stead, D.; Walker, J.; Selway, L.; Smith, D.; Brown, A.J.P.; Quinn, J. A proteomic analysis of the salt, cadmium and peroxide stress responses in *Candida albicans* and the role of the Hog1 Sapk in regulating the stress-induced proteome. *Proteomics* **2009**, *9*, 4680–4703.
89. Alonso-Monge, R.; Carvaihlo, S.; Nombela, C.; Rial, E.; Pla, J. The Hog1 MAP kinase controls respiratory metabolism in the fungal pathogen *Candida albicans*. *Microbiology* **2009**, *155*, 413–423.
90. Navarro-Garcia, F.; Eisman, B.; Fiuza, S.M.; Nombela, C.; Pla, J. The MAP kinase Mkc1p is activated under different stress conditions in *Candida albicans*. *Microbiology* **2005**, *151*, 2737–2749.
91. Rauceo, J.M.; Blankenship, J.R.; Fanning, S.; Hamaker, J.J.; Deneault, J.S.; Smith, F.J.; Nantel, A.; Mitchell, A.P. Regulation of the *Candida albicans* cell wall damage response by transcription factor Sko1 and PAS kinase Psk1. *Mol. Biol. Cell* **2008**, *19*, 2741–2751.
92. Alonso-Monge, R.; Roman, E.; Arana, D.M.; Prieto, D.; Urrialde, V.; Nombela, C.; Pla, J. The sko1 protein represses the yeast-to-hypha transition and regulates the oxidative stress response in *Candida albicans*. *Fungal Genet. Biol.* **2010**, *47*, 587–601.
93. Arana, D.M.; Nombela, C.; Alonso-Monge, R.; Pla, J. The Pbs2 MAP kinase kinase is essential for the oxidative-stress response in the fungal pathogen *Candida albicans*. *Microbiology* **2005**, *151*, 1033–1049.
94. Cheetham, J.; Smith, D.A.; da Silva Dantas, A.; Doris, K.S.; Patterson, M.J.; Bruce, C.R.; Quinn, J. A single MAPKKK regulates the Hog1 MAPK pathway in the pathogenic fungus *Candida albicans*. *Mol. Biol. Cell* **2007** *18*, 4603–4614.
95. Thomas, E.; Roman, E.; Claypool, S.; Manzoor, N.; Pla, J.; Panwar, S.L. Mitochondria influence Cdr1 efflux pump activity, Hog1-mediated oxidative stress pathway, iron homeostasis, and ergosterol levels in *Candida albicans*. *Antimicrob. Agents Chemother.* **2013**, *57*, 5580–5599.
96. Horie, T.; Tatebayashi, K.; Yamada, R.; Saito, H. Phosphorylated Ssk1 prevents unphosphorylated

Ssk1 from activating the Ssk2 mitogen-activated protein kinase kinase kinase in the yeast high-osmolarity glycerol osmoregulatory pathway. *Mol. Cell. Biol.* **2008**, *28*, 5172–5183.

97. Posas, F.; Saito, H. Activation of the yeast Ssk2 map kinase kinase kinase by the Ssk1 two-component response regulator. *EMBO J.* **1998**, *17*, 1385–1394.
98. Chauhan, N.; Inglis, D.; Roman, E.; Pla, J.; Li, D.; Calera, J.A.; Calderone, R. *Candida albicans* response regulator gene Ssk1 regulates a subset of genes whose functions are associated with cell wall biosynthesis and adaptation to oxidative stress. *Eukaryot. Cell* **2003**, *2*, 1018–1024.
99. Li, D.; Gurkovska, V.; Sheridan, M.; Calderone, R.; Chauhan, N. Studies on the regulation of the two-component histidine kinase gene *CHK1* in *Candida albicans* using the heterologous lacz reporter gene. *Microbiology* **2004**, *150*, 3305–3313.
100. Roman, E.; Nombela, C.; Pla, J. The *SHO1* adaptor protein links oxidative stress to morphogenesis and cell wall biosynthesis in the fungal pathogen *Candida albicans*. *Mol. Cell. Biol.* **2005**, *25*, 10611–10627.
101. Buck, V.; Quinn, J.; Soto Pino, T.; Martin, H.; Saldanha, J.; Makino, K.; Morgan, B.A.; Millar, J.B. Peroxide sensors for the fission yeast stress-activated mitogen-activated protein kinase pathway. *Mol. Biol. Cell* **2001**, *12*, 407–419.
102. Quinn, J.; Malakasi, P.; Smith, D.A.; Cheetham, J.; Buck, V.; Millar, J.B.; Morgan, B.A. Two-component mediated peroxide sensing and signal transduction in fission yeast. *Antioxid. Redox Signal.* **2011**, *15*, 153–165.
103. Menon, V.; Li, D.; Chauhan, N.; Rajnarayanan, R.; Dubrovska, A.; West, A.H.; Calderone, R. Functional studies of the Ssk1p response regulator protein of *Candida albicans* as determined by phenotypic analysis of receiver domain point mutants. *Mol. Microbiol.* **2006**, *62*, 997–1013.
104. Mavrianos, J.; Desai, C.; Chauhan, N. Two-component histidine phosphotransfer protein Ypd1 is not essential for viability in *Candida albicans*. *Eukaryot. Cell* **2014**, *13*, 452–460.
105. Bruce, C.R.; Smith, D.A.; Rodgers, D.; da Silva Dantas, A.; MacCallum, D.M.; Morgan, B.A.; Quinn, J. Identification of a novel response regulator, Crr1, that is required for hydrogen peroxide resistance in *Candida albicans*. *PLOS ONE* **2011**, *6*, e27979.
106. Desai, C.; Mavrianos, J.; Chauhan, N. *Candida albicans* Srr1, a putative two-component response regulator gene, is required for stress adaptation, morphogenesis, and virulence. *Eukaryot. Cell* **2011**, *10*, 1370–1374.
107. O'Rourke, S.M.; Herskowitz, I. A third osmosensing branch in *Saccharomyces cerevisiae* requires the Msb2 protein and functions in parallel with the Sho1 branch. *Mol. Cell. Biol.* **2002**, *22*, 4739–4749.
108. Veal, E.A.; Day, A.M.; Morgan, B.A. Hydrogen peroxide sensing and signaling. *Mol. Cell* **2007**, *26*, 1–14.
109. Veal, E.A.; Findlay, V.J.; Day, A.M.; Bozonet, S.M.; Evans, J.M.; Quinn, J.; Morgan, B.A. A 2-cys peroxiredoxin regulates peroxide-induced oxidation and activation of a stress-activated MAP kinase. *Mol. Cell* **2004**, *15*, 129–139.
110. Saitoh, M.; Nishitoh, H.; Fujii, M.; Takeda, K.; Tobiume, K.; Sawada, Y.; Kawabata, M.; Miyazono, K.; Ichijo, H. Mammalian thioredoxin is a direct inhibitor of apoptosis signal-regulating kinase (ASK) 1. *EMBO J.* **1998**, *17*, 2596–2606.
111. Nadeau, P.J.; Charette, S.J.; Landry, J. Redox reaction at ASK1-Cys250 is essential for activation

of jnk and induction of apoptosis. *Mol. Biol. Cell* **2009**, *20*, 3628–3637.

112. Nadeau, P.J.; Charette, S.J.; Toledano, M.B.; Landry, J. Disulfide bond-mediated multimerization of ASK1 and its reduction by thioredoxin-1 regulate H_2O_2-induced c-Jun NH_2-terminal kinase activation and apoptosis. *Mol. Biol. Cell* **2007**, *18*, 3903–3913.
113. Brandes, N.; Schmitt, S.; Jakob, U. Thiol-based redox switches in eukaryotic proteins. *Antioxid. Redox Signal.* **2009**, *11*, 997–1014.
114. Alonso-Monge, R.; Navarro-Garcia, F.; Molero, G.; Diez-Orejas, R.; Gustin, M.; Pla, J.; Sanchez, M.; Nombela, C. Role of the mitogen-activated protein kinase Hog1p in morphogenesis and virulence of *Candida albicans*. *J. Bacteriol.* **1999**, *181*, 3058–3068.
115. Cheetham, J.; MacCallum, D.M.; Doris, K.S.; da Silva Dantas, A.; Scorfield, S.; Odds, F.; Smith, D.A.; Quinn, J. MAPKKK-independent regulation of the Hog1 stress-activated protein kinase in *Candida albicans*. *J. Biol. Chem.* **2011**, *286*, 42002–42016.
116. Prieto, D.; Roman, E.; Correia, I.; Pla, J. The Hog pathway is critical for the colonization of the mouse gastrointestinal tract by *Candida albicans*. *PLOS ONE* **2014**, *9*, e87128.
117. Arana, D.M.; Alonso-Monge, R.; Du, C.; Calderone, R.; Pla, J. Differential susceptibility of mitogen-activated protein kinase pathway mutants to oxidative-mediated killing by phagocytes in the fungal pathogen *Candida albicans*. *Cell. Microbiol.* **2007**, *9*, 1647–1659.
118. Loll-Krippleber, R.; d'Enfert, C.; Feri, A.; Diogo, D.; Perin, A.; Marcet-Houben, M.; Bougnoux, M.E.; Legrand, M. A study of the DNA damage checkpoint in *Candida albicans*: Uncoupling of the functions of Rad53 in DNA repair, cell cycle regulation and genotoxic stress-induced polarized growth. *Mol. Microbiol.* **2014**, *91*, 452–471.
119. Shockley, A.H.; Doo, D.W.; Rodriguez, G.P.; Crouse, G.F. Oxidative damage and mutagenesis in *Saccharomyces cerevisiae*: Genetic studies of pathways affecting replication fidelity of 8-oxoguanine. *Genetics* **2013**, *195*, 359–367.
120. Leroy, C.; Mann, C.; Marsolier, M.C. Silent repair accounts for cell cycle specificity in the signaling of oxidative DNA lesions. *EMBO J.* **2001**, *20*, 2896–2906.
121. Muniyappa, H.; Song, S.; Mathews, C.K.; Das, K.C. Reactive oxygen species-independent oxidation of thioredoxin in hypoxia: Inactivation of ribonucleotide reductase and redox-mediated checkpoint control. *J. Biol. Chem.* **2009**, *284*, 17069–17081.
122. Guo, Z.; Kozlov, S.; Lavin, M.F.; Person, M.D.; Paull, T.T. ATM activation by oxidative stress. *Science* **2010**, *330*, 517–521.
123. Wilson, D.; Tutulan-Cunita, A.; Jung, W.; Hauser, N.C.; Hernandez, R.; Williamson, T.; Piekarska, K.; Rupp, S.; Young, T.; Stateva, L. Deletion of the high-affinity camp phosphodiesterase encoded by Pde2 affects stress responses and virulence in *Candida albicans*. *Mol. Microbiol.* **2007**, *65*, 841–856.
124. Deveau, A.; Piispanen, A.E.; Jackson, A.A.; Hogan, D.A. Farnesol induces hydrogen peroxide resistance in *Candida albicans* yeast by inhibiting the Ras-cyclic AMP signaling pathway. *Eukaryot. Cell* **2010**, *9*, 569–577.
125. Kamthan, M.; Nalla, V.K.; Ruhela, D.; Kamthan, A.; Maiti, P.; Datta, A. Characterization of a putative spindle assembly checkpoint kinase Mps1, suggests its involvement in cell division, morphogenesis and oxidative stress tolerance in *Candida albicans*. *PLOS ONE* **2014**, *9*, e101517.
126. Bai, C.; Ramanan, N.; Wang, Y.M.; Wang, Y. Spindle assembly checkpoint component CaMad2p

is indispensable for *Candida albicans* survival and virulence in mice. *Mol. Microbiol.* **2002**, *45*, 31–44.

127. Du, C.; Calderone, R.; Richert, J.; Li, D. Deletion of the SSK1 response regulator gene in *Candida albicans* contributes to enhanced killing by human polymorphonuclear neutrophils. *Infect. Immun.* **2005,** *73*, 865–871.
128. Calera, J.A.; Zhao, X.J.; Calderone, R. Defective hyphal development and avirulence caused by a deletion of the *SSK1* response regulator gene in *Candida albicans. Infect. Immun.* **2000**, *68*, 518–525.

5

Anti-Inflammatory Activity of Haskap Cultivars is Polyphenols-Dependent

H. P. Vasantha Rupasinghe [1,*], Mannfred M. A. Boehm [1], Satvir Sekhon-Loodu [1], Indu Parmar [1], Bob Bors [2] and Andrew R. Jamieson [3]

[1] Department of Environmental Sciences, Faculty of Agriculture, Dalhousie University, Truro, NS B2N 5E3, Canada; E-Mails: mboehm@dal.ca (M.M.A.B.); Satvir.sekhon@dal.ca (S.S-L.); induparmar@Dal.Ca (I.P.)

[2] Department of Plant Sciences, University of Saskatchewan, Saskatoon, SK S7N 5A8, Canada; E-Mail: Bob.bors@usask.ca

[3] Agriculture and Agri-Food Canada, Kentville, NS B4N 1J5, Canada; E-Mail: andrew.jamieson@agr.gc.ca

* Author to whom correspondence should be addressed; E-Mail: vrupasinghe@dal.ca;

Academic Editors: Michael Breitenbach and Peter Eckl

Abstract: Haskap (*Lonicera caerulea* L.) berries have long been used for their health promoting properties against chronic conditions. The current study investigated the effect of Canadian haskap berry extracts on pro-inflammatory cytokines using a human monocytic cell line THP-1 derived macrophages stimulated by lipopolysaccharide. Methanol extracts of haskap from different growing locations in Canada were prepared and characterized for their total phenolic profile using colorimetric assays and liquid chromatography—Mass spectrometry (UPLC-MS/MS). Human THP-1 monocytes were seeded in 24-well plates (5×10^5/well) and treated with phorbol 12-myristate 13-acetate (PMA, 0.1 μg/mL) for 48 h to induce macrophage differentiation. After 48 h, the differentiated macrophages were washed with Hank's buffer and treated with various concentrations of test compounds for 4 h, followed by the lipopolysaccharide (LPS)-stimulation (18 h). Borealis cultivar showed the highest phenolic content, flavonoid content and anthocyanin content ($p < 0.05$). A negative correlation existed between the polyphenol concentration of the extracts and pro-inflammatory cytokines: Interleukin-6 (IL-6), tumour necrosis factor-alpha (TNF-α), prostaglandin (PGE_2), and cyclooxygenase-2 (COX-2) enzyme. Borealis exhibited comparable anti-inflammatory effects to COX inhibitory drug, diclofenac. The results showed that haskap berry polyphenols has the potential to act as an effective inflammation inhibitor.

Keywords: haskap; inflammation; cytokines; polyphenols; macrophages; functional food

---

1. Introduction

Free radicals, produced as metabolic process mediators have a tendency to attack biomolecules like DNA, RNA, proteins and lipids [1]. The damage caused by these reactive oxygen species can vary from loss of enzyme function, increased cell permeability, and affected cell signalling causing inflammation, diabetes, and cancer [2–4]. Chronic inflammation is known to contribute to the risk of developing metabolic disease, which has become especially prevalent in recent years [5,6]. Inflammation is a pathophysiological response of living tissue to injuries, mediated by macrophages, leukocytes and neutrophils through oxidative species. The inflammation cascade involves various factors and enzymes, mainly cyclooxygenase (COX)-2, tumour necrosis factor-alpha (TNF-α), interferon-gamma (IFN-γ) among others. However, metabolic syndrome can be amended through addressing obesity, elevated blood sugar and cholesterol, physical exercise, and improving diet [7].

The commonly prescribed non-steroidal anti-inflammatory drugs (NSAIDs) such as aspirin (acetyl salicylic acid), diclofenac (dichloranilino phenylacetic acid) (non-specific COX inhibitor), nimesulide [N-(2-phenyloxy-4-nitrophenyl)methanesulphonamide] (COX-2 specific inhibitor) have the ability to inhibit COX enzyme which is responsible for the production of pro-inflammatory prostaglandins and prostacyclins. However, there are different undesirable effects related with these NSAID, mainly causing bleeding and ulceration in gastrointestinal tract and platelet dysfunction by blocking COX-1 derived prostanoids [8]. Hence, there is a growing interest in finding alternative food therapies, as well as new anti-inflammatory nutraceuticals derived from plant-based foods, such as fruit crops.

Phytochemicals present in functional foods such as cool climate berries offer a great hope as an alternative therapy for chronic disorders. Polyphenols in fruits have demonstrated the potential to terminate free radical reactions and many antioxidant and physiological benefits in biological systems [9]. Haskap (*Lonicera caerulea* L.), also known as blue honeysuckle, honeyberry, or sweet berry honeysuckle, is native to Siberia, China, and Japan. Haskap is fairly new to Canada, with only three major varieties in production, namely Borealis, Indigo Gem, and Tundra [10]. Along with a flavour similar to raspberries, black currants, and blueberries, haskap polyphenols also attenuates nuclear factor (NF)-kappaB dependent signaling pathway and subsequent production of proinflammatory mediators [11]. Our previous study has demonstrated the antioxidant properties and total phenolic content of the Borealis cultivar of haskap to be competitive with other berries [12], thereby suggesting its health promoting potential. Although various *in vitro* studies have evaluated blue honeysuckle extracts for their antimicrobial, anti-adherence, antioxidant effects, protective activity against ultraviolet B (UVB)-caused injury of keratinocytes (HaCaT cells), lipopolysaccharide-induced inflammation, and gingival fibroblast oxidative damage by lipopolysaccharide [11,13,14], the information on phenolic characterization and biological activity of recently introduced cultivars of Canadian haskap is limited.

Considering the previous reports, the present study was conducted to investigate the sugar, organic acid and polyphenol profiles of different haskap cultivars grown in Canada and to study their anti-inflammatory potential. In this study, human monocytes (THP-1 cells) differentiated macrophages were employed to investigate the anti-inflammatory properties of haskap berry extracts, as macrophages are predominately involved at the initial stage of inflammation process and secreting cell-signalling molecules [15].

2. Materials and Methods

2.1. Plant Material

Fruits of four haskap cultivars, Berry Blue (BL), Borealis (BR), Tundra (TN) and Indigo Gem (IG) were obtained from LaHave Forests Farm, Blockhouse, NS, Canada; two cultivars Indigo Gem (SAS-IG) and Tundra (SAS-TN) were obtained from the University of Saskatchewan, Saskatoon, SK, Canada and five genotypes, LC12, LC13, LC16, LC23, and LC47, were obtained from Agriculture and Agric-Food Canada, Kentville, NS, Canada. Berry Blue (also known as Czech No. 17) is a Czech cultivar of Russian descent with early maturity and tall plants, often used as a polliniser. The other cultivars (BR, TN, and IG) are sister seedlings bred in Saskatchewan from Blue Velvet (Kiev #3) × Blue Belle (also known as Tomichka). Kiev #3 was derived from an open pollinated plant of *L. caerulea* var. *kamtschatica* Sevast, gathered from the Kurile Islands. Czech No. 17 and Tomichka, like most cultivars of Russian descent, were derived mostly from four varieties of *L. caerulea* L., but were not derived from varieties from Japan [16].

The Kentville haskap samples (*L. caerulea* L.) were selected from seedlings planted in 2007. The seed came from open pollenated fruit grown in BC, Canada in 2006 from plants originally developed by Dr. Maxine Thompson of the National Clonal Germplasm Repository in Corvallis, Oregon. The BC plants were derived from *L. caerulea* var. *emphyllocalyx* Nakai from Hokkaido. The Kentville selections flower and their fruit ripen later than the listed cultivars. The soil in Kentville was a sandy loam of the Berwick series with pH 6.8 and 2.8% organic matter. Plants grew with no pruning and minimal fertilization and 5 plants were selected based on positive fruit characteristics and designated LC-12, LC-13, LC-16, LC-23, and LC-47. Fruit from these five plants was hand-harvested for analysis on 10 July 2013. The samples were frozen at −20 °C until use.

2.2. Chemicals

HPLC grade methanol, acetonitrile, formic acid, Folin-Ciocalteu reagent and dimethylcinnamaldehyde were purchased from Sigma-Aldrich (Oakville, ON, Canada). The liquid chromatography standards were purchased as follows: Cyanidin-3-*O*-glucoside from Extrasynthese (Genay Cedex, France); phloridzin, phloretin, chlorogenic acid, ferulic acid and caffeic acid from Sigma-Aldrich; catechin, epicatechin, quercetin, quercetin-3-*O*-galactoside and quercitin-3-*O*-glucoside from ChromaDex, Inc. (Santa Ana, CA, USA); quercitin-3-*O*-rhamnoside, quercitin-3-*O*-galactoside and anthocyanin standards from Indofine Chemical Company (Hillsborough, NJ, USA). Hydrochloric acid, sulfuric acid, and 96-well microplates were purchased from Fisher Scientific (Ottawa, ON, Canada). Phorbol 12-myristate 13-acetate, diclofenac sodium salt, nimesulide and lipopolysaccharide were obtained from Sigma-Aldrich. COX-2 human ELISA kit was purchased from Enzo Life Sciences, Inc. (Farmingdale, NY, USA).

Prostaglandin E_2 EIA kit and nitric oxide quantification kit was purchased from Cayman Chemical Company (Ann Arbor, MI, USA). The TNF-α and IL-6 ELISA kits were purchased from BD Biosciences (San Diego, CA, USA).

2.3. Extraction

Frozen fruit samples (50 g) were ground and extracted with 100% methanol (250 mL) using a blender (Model HBB909, Hamilton Beach Brands Inc., Glen Allen, VA, USA) under semi-dark conditions. The extract was filtered through eight layers of cheese cloth and centrifuged at 4900× *g* for 10 min and supernatant was stored at −20 °C. Prior to performing the assays the berry extracts were evaporated of methanol under a nitrogen evaporator and stored at −80 °C until assays were performed.

2.3.1. Total Phenolic Content

Total phenolic content was determined by using the modified Folin-Ciocalteu assay as described elsewhere [12]. Total phenolic content was calculated against a gallic acid calibration curve, and samples diluted when necessary to fit range. Total phenolics were expressed as gallic acid equivalents (GAE) 100 g^{-1} fresh weight (FW).

2.3.2. Total Flavonoid Content

The total flavonoid assay was based on the aluminum chloride colorimetric method as described by Marinova *et al.* [17] and modified by Rupasinghe *et al.* [12]. Results were expressed as mmole quercetin equivalents (QE) 100 g^{-1} FW. The flavonoid quercetin was used as a standard to produce a calibration curve (range 50–500 mM).

2.3.3. Total Anthocyanin Content

Total anthocyanin concentration (TAC) in the samples was based on the pH-differential method (AOAC method 2005.02) as previously described [18] and their concentration expressed as mg cyanidin-3-*O*-glucoside (C3G) per 100 gram FW using a molar extinction coefficient (ε) 28,000, molecular weight (MW) 484.8 for C3G.

2.3.4. Total Proanthocyanidin Content

Total proanthocyanidin content was determined by the 4-dimethylaminocinnamaldehyde (DMAC) assay, as reported by Prior *et al.* [19] with modifications. DMAC reagent was prepared by adding 10 mL of acidified methanol to 0.01 g of DMAC. Then, 150 μL of DMAC reagent was added to a 96-well plate with 50 μL of berry extract and read at 640 nm. Samples were standardized against a 1000 ppm catechin stock in 5 dilutions with methanol. Total proanthocyanidin concentration was expressed as mg catechin equivalents (CE) 100 g^{-1} FW.

2.3.5. LC-MS/MS Analysis of Specific Polyphenols

Total monomeric polyphenols were identified and massed by liquid chromatography—Mass spectrometry (LC-MS/MS) analysis as described elsewhere [20]. Analysis was carried out using a Waters H-class UPLC separation module (Waters, Milford, MA, USA) coupled with a Micromass Quattro micro API MS/MS system and controlled with MassLynx V4.0 data analysis system (Micromass, Cary, NC, USA). An Aquity BEH C18 (100 mm × 2.1 mm, 1.7 µm) column (Waters, Milford, MA, USA) was used.

The analysis of flavonol, flavan-3-ol, phenolic acid, and dihydrochalcone compounds was done by electrospray ionization in negative ion mode (ESI−), with a capillary voltage of 3000 V, nebulizer gas (N_2) temperature of 375 °C, and flow rate of 0.35 mL·min^{-1}. Anthocyanin compounds were analysed by electrospray ionization in positive ion mode (ESI+), with capillary voltage 3500 V, nebulizer gas at 375 °C, and flow rate of 0.35 mL·min^{-1}. The cone voltage (25–50 V) was optimized for individual compounds.

2.3.6. Sugars and Organic Acid Analyses

Samples were evaporated to remove methanol and dissolved in 100% deionised water prior to their filtration through a 0.45 µm nylon filter (Chromaspec, Brockville, ON, Canada) for analyses of sugars (glucose, fructose and sucrose) and organic acids (lactic, malic, quinic, citric acid) by HPLC using a Waters Alliance 2695 HPLC system connected to a Waters 2414 refractive index (RI) detector. A Rezex ROA column (250 × 4.6 mm; 8 µm; Phenomenex, Torrance, CA, USA) was used and the analysis conditions were: Temperature of 40 °C and 30 °C for column and detector, respectively; mobile phase of 0.005 N sulfuric acid in deionised water; run time of 30 min and isocratic flow rate 0.6 mL/min. An injector volume of 10 µL was used for each run. Four point calibration curves were obtained by using mixed standard solutions of sugars and organic acids at concentration range between 100 and 1000 mg/L. The results are expressed in mg 100 g^{-1} FW.

2.4. Cell Culture

The human leukemia monocytes THP-1 (ATCC® TIB202™) cell line was cultured at 37 °C with 5% humidified CO_2 in Roswell Park Memorial Institute (RPMI) 1640 media with 0.05 mM of 2-mercaptoethanol and 10% fetal bovine serum. The cells were seeded in 24-well plates (5 × 10^5/well) and treated with phorbol 12-myristate 13-acetate (PMA, 0.1 µg/ mL) to induce macrophage differentiation. Differentiated macrophages were verified by observing the cell morphology through an inverted phase contrast microscope (Figure A1) (Nikon Eclipse E 100, Nikon, Mississauga, ON, Canada). The differentiated macrophages were washed and treated with various concentrations of extracts for 4 h, followed by LPS (18 h). The supernatants were collected and stored at −20 °C for further analysis. Extracts that were used in cell culture studies were completely dried to remove solvents and reconstituted in DMSO. The controls contain the DMSO concentration equivalent to all the treatments.

2.4.1. Measurement of Cell Viability

Cell viability was determined using the 3-(4,5-dimethylthiazol-2-yl)-5-(3-carboxymethoxyphenyl)-2-(4-sulfophenyl)-2H-tetrazolium (MTS) assay. THP-1 monocytes differentiated macrophages were cultured in 96-well tissue culture plate (5 × 10^4 cells/well). Cells were treated with 50 and 100 µg/mL of haskap extract and incubated with MTS reagent for 4 h. Optical density was measured at 490 nm with a microplate reader (FLUOstar OPTIMA, BMG Labtech, Durham, NC, USA).

2.4.2. Measurement of Nitric Oxide

Products of nitric oxide synthase activity were measured using Cayman's nitrate/nitrite colorimetric assay kit (Cayman Chemical Co., Ann Arbor, MI, USA). Nitrates were converted to nitrites by nitrate reductase and total accumulated nitrites were converted to an azo compound using Griess reagent. The absorbance was read at 540 nm using the microplate reader.

2.4.3. Measurement of COX-2 Activity

The COX-2 concentration was measured using human COX-2 ELISA kit provided by Enzo Life Sciences. Samples were prepared by extracting COX-2 from the cells and standards were prepared, according to instructions of the manufacturer. The plate was incubated at 37 °C for 1 h, followed by washing steps and labeled antibody was added. The plate was incubated at 4 °C for 30 min, followed by washing steps and addition of substrate solution. The reaction was stopped and absorbance was measured at 450 nm using a FLUOstar OPTIMA plate reader (BMG Labtech). The concentrations were calculated in reference to the standard curve and presented as the percentage of the inflammation control.

2.4.4. Measurement of IL-6 and TNF-α

The concentrations of pro-inflammatory cytokines (TNF-α and IL-6) were measured from culture medium of control and treated cells by enzyme linked immunosorbant assay (ELISA) kit provided by BD Biosciences (Mississauga, ON, Canada). Anti-human monoclonal antibodies coated plates were developed by using detection antibodies and streptavidin-horseradish peroxidase conjugate provided by the manufacturer with each kit, according to the provided instructions. The absorbance was read at 450 nm using a FLUOstar OPTIMA plate reader (BMG Labtech). The concentrations were calculated against the standard curve and presented as the percentage of the inflammation control.

2.4.5. Measurement of PGE_2

The concentration of PGE_2 released by LPS-stimulated macrophages was determined by acetylcholinesterase™ (ACE™) competitive enzyme immunoassay (EIA) kit purchased from Cayman Chemicals (Burlington, ON, Canada). The standards were prepared according to the manufacturer's instructions. The samples (50 µL) were added to the designated wells, followed by PGE_2-ACE tracer (50 µL) and PGE_2-monoclonal antibody. The plates were covered for 18 h at 4 °C. The plates were developed according to the manufacturer's instructions. The absorbance was read at 420 nm using a FLUOstar OPTIMA plate reader (BMG Labtech).

2.5. Statistical Analysis

All measurements were conducted in triplicate with the mean and standard deviation calculated. The normal distribution of the residuals was tested using the Anderson-Darling test. The data was analysed using ANOVA and multiple mean comparison was done using Tukey's student range test (*t*-test) on SAS version 9.3 for Windows (SAS Institute, Cary, NC, USA). Significant differences were compared using a p value ≤ 0.05 for all the parameters and correlations recorded with Pearson correlation coefficients.

3. Results and Discussion

3.1. Qualitative Phenolic Composition

The total phenolic content determined by the Folin-Ciocalteu assay ranged from 634 to 1154 mg GAE 100 g^{-1} FW with the mean value of 832 mg GAE 100 g^{-1} FW (Table 1). Cultivar Borealis (BR) had the highest phenolic content and was consistent with the previous findings [12]. Tundra collected from both locations (TN and SAS-TN) had the second highest levels of phenolics (953–1015 mg GAE 100 g^{-1} FW, $p > 0.05$), while the LCs contained the lowest total phenolic content (634–849 mg GAE 100 g^{-1}). Previously reported values of phenolic content of *L. caerulea* have been 428.1–622.5 mg GAE 100 g^{-1} FW by Rupasinghe *et al.* [12] and 575 to 903 mg GAE 100 g^{-1} FW by Rop *et al.* [21] that support the current data.

Table 1. Effect of location and cultivars on total phenolic, total flavonoid, total proanthocyanidins and total anthocyanidin content of haskap berry.

Growing Location	Cultivar	Total Phenolics (mg GAE/100 g FW)	Total Flavonoids (mg QE/100 g FW)	Total Proanthocyanidins (mg CE/100 g FW)	Total Anthocyanins (mg CGE/100 g FW)
LaHave farm	BL	755.9 ± 9.4 [d,e,f]	1156.6 ± 121.7 [b,c]	13.2 ± 1.0 [c]	163.0 ± 10.1 [c]
	BR	1154.1 ± 59.7 [a]	1582.8 ± 140.5 [a]	16.3 ± 0.9 [c]	314.0 ± 2.7 [a]
	TN	952.9 ± 18.7 [b,c]	1260.3 ± 69.0 [b,c]	16.0 ± 1.0 [c]	234.4 ± 2.6 [b]
	IG	884.3 ± 25.0 [c,d]	1327.0 ± 12.1 [b,c]	14.4 ± 0.5 [c]	246.9 ± 13.7 [b]
Kentville	LC-12	849.2 ± 28.3 [c,d]	1035.5 ± 86.7 [b,c]	37.2 ± 0.9 [b]	164.9 ± 5.4 [c]
	LC-13	796.8 ± 5.2 [d]	997.0 ± 32.0 [b,c]	41.0 ± 3.6 [b]	142.5 ± 9.1 [c]
	LC-16	664.9 ± 51.4 [e,f]	900.7 ± 16.6 [c,d]	52.3 ± 2.3 [a]	70.2 ± 2.2 [d]
	LC-23	658.1 ± 0.4 [e,f]	956.3 ± 11.0 [b,c]	34.2 ± 0.8 [b]	120.2 ± 1.5 [c,d]
	LC-47	634.4 ± 29.9 [f]	916.5 ± 65.2 [c,d]	47.0 ± 0.4 [a,b]	133.4 ± 1.2 [c]
Saskatchewan	SAS-IG	790.3 ± 68.4 [d]	1128.5 ± 54.2 [c,d]	19.6 ± 2.0 [c]	246.3 ± 2.8 [b]
	SAS-TN	1015.3 ± 78.2 [a,b]	1428.4 ± 35.1 [a,b]	38.5 ± 0.9 [b]	303.2 ± 5.5 [a]

Results represent the mean ± SD ($n = 3$), Tukey's test, $p < 0.05$. BL, Berry Blue; BR, Borealis; TN, Tundra; IG, Indigo Gem; SAS-TN, Saskatoon Tundra; SAS-IG, Saskatoon Indigo GemGAE, gallic acid equivalents; QE, quercetin equivalents; CE, catechin equivalents; CGE, cyanidin-3-glucoside equivalents. Significant differences in each column were compared using the p value ≤ 0.05. [a–f] Different letters within rows denote significant differences between values as determined by one way ANOVA analysis.

3.1.1. Total Flavonoid Content

The total flavonoid contents of the four cultivars varied from 901 to 1583 mg QE 100 g^{-1} FW with the mean value of 1154 mg QE 100 g^{-1} FW (Table 1). BR was again significantly higher in flavonoids as it was in total phenolics, which was followed by Tundra (SAS-TN and TN) and Indigo Gem (IG). Similar to the total phenolics results, the LC breeding lines had the least amount of flavonoids. Additionally, Berry Blue (BL) demonstrated a low flavonoid content of merely 1156 mg QE 100 g^{-1} FW. The present data shows about two times higher values of TFC for *L. caerulea* than one of the previous reports ranging between 594 to 699 mg QE g^{-1} FW, with Borealis containing the greatest amounts [12]. However, another finding reported much less total flavonoids (30 to 40 mg QE 100 g^{-1} FW) in blue honeysuckle [21]. This, the large variation in flavonoid content may be attributed to differences in cultivars and growing locations and conditions.

3.1.2. Total Anthocyanin Content

The total anthocyanin content determined by the pH-differential method was found to have a large variation between 70 to 314 mg C3G 100 g^{-1} FW (Table 1). BR and SAS-TN were attributed with the highest levels of anthocyanins ($p < 0.05$), while the LCs was found to be significantly lower in these compounds. However, these results were lower than the previously reported values of 1470 mg C3G 100 g^{-1} for *L. caerulea* [22].

3.1.3. Total Proanthocyanidin Content

The total proanthocyanidin content was found to be between 13 and 52 mg CE 100 g^{-1} FW (Table 1). Interestingly, all the LC samples showed more than twice the higher proanthocyanidin concentration than the other cultivars. While LC47 and LC16 exhibited the highest level of total proanthocyanidins, on the other hand, BL followed by BR, TN and IG showed the least content. Proanthocyanidins are considered as one of the major phenolic groups found in *L. caerulea* species [23]. However, in contrast to the previous studies [24,25] showing edible honeysuckle with very high proanthocyanidin content (195–772 mg 100 g^{-1} FW), the present results demonstrated much lower values. This discrepancy could be attributed to the difference in standards used for calibration, cultivars and geographical locations. At the time of writing, no previously published data existed stating the concentration of proanthocyanidins (for berries of genus *Lonicera*) expressed as catechin equivalents.

3.1.4. LC-MS/MS Composition of Haskap Berry Extract

The phenolic characterization of haskap by LC-MS/MS is given in Table 2. In line with previous studies, cyanidin-3-*O*-glucoside followed by cyanidin-*3-O*-rutinoside were the most abundant anthocyanins observed in selected cultivars of haskap [26,27]. BR exhibited the highest cyanidin-3-*O*-glucoside concentration (76% of the total anthocyanins determined by LCMS), followed by SAS-TN (59% of the total anthocyanins). On the other hand, TN demonstrated the highest cyanidin-3-*O*-rutinoside concentration (37% of the total anthocyanins) among all the investigated cultivars. Consistent with [28], the current study also found trace amounts of petunidin and malvidin. Except for LC samples, quercetin (Q) rutinoside (also called rutin) represented the highest concentrations of 27%–88% of the total flavonols analysed by LCMS. BR presented the highest amount of 24.3 mg 100 g^{-1} FW; while LC-13 exhibited the least amount of 6.2 mg 100 g^{-1} FW. These results are in agreement with previous studies, which reported a range of 7–17 mg 100 g^{-1} FW for rutin [23,24]. This was followed by the presence of Q. arabinoside and Q. glucoside; while Q. galactoside was the least present flavonol in the selected haskap cultivars. The Kentville genotypes had lower Q. rutinoside but markedly higher Q. arabinoside (Table 2). This may prove to be a characteristic of *L. caerulea* var. *emphyllocalyx*. Among flavan-3-ol monomers, epicatechin and catechin were found most abundant in the selected haskap cultivars representing around 25%–77% and 15%–70% of the total flavan-3-ols, respectively. LC-23 and LC-16 exhibited the highest concentrations of catechin and epicatechin; while BL represented their least amount. Chlorogenic acid was the major phenolic acid observed in the Canadian haskap cultivars with amounts varying between 201–234 mg 100 g^{-1} FW. Overall, the phenolic content was found to be consistent with a previous report comparing haskap to other berries [12].

Cyanidin-3-*O*-glucoside has been reported to down-regulate the expression of inducible nitric oxide synthase (*i*NOS) in mice by suppressing the levels of proinflammatory cytokines (TNF-α, IL-6, and IL-1 β) [29], while both rutin and cyanidin-3-*O*-glucoside have shown to decrease the activity of cytokine mediated transcription factors c-Jun and nuclear factor-κB (NF-κB) [30]. Chlorogenic acid is a phenylpropanoid, produced by plants as a stress response in pathogenesis [31] and is an intermediate produced in lignin biosynthesis [32]. This compound has demonstrated an ability to reduce levels of NO, expression of COX-2 and *i*NOS, cytokines TNF-α, IL-6, IL-1 β as well as impeding the nuclear translocation of NF-κB [33]. Rutin has been reported to reduce TNF-α production and a range of interleukins in mice model [34]. It has also been reported to maintain glucose sensitivity in mice while blocking the development of insulin resistance and preventing macrophage activation [35].

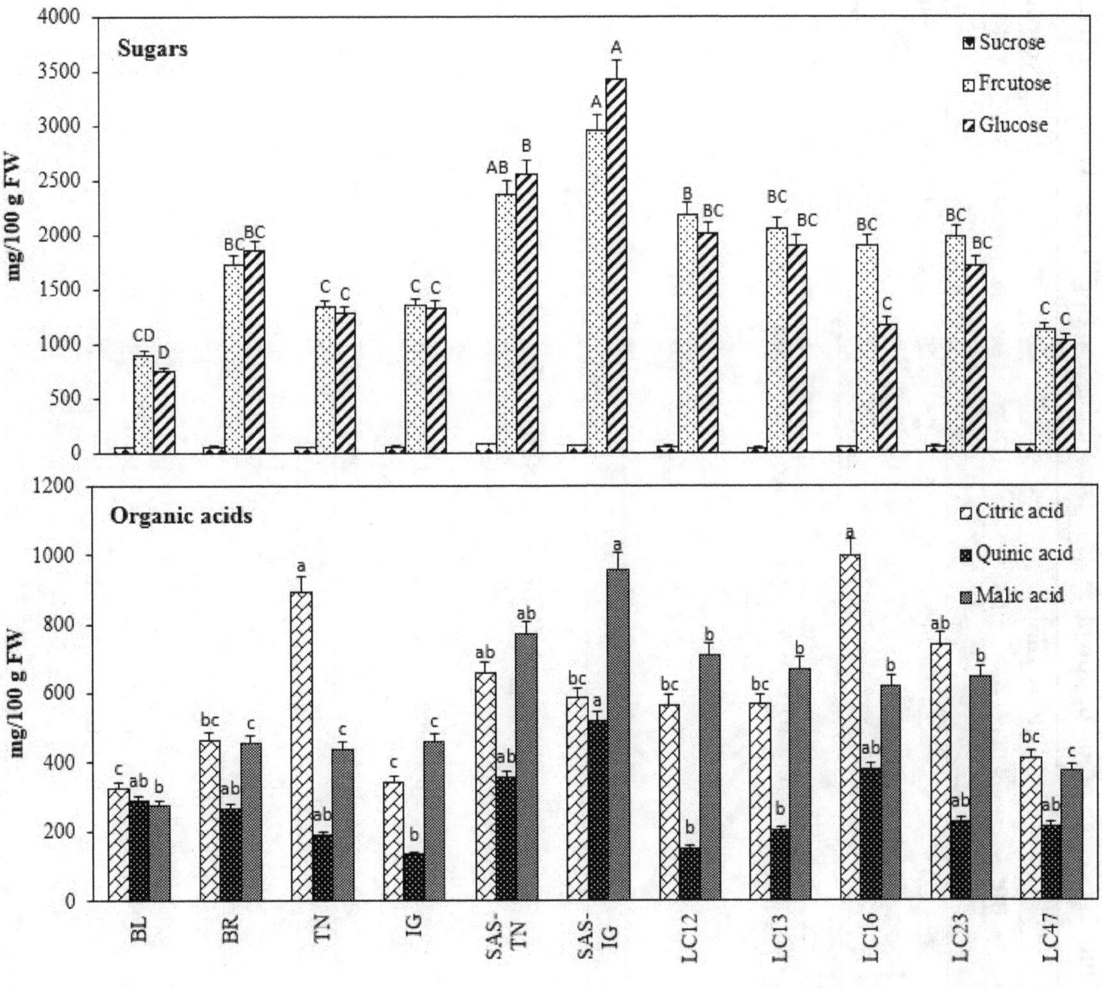

Figure 1. Sugar and organic acid profile of different haskap cultivars. Results represent the mean ± SD ($n = 3$), Tukey's test, $p < 0.05$. BL, Berry Blue; BR, Borealis; TN, Tundra; IG, Indigo Gem; SAS-TN, Saskatoon Tundra; SAS-IG, Saskatoon Indigo Gem. Significance mean values at p value ≤ 0.05: Capital letter grouping (A–D) is for sugars and small letter grouping (a–c) is for organic acids.

Table 2. Concentration of polyphenols present in haskap (mg/100 g FW) as affected by cultivars and growing location.

	LaHave Farm				Kentville					Saskatchewan	
Compounds	**BR**	**BL**	**TN**	**IG**	**LC-12**	**LC-13**	**LC-16**	**LC-23**	**LC-47**	**SAS-IG**	**SAS-TN**
Phenolic acids											
Chlorogenic acid	25.6 ± 3.6	23.1 ± 2.4	26.0 ± 2.4	22.4 ± 2.5	29.7 ± 0.6	32.7 ± 3.2	23.0 ± 0.2	23.9 ± 1.3	28.6 ± 0.1	20.7 ± 1.1	33.8 ± 0.7
Caffeic acid	0.2 ± 0.0	0.1 ± 0.0	0.2 ± 0.0	0.1 ± 0.0	0.1 ± 0.0	0.1 ± 0.0	ND	0.1 ± 0.0	0.1 ± 0.0	0.1 ± 0.0	0.1 ± 0.0
Total	25.8	23.2	26.2	22.5	29.8	32.8	23.0	24.0	28.7	20.8	33.9
Flavan-3-ols											
EGC	0.1 ± 0.0	0.1 ± 0.0	0.1 ± 0.0	0.6 ± 0.0	0.1 ± 0.0	0.1 ± 0.0	0.1 ± 0.0	0.1 ± 0.0	0.1 ± 0.0	0.1 ± 0.0	0.1 ± 0.0
Catechin	2.5 ± 0.2	1.7 ± 0.3	3.5 ± 0.2	2.9 ± 0.3	2.1 ± 0.1	2.1 ± 0.3	3.9 ± 0.1	5.4 ± 0.0	2.1 ± 0.1	2.2 ± 0.1	3.4 ± 0.2
Epicatechin	1.2 ± 0.1	1.7 ± 0.4	0.7 ± 0.1	1.5 ± 0.1	4.5 ± 0.1	5.8 ± 0.5	7.1 ± 0.1	6.2 ± 0.1	1.5 ± 0.1	0.9 ± 0.0	0.9 ± 0.0
EGCG	0.1 ± 0.0	0.2 ± 0.0	0.2 ± 0.0	0.2 ± 0.0	0.1 ± 0.0	0.3 ± 0.0	0.1 ± 0.0	0.3 ± 0.0	0.2 ± 0.0	0.2 ± 0.0	0.2 ± 0.0
Total	3.9	3.7	4.5	5.1	6.8	8.3	11.2	12.0	4.0	3.4	4.6
Flavonols											
Q. galactoside	ND	0.1 ± 0.0	ND	ND	0.1 ± 0.0	0.1 ± 0.0	ND	ND	ND	ND	ND
Q. glucoside	3.6 ± 0.2	4.2 ± 0.8	1.4 ± 0.2	2.8 ± 0.2	7.3 ± 0.8	4.4 ± 0.6	3.7 ± 0.1	3.8 ± 0.1	3.7 ± 0.1	2.7 ± 0.1	4.0 ± 0.2
Q. arabinoside	2.9 ± 0.3	2.0 ± 0.3	1.4 ± 0.2	1.1 ± 0.1	11.9 ± 0.8	9.2 ± 0.8	11.4 ± 0.3	7.3 ± 0.4	10.0 ± 0.0	1.0 ± 0.0	2.9 ± 0.1
Q. rhamnoside	0.1 ± 0.0	0.2 ± 0.0	0.0 ± 0.0	0.1 ± 0.0	2.2 ± 0.2	1.2 ± 0.2	0.4 ± 0.0	0.3 ± 0.0	0.4 ± 0.0	0.1 ± 0.0	0.1 ± 0.0
Q	0.2 ± 0.0	0.3 ± 0.0	0.2 ± 0.0	0.2 ± 0.0	0.1 ± 0.0	0.1 ± 0.0	0.1 ± 0.0	0.1 ± 0.0	0.1 ± 0.0	0.2 ± 0.0	0.1 ± 0.0
Q. rutinoside	24.3 ± 1.2	16.7 ± 6.0	19.9 ± 3.0	21.6 ± 0.6	8.5 ± 0.3	6.2 ± 0.2	7.7 ± 0.6	10.9 ± 1.1	6.7 ± 0.1	19.5 ± 1.3	20.4 ± 1.3
Total	31.1	23.5	22.9	25.8	30.1	21.2	23.3	22.4	20.9	23.5	27.5
Dihydrochalcones											
Phloridzin	0.4 ± 0.0	0.3 ± 0.0	0.2 ± 0.0	0.3 ± 0.0	0.25 ± 0.0	0.2 ± 0.0	0.1 ± 0.0	0.2 ± 0.0	0.1 ± 0.0	0.3 ± 0.0	0.1 ± 0.0
Anthocyanins											
C-3-gluc	170.0 ± 10.1	140.8 ± 8.3	104.7 ± 10.3	143.9 ± 1.5	147.2 ± 5.6	107.7 ± 2.1	67.7 ± 2.7	103.4 ± 3.1	107.2 ± 0.1	138.8 ± 8.2	164.3 ± 6.2
D-3-glu	0.4 ± 0.0	0.4 ± 0.0	0.3 ± 0.0	0.4 ± 0.0	0.1 ± 0.0	0.2 ± 0.0	0.2 ± 0.0	0.2 ± 0.0	0.2 ± 0.0	0.3 ± 0.0	0.4 ± 0.0
P-3-gluc	13.7 ± 0.9	9.5 ± 0.5	6.7 ± 0.7	8.2 ± 0.3	9.9 ± 0.1	4.3 ± 0.2	3.9 ± 0.3	7.6 ± 0.3	4.9 ± 0.3	7.8 ± 0.3	14.7 ± 0.4
D-3-rutin	0.2 ± 0.0	0.1 ± 0.0	0.1 ± 0.0	0.1 ± 0.0	0.4 ± 0.0	0.2 ± 0.0	0.1 ± 0.0	0.0 ± 0.0	0.1 ± 0.0	0.0 ± 0.0	0.1 ± 0.0
C-3-rutin	39.2 ± 2.3	26.1 ± 7.5	64.9 ± 9.5	38.6 ± 2.5	19.5 ± 1.0	17.4 ± 0.6	11.8 ± 1.3	32.0 ± 0.2	22.7 ± 0.2	31.9 ± 1.3	34.6 ± 1.1
C-3-galact	0.4 ± 0.0	0.2 ± 0.0	0.8 ± 0.1	0.0 ± 0.0	0.1 ± 0.0	0.1 ± 0.0	0.0 ± 0.0	0.0 ± 0.0	0.1 ± 0.0	0.1 ± 0.0	0.4 ± 0.0

Table 2. *Cont.*

	LaHave Farm				Kentville					Saskatchewan	
Compounds	**BR**	**BL**	**TN**	**IG**	**LC-12**	**LC-13**	**LC-16**	**LC-23**	**LC-47**	**SAS-IG**	**SAS-TN**
Total	223.9	177.1	177.5	191.2	177.2	129.9	83.7	143.2	135.2	178.9	214.5
Total phenolics by LCMS	285.1 ^{a}	227.8 ^{c,d}	231.29 ^{c,d}	244.87 ^{b,c}	244.25 ^{b,c}	192.5 ^{f}	209.12 ^{c,d,e}	201.92 ^{d,e}	188.9 ^{f}	226.9 ^{c,d}	280.6 ^{a}

EGC, (−)-epigallocatechin; EGCG, (−)-epigallocatechin gallate; ECG, (−)-epicatechin gallate; Q, quercetin; C, cyanidin; D, delphinidin; P, peonidin; Results represent the mean ± SD ($n = 3$), Tukey's test, $p < 0.05$. BL, Berry Blue; BR, Borealis; TN, Tundra; IG, Indigo Gem; SAS-TN, Saskatoon Tundra; SAS-IG, Saskatoon Indigo Gem; ND, Not detectable at the limit of detection of 50 ppb. Significant differences of total phenolics were compared using the p value ≤ 0.05. $^{a–f}$ Different letters within rows denote significant differences between values as determined by one way ANOVA analysis.

3.1.5. Sugar and Organic Acid Profile

Citric, quinic, malic and lactic acid were quantified in the selected haskap cultivars (Figure 1). Citric acid was the predominant organic acid, accounting for 30%–58% of the total organic acid content. These results supported a recent report on *L. caerulea* that showed citric acid representing around 47% of the total organic acid content in haskap samples [23]. This was closely followed by malic acid content, which ranged between 28%–50% of the total organic acid content. The present study showed the presence of 10%–32% quinic acid with respect to the total organic acid, which was higher than reported by the previous study [23]. In contrast, lactic acid was not detected in all Canadian haskap cultivars. SAS-IG and LC-16 exhibited the highest concentration of organic acid.

The monosaccharides, glucose and fructose, predominated the haskap berries and together accounted for more than 95% of the total sugars analysed (Figure 1). Trace amounts of sucrose were also identified in the analysed fruits. Similar to the organic acid content, the highest sugar content was found in SAS-IG (6.5 g 100 g^{-1} FW; $p < 0.05$), while the lowest levels were in BL (1.7 g 100 g^{-1} FW). The content of glucose varied from 0.8–2 and 2.5–3.4 g 100 g^{-1} FW for Nova Scotia and Saskatoon berries, respectively. Similarly fructose content ranged from 0.9–2.1 and 2.3–2.9 g 100 g^{-1} FW for Nova Scotia and Saskatoon fruits, respectively. Consistent with the previously described pattern, the results showed that the cultivars with a high fructose content also had a high glucose content [23].

3.2. Inhibition of Inflammatory Markers by Haskap Berry Extract

The working concentrations did not reduce the cell viability significantly. Following LPS stimulation, five markers of inflammation were measured and compared to two common COX inhibitors, diclofenac and nimesulide (Figures 2 and 3). Cultivar BR exhibited the most consistent inhibitory effects. At a concentration of 100 μg/mL, BR extract was found to have dose dependent effects on inflammation by inhibiting cytokines TNF-α, IL-6, PGE_2 and COX-2 to 55%, 50%, 52% and 38%, respectively (Figure 2). Furthermore, the concentrations of all five measured inflammatory markers in response to 100 μg/mL haskap extracts were comparable to the performance of diclofenac. Considering that BR was found to have the highest concentration of total phenolics (Tables 1 and 2), the inhibition of inflammatory markers was significantly correlated with total flavonols, total anthocyanins and total phenolics (Table 3). Additionally, Borealis had the highest levels of cyanidin-3-*O*-glucoside and rutin (Table 2).

Extract of TN cultivar berries also showed significant dose dependent effects; at 10 μg/mL all markers except PGE_2 were significantly reduced and at 100 μg/mL all parameters, TNF-α and IL-6, PGE_2 and COX-2 were further reduced to 68%, 70% and 74%, respectively, and was as effective as diclofenac in inhibiting COX-2 (47%) and PGE_2 (53%) (Figure 2). Cultivar IG was not an effective inhibitor for TNF-α at 10 μg/mL concentration, however, was comparable to the other three cultivars at 100 μg/mL concentration in reducing the levels of markers (Figure 2). Extract of cultivar BL was consistently the least effective at a concentration of 10 μg/mL and was not able to effectively reduce levels of COX-2 (81%), TNF-α (91%), IL-6 (99%) ($p < 0.05$) (Figure 2). Furthermore, its inhibitory effects on COX-2, TNF-α, and IL-6 at 100 μg/mL concentration were comparable to that of the other three cultivars at 10 μg/mL concentration. The dose dependent effect was not observed on NO inhibition in BR, TN extracts and diclofenac treatments (Figure 3). The extract of BL cultivar reduced NO level by 37%, followed by BR (32%), IG (30%), TN (27%), respectively, compared to LPS treatment at 100 μg/mL. Nimesulide was most effective and reduced NO level by 48% at 100 μg/mL. All extracts were comparable to diclofenac in reducing NO level (Figure 3).

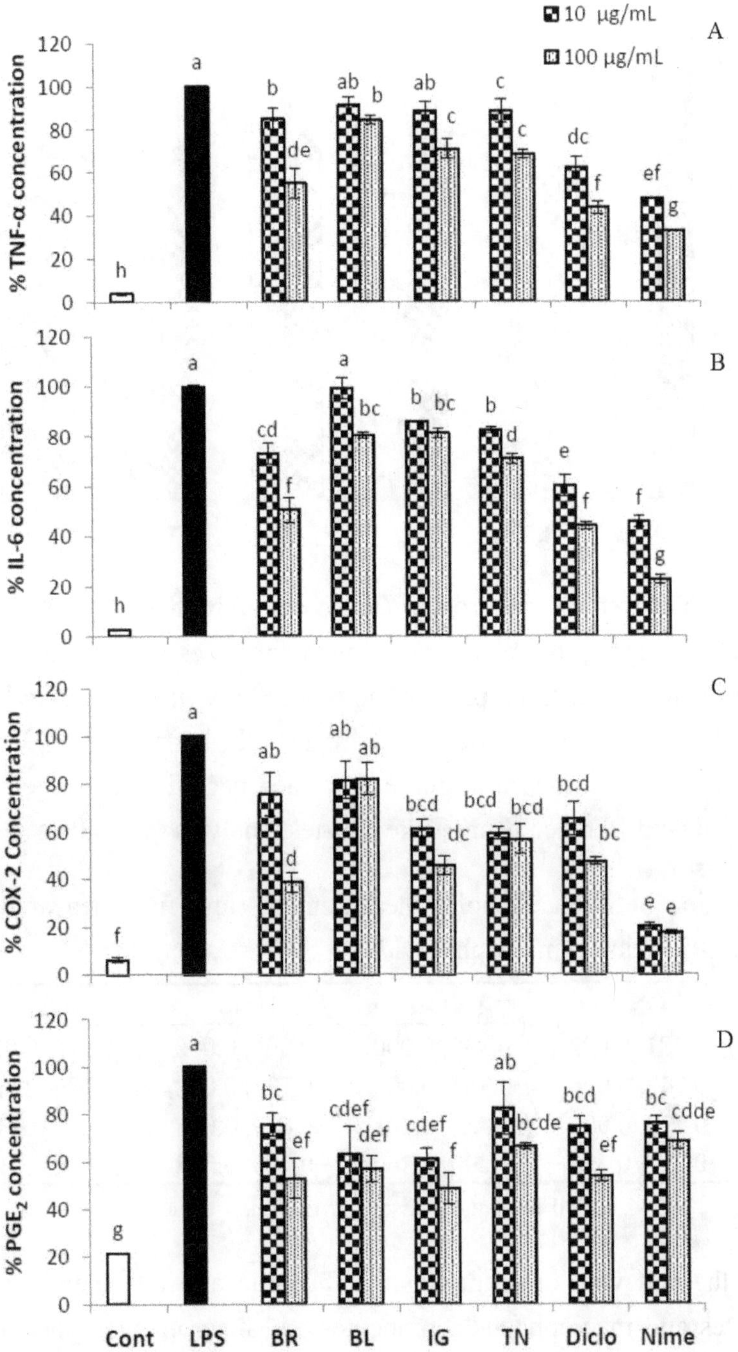

Figure 2. Quantification of TNF-α (**A**), IL-6 (**B**), COX-2 (**C**) and PGE_2 (**D**) in THP-1 differentiated macrophages. The cells were pretreated with test compounds for 4 h, followed by 18 h LPS-stimulation. Results represent the mean ± SD ($n = 3$), Tukey's test, $p < 0.05$. Cont, control; LPS, lipopolysaccharide; BR, borealis; BL, berry blue; IG, indigo gem; TN, tundra; Diclo, diclofenac; Nime, nimesulide. For each bar, different letters a-g represents significance statistical difference using Tukey's *t*-test analysis ($p \leq 0.05$). The bars with one or more same letter are not statistically different.

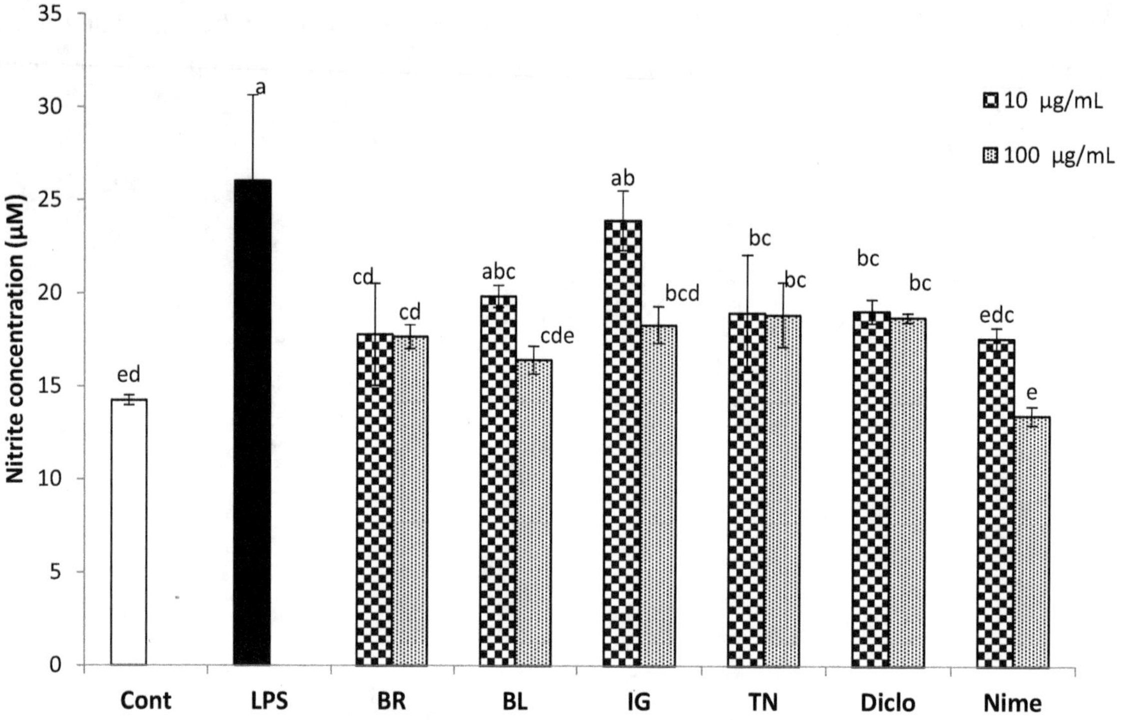

Figure 3. Inhibition of free-radical product (nitric oxide) by haskap berry extract in THP-1 differentiated macrophages. The cells were pretreated with test compounds for 4 h, followed by 18 h LPS-stimulation. Results represent the mean ± SD ($n = 3$), Tukey's test, $p < 0.05$. Cont, control; LPS, lipopolysaccharide; BR, borealis; BL, berry blue; IG, indigo gem; TN, tundra; Diclo, diclofenac; Nime, nimesulide. For each bar, different letters a–e represents significance statistical difference using Tukey's *t*-test analysis ($p \leq 0.05$).

Table 3. Correlation between phenolic measurements and five measured parameters of inflammation in LPS-induced macrophages.

Parameters	COX-2	TNF-α	IL-6	PGE_2	NO
Phenolics	−0.781, 0.003	−0.935, 0.000	−0.896, 0.000	−0.026, 0.936	0.324, 0.304
Flavonoids	−0.742, 0.006	−0.781, 0.003	−0.723, 0.008	−0.186, 0.563	0.264, 0.406
Anthocyanins	−0.728, 0.007	−0.831, 0.001	−0.629, 0.028	0.005, 0.988	0.327, 0.300
Proanthocyanidins	−0.211, 0.511	−0.290, 0.360	−0.298, 0.348	0.459, 0.134	0.084, 0.796

Cell contents: Pearson correlation, *p*-value.

LPS is present in the cell wall of gram-negative bacteria and considered to be the most potent stimulator of monocytes and macrophages. Monocytes and macrophages play important roles at the initiation of inflammation and mobilization of the host defence against bacterial infection. LPS-stimulated macrophage activation leads to the secretion of pro-inflammatory signalling molecules known as cytokines [36]. Two common cytokines associated with inflammation are TNF-α and IL-6. TNF-α is an inflammation response protein produced in the body by macrophages, monocytes and adipocytes and endothelial cells. Long term elevated levels of TNF-α have been linked to both Type II diabetes and metabolic syndrome [37]. Reactive oxygen species can activate NF-κB by releasing inhibitor IκB-α and activated NF-κB induces the production of various cytokines [38]. The elevation of TNF-α and IL-6 levels are linked to diseases induced by inflammation and are known to stimulate endothelial cells to produce vascular cell adhesion molecules, leading to blood clots and increasing the chances of atherosclerosis [39]. Inflammatory cytokines are known to be associated with COX-2 induction, which plays a role in inflammation, as well as cell growth [40]. COX-2 is a key enzyme attributed to inducing an inflammatory response. It is known to be responsible for the production of pro-inflammatory prostaglandins (PGE_2) at sites of injury [41]. COX-2 has been associated with a range of ailments such as heart disease, joint and muscle pain, and minor headaches. NSAIDS research to treat such symptoms accounts for a multi-billion dollar industry in the United States [42]. The inducible nitric oxide, produced by macrophages induced by TNF-α, IL-1 and IL-6, has pro-inflammatory role, whereas e-NO regulates vasodilation is anti-inflammatory in nature. Based on the growing body of evidences, the inhibition of major cytokines like TNF-α and IL-6 could be the best solution for preventing inflammation related diseases. Inhibition of inflammation appears to be most effective at the level of the two measured cytokines and COX-2 or in other words, the initial steps in the inflammation cascade. This is particularly relevant as it indicates that the haskap extract may be more effective as a preventive rather than responsive treatment to inflammation.

3.3. Correlation between Phenolics and Inflammatory Parameters

Overall, the correlation between levels of phenolics of the haskap and the concentrations of COX-2, IL-6, and TNF-α was found to be negative by Pearson's correlation (Table 3). Furthermore, the flavonoids and specifically the anthocyanins showed significant negative correlation with levels of IL-6 and no significant effects were measured on levels of PGE_2 or NO. Overall, proanthocyanidins demonstrated no statistically significant effects on the five measured markers of inflammation.

The various epidemiological studies have inversely correlated flavonoid consumption with incidence of stroke, CVD and cancer [42]. Flavonoids are renowned for their free radical scavenging abilities, which are attributed to both their physical structure and chemical activity. As an antioxidant, their primary mode of action is hydrogen atom donation by a hydroxyl group, which stabilizes the target radical to a flavonoid phenoxyl radical [43]. Quercetin, a commonly occurring flavonoid in fruits and vegetables, has been found to inhibit pro-inflammatory cytokine TNF-α by modulating the NF-κB transcription factor associated with cytokine expression [44]. Quercetin has also been reported for its benefits in diabetes [45] and hyperglycemia [46].

Anthocyanins are a sub-class of flavonoids and also function as free radical scavengers. Anthocyanins have been reported to inhibit the expression of *i*NOS responsible for producing nitric oxide [29] by suppressing the levels of pro-inflammatory cytokines earlier in the inflammatory cascade. Similarly, anthocyanins have been shown to modulate transcription factors (c-Jun and NF-κB) responsible for expressing cytokines [30]. This is an exciting prospect because the anti-inflammatory properties of haskap might not be limited to free-radical scavenging, could also be involved in gene expression of the cytokines and enzymes responsible for initiating the inflammatory cascade.

In summary, we characterized various haskap cultivars for their phenolic composition, sugar and organic profile along with anti-inflammatory properties. Among the selected haskap cultivars, Borealis was found to be the most rich in phenolic content, flavonoid content and anthocyanin content ($p < 0.05$). Cyanidin-3-*O*-glucoside, cyanidin-3-*O*-rutinoside, chlorogenic acid and rutin were polyphenols found in the highest concentrations and have been implicated in inhibiting inflammatory transcription factors in previous reports [13,14]. The phenolic profile was consistent with a previously published study of haskap [12]. We demonstrated that the extracts of haskap berry, specifically Borealis and Tundra, have markedly suppressed LPS-induced inflammation in an *in vitro* model of THP-1 derived human macrophages, when compared that with the non-specific COX inhibitor drug, diclofenac. A negative correlation was found between the polyphenols and the major inflammatory mediators. These inflammatory mediators are predominately involved at the initial stages of the inflammatory cascade and, therefore, results suggest that anti-inflammatory compounds could effectively control onset of inflammation. Overall, the results suggest that polyphenols-rich haskap berry has a potential to use as an effective functional food to control inflammation. Recently, Weidinger and colleagues [47] have demonstrated that intracellular signalling pathways mediated by NO and ROS are linked to each other via mitochondrial ROS (mtROS) and form an inducible nitric oxide synthase (iNOS)-mtROS feed-forward loop which intensifies liver failure upon acute inflammation. In general, prospective trials have shown that habitual intake of dietary polyphenols reduces the risk of dementia, stroke and neurodegenerative disorders through a wide spectrum of activities including free radical scavenging, transition metal chelation, activation of survival genes and signalling pathways, regulation of mitochondrial function and modulation of neuroinflammation [48]. Therefore, investigations will be continued to understand the possible mechanisms of haskap polyphenols in regulation of mtROS and other specific targets associated with inflammation. Further investigations are also required to find out the effectiveness of haskap berry polyphenol in animal model systems.

4. Conclusions

In conclusion, among the Canadian haskap cultivars assessed, Borealis exhibited the highest phenolic, flavonoid and anthocyanin contents. The polyphenol extracts of haskap could able to suppress the major pro-inflammatory cytokines including IL-6, TNF-α and PGE2 as well as COX-2 enzyme in LPS-stimulated human macrophages *in vitro*. The results suggest that haskap berry polyphenols need to be further assessed as a natural health product to control inflammation.

Acknowledgements

This research was funded by Discovery Grant program of the Natural Sciences and Engineering Research Council (NSERC) of Canada. We thank LaHave Forest Farm Inc., Blockhouse, NS, Canada for providing haskap samples.

Author Contributions

H. P. Vasantha Rupasinghe is the principal investigator who designed and executed the study. Bob Bors and Andrew R. Jamieson provided the plant materials. Indu Parmar and H. P. Vasantha Rupasinghe performed antioxidant and phenolic characterization studies. Mannfred M. A. Boehm and Satvir Sekhon-Loodu performed the cell culture studies of inflammation. Mannfred M. A. Boehm, Indu Parmar, and Satvir Sekhon-Loodu carried out data analysis and interpretation of the results. All the authors contributed to writing of this manuscript, read and approved the final manuscript.

Appendix

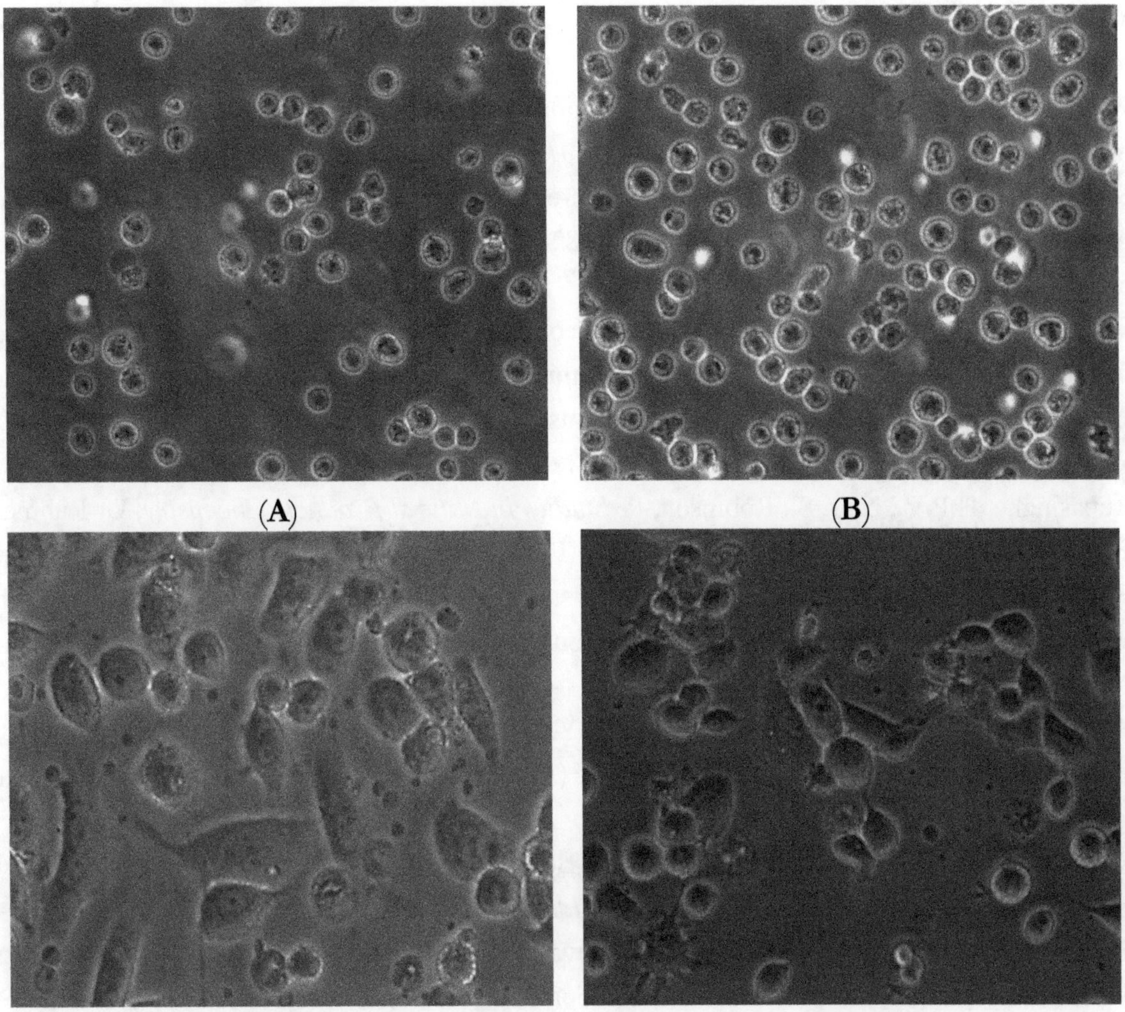

(A) (B) (C) (D)

Figure A1. Morphology of human monocytes (THP-1) and LPS-induced macrophages. Monocytes exist in suspension phase and have spherical shape. Monocytes upon differentiation take adherent phase and form various cell shapes. The undifferentiated monocytes (**A**), (**B**) and differentiated macrophages (**C**), (**D**) are shown in the figure.

References

1. Diplock, A.T.; Charleux, J.L.; Crozier-Willi, G.; Kok, F.J.; Rice-Evans, C.; Roberfroid, M.; Stahl, W.; Vina-Ribes, J. Functional food science and defence against reactive oxidative species. *Br. J. Nutr.* **1998**, *80*, S77–S112.
2. Antonicelli, F.; Parmentier, M.; Hirani, N.; Drost, E.; Rahman, I.; Donaldson, K.; MacNee W. LPS stimulation of IL-8 release is inhibited by thiol antioxidant at the transcriptional level in THP-1 macrophage cells. *Am. J. Respir. Crit. Care Med.* **2000**, *161*, 1319–1327.

3. Mehta, J.L.; Rasouli, N.; Sinha, A.K.; Molavi, B. Oxidative stress in diabetes, a mechanistic overview of its effects on atherogenesis and myocardial dysfunction. *Int. J. Biochem. Cell. Biol.* **2006**, *38*, 794–803.
4. Valko, M.; Rhodes, C.J.; Moncol, J.; Izakovic, M.; Mazur, M. Free radicals; metals and antioxidants in oxidative stress-induced cancer. *Chem. Biol. Interact.* **2006**, *160*, 1–40.
5. Prescott, S.L. Early-life environmental determinants of allergic diseases and the wider pandemic of inflammatory non-communicable diseases. *J. Allergy Clin. Immunol.* **2013**, *131*, 23–30.
6. Statistics Canada. Canadian Health Measures Survey: Metabolic Syndrome in Canadians. Available online: http://www.statcan.gc.ca/pub/82-625-x/2012001/article/11735-eng.htm#n1 (accessed on 13 February 2014).
7. Ford, E.S.; Giles, W.H.; Dietz, W.H. Prevalence of the metabolic syndrome among us adults: Findings from the third national health and nutrition examination survey. *JAMA* **2002**, *287*, 356–359.
8. Gautam, R.; Jachak, S.M. Recent developments in anti-inflammatory natural products. *Med. Res. Rev.* **2009**, *29*, 767–820.
9. Rupasinghe, H.P.V.; Nair, S.; Robinson, R. *Studies in Natural Products Chemistry*; Ur Rahman, A., Ed.; Elsevier Science Publishers: Amsterdam, The Netherlands, 2014; Volume 42, pp. 229–266.
10. Bors, B. Breeding of *Lonicera caerulea* L. for saskatchewan and Canada. In Proceedings of the 1st Virtual International Scientific Conference on *Lonicera caerulea* L., Saskatoon, SK, Canada, 23 March–23 April 2009; pp. 88–98.
11. Jin, X.H.; Ohgami, K.; Shiratori, K.; Suzuki, Y.; Koyama, Y.; Yoshida, K.; Ilieva, I.; Tanaka, T.; Onoe, K.; Ohno, S. Effects of blue honeysuckle (*Lonicera caerulea* L.) extract on lipopolysaccharide induced inflammation *in vitro* and *in vivo*. *Exp. Eye. Res.* **2006**, *82*, 860−867.
12. Rupasinghe, H.P.V.; Yu, L.J.; Bhullar, K.S.; Bors, B. Haskap (*Lonicera caerulea*): A new berry crop with high antioxidant capacity. *Can. J. Plant Sci.* **2012**, *92*, 1311−1317.
13. Palikova, I.; Valentova, K.; Oborna, I.; Ulrichova, J. Protectivity of blue honeysuckle extract against oxidative human endothelial cells and rat hepatocyte damage. *J. Agric. Food Chem.* **2009**, *57*, 6584−6589.
14. Zdarilova, A.; Svobodova, A.R.; Chytilova, K.; Simanek, V.; Ulrichova, J. Polyphenolic fraction of *Lonicera caerulea* L. fruits reduced oxidative stress and inflammatory markers induced by lipopolysaccharide in gingival fibroblasts. *Food Chem. Toxicol.* **2010**, *48*, 1555−1561.
15. Olefsky, J.M.; Glass, C.K. Macrophages, inflammation, and insulin resistance. *Annu. Rev. Physiol.* **2010**, *72*, 219−246.
16. Plekhanova, M.N. Blue honeysuckle (*Lonicera caerulea* L.)—A new commercial berry crop for temperate climate: Genetic resources and breeding. *Acta Hortic.* **2000**, *538*, 159–164.
17. Marinova, D.; Ribarova, F.; Atanassova, M. Total phenolics and total flavonoids in Bulgarian fruits and vegetables. *J. Univ. Chem. Tech. Metall.* **2005**, *40*, 255−260.
18. Ratnasooriya, C.; Rupasinghe, H.P.V.; Jamieson, A. Juice quality and polyphenol concentration of fresh fruits and pomace of selected Nova Scotia-grown grape cultivars. *Can. J. Plant Sci.* **2010**, *90*, 193−205.
19. Prior, R.L.; Fan, E.; Ji, H.; Howell, A.; Nio, C.; Payne, M.J.; Reed, J. Multi-laboratory validation of a standard method for quantifying proanthocyanidins in cranberry powders. *J. Sci. Food Agric.* **2010**, *90*, 1473−1478.

20. Rupasinghe, H.P.V.; Erkan, N.; Yasmin, A. Antioxidant protection of eicosapentaenoic acid and fish oil oxidation by polyphenolic-enriched apple skin extract. *J. Agric. Food Chem.* **2010**, *58*, 1233–1239.
21. Rop, O.; Reznicek, V.; Mlcek, J.; Jurikova, T.; Balik, J.; Sochor, J.; Kramarova, D. Antioxidant and radical oxygen species scavenging activities of 12 cultivars of blue honeysuckle fruit. *Hortic. Sci.* **2011**, *38*, 63–70.
22. Petrova, V.P. *Biochimija Dikorastuščich Plodovo—Jagodnych Rastenij* (In Ukraine), 1st ed.; Golovnoe Izdatel'stvo Objedenija, Kijev, Ukraine, 1986; pp. 260–266.
23. Wojdylo, A.; Jauregui, P.N.N.; Carbonell-Barrachina, A.; Oszmianski, J.; Golis, T. Variability of phytochemical properties and content of bioactive compounds in *Lonicera caerulea* L. var. *kamtschatica* berries. *J. Agric. Food Chem.* **2013**, *61*, 12072–12084.
24. Orincak, J.; Matuskovic, J.; Jurcak, S. *Possibilities of Species Lonicera caerulea in Utilization of the Secondary Metabolism in Food and Pharmaceutical Processing*, 1st ed.; SPU: Nitra, Slovak, 2003; pp. 210–219.
25. Plekhanova, M.N.; Streltsyna, S.A.; Rostova, N.S. Phenolic compounds in berries of *Lonicera* subsect. *Caerulea* species. *Plant Res.* **1993**, *29*, 16–25.
26. Andersen, O.M.; Jordheim, M. The anthocyanins. *Flavonoids Chemistry, Biochemistry and Applications*, 4th ed.; Andersen, O.M., Markham, K.R., Eds.; CRC Press: Boca Raton, FL, USA, 2006; pp. 471–552.
27. Gazdik, Z.; Krska, B.; Adam, V.; Saloun, J.; Jurikova, T.; Reznicek, V.; Horna, A.; Kizek, R. Electrochemical determination of antioxidant potential of some less common fruit species. *Sensors* **2008**, *8*, 7564–7570.
28. Bakowska, A.M.; Marianchuk, M.; Kolodziejczyk, P. Survey of bioactive components in Western Canadian berries. *Can. J. Physiol. Pharmacol.* **2007**, *85*, 1139–1152.
29. Tsuda, T.; Horia, F.; Osawa, T. Cyanidin 3-*O*-β-D-glucoside suppresses nitric oxide production during a zymosan treatment in rats. *J. Nutr. Sci. Vitaminol.* **2002**, *48*, 305–310.
30. Karlsen, A.; Retterstol, L.; Laake, P.; Paur, I.; Kjolsrud-Bohn, S.; Sandvik, L.; Blomhoff, R. Anthocyanins inhibit nuclear factor-κB activation in monocytes and reduce plasma concentrations of pro-inflammatory mediators in healthy adults. *J. Nutr.* **2007**, *137*, 1951–1954.
31. Leiss, K.; Maltese, F.; Choi, Y.H.; Verpoorte, R.; Klinkhamer, P.G.L. Identification of chlorogenic acid as a resistance factor for thrips in chrysanthemum. *Plant Physiol.* **2009**, *150*, 1567–1575.
32. Boerjan, W.; Ralph, J.; Baucher, M. Lignin biosynthesis. *Annu. Rev. Plant Biol.* **2003**, *54*, 519–546.
33. Hwang, S.J.; Kim, Y.W.; Park, Y.; Lee, H.J.; Kim, K.W. Anti-inflammatory effects of chlorogenic acid in lipopolysaccharide-stimulated RAW 264.7 cells. *Inflamm. Res.* **2014**, *63*, 81–90.
34. Choi, J.K.; Kim, S.H. Rutin suppresses atopic dermatitis and allergic contact dermatitis. *Exp. Biol. Med.* **2012**, *238*, 410–417.
35. Gao, M.; Ma, Y.; Liu, D. Rutin suppresses palmitic acids-triggered inflammation in macrophages and blocks high fat diet-induced obesity and fatty liver in mice. *Pharm. Res.* **2013**, *30*, 2940–2950.
36. Kim, Y.; So, H.S.; Moon, B.S.; Youn, M.J.; Kim, H.J.; Shin, Y.I.; Moon, S.K.; Song, M.S.; Choi, K.Y.; Song, J.; *et al.* Sasim attenuates LPS-induced TNF-alpha production through the induction of HO-1 in THP-1 differentiated macrophage-like cells. *J. Ethnopharmacol.* **2008**, *119*, 122–128.

37. Yang, J.; Park, Y.; Zhang, H.; Gao, X.; Wilson, E.; Zimmer, W.; Abbott, L.; Zhang, C. Role of MCP-1 in tumor necrosis factor-α-induced endothelial dysfunction in type 2 diabetic mice. *Am. J. Physiol. Heart Circ. Physiol.* **2009**, *297*, H1208–H1216.
38. Rahman, I.; Gilmour, P.S.; Jimenez, L.A.; MacNee, W. Oxidative stress and TNF alpha induce histone acetylation and NF-kappaB/AP-1 activation in alveolar epithelial cells: Potential mechanism in gene transcription in lung inflammation. *Mol. Cell Biochem.* **2002**, *234–235*, 239–248.
39. Thilakarathna, S.H.; Rupasinghe, H.P.V. Anti-atherosclerotic effects of fruit bioactive compounds: A review of current scientific evidence. *Can. J. Plant Sci.* **2012**, *92*, 407–419.
40. Vane, J.R.; Bakhle, Y.S.; Botting, R.M. Cyclooxygenases 1 and 2. *Annu. Rev. Pharmacol.* **1998**, *38*, 97–120.
41. Das, U. Is obesity an inflammatory condition? *Nutrition* **2001**, *17*, 953–966.
42. Knekt, P.; Kumpulainen, J.; Jarvinen, R.; Rissanen, H.; Heliovaara, M.; Reunanen, A.; Hakulinen, T.; Aromaa, A. Flavonoid intake and risk of chronic diseases. *Am. J. Clin. Nutr.* **2002**, *76*, 560–568.
43. Amic, D.; Davidovic-Amic, D.; Beslo, D.; Rastija, V.; Lucic, B.; Trinajstic, N. SAR and QSAR of the antioxidant activity of flavonoids. *Curr. Med. Chem.* **2007**, *14*, 827–845.
44. Dias, A.S.; Porawski, M.; Alonso, M.; Marroni, N.; Collado, P.S.; Gonzalez-Gallego, J. Quercetin decreases oxidative stress, NF-kappa β activation, and iNOS overexpression in liver of streptozotocin-induced diabetic rats. *J. Nutr.* **2005**, *135*, 2299–2304.
45. Nair, M.P.; Mahajan, S.; Reynolds, J.L.; Aalinkeel, R; Nair, H.; Schwartz, S.A.; Kandaswami, C. The flavonoid quercetin inhibits proinflammatory cytokine (tumor necrosis factor alpha) gene expression in normal peripheral blood mononuclear cells via modulation of the NF-kappa beta system. *Clin. Vaccine Immunol.* **2006**, *13*, 319–328.
46. Jung, M.; Triebel, S.; Anke, T.; Richling, E.; Erkel, G. Influence of apple polyphenols on inflammatory gene expression. *Mol. Nutr. Food Res.* **2009**, *53*, 1263–1280.
47. Weidinger, A.; Mullebner, A.; Paier-Pourani, J.; Banerjee, A.; Miller, I.; Lauterbock, L.; Duvigneau, J.C.; Skulachev, V.P.; Redl, H.; Kozlov, A.V. Vicious inducible nitric oxide synthase-mitochondrial reactive oxygen species cycle accelerates inflammatory response and causes liver injury in rats. *Antioxid. Redox Signal.* **2015**, *22*, 572–586.
48. Jones, Q.R.D.; Warford, J.; Rupasinghe, H.P.V.; Robertson, G.S. Target-based selection of flavonoids for neurodegenerative disorders. *Trends Pharmacol. Sci.* **2012**, *33*, 602–610.

6

The Impact of Non-Enzymatic Reactions and Enzyme Promiscuity on Cellular Metabolism during (Oxidative) Stress Conditions

Gabriel Piedrafita [1,†], Markus A Keller [1,†] and Markus Ralser [1,2,*]

[1] Department of Biochemistry, University of Cambridge, 80 Tennis Court Rd, Cambridge CB2 1GA, UK; E-Mails: gp394@cam.ac.uk (G.P.); mk747@cam.ac.uk (M.A.K.)

[2] The Francis Crick Institute, Mill Hill Laboratory, The Ridgeway, London NW1 7AA, UK

† These authors contributed equally to this work.

* Author to whom correspondence should be addressed; E-Mail: markus.ralser@crick.ac.uk;

Academic Editor: Michael Breitenbach

Abstract: Cellular metabolism assembles in a structurally highly conserved, but functionally dynamic system, known as the metabolic network. This network involves highly active, enzyme-catalyzed metabolic pathways that provide the building blocks for cell growth. In parallel, however, chemical reactivity of metabolites and unspecific enzyme function give rise to a number of side products that are not part of canonical metabolic pathways. It is increasingly acknowledged that these molecules are important for the evolution of metabolism, affect metabolic efficiency, and that they play a potential role in human disease—age-related disorders and cancer in particular. In this review we discuss the impact of oxidative and other cellular stressors on the formation of metabolic side products, which originate as a consequence of: (i) chemical reactivity or modification of regular metabolites; (ii) through modifications in substrate specificity of damaged enzymes; and (iii) through altered metabolic flux that protects cells in stress conditions. In particular, oxidative and heat stress conditions are causative of metabolite and enzymatic damage and thus promote the non-canonical metabolic activity of the cells through an increased repertoire of side products. On the basis of selected examples, we discuss the consequences of non-canonical metabolic reactivity on evolution, function and repair of the metabolic network.

Keywords: oxidative stress; reactive oxygen species; metabolic damage; metabolite repair; enzyme promiscuity; underground metabolism

1. Introduction

During its lifespan, a cell is exposed to many hazards and changes in the environment. These include varying pH and temperature, osmotic imbalances, radiation, drought, starvation, contact with deleterious or toxic chemicals, or biotic interactions with other species. These stress situations can manifest as an acute pulse, or be sustained over time, and can affect cellular functioning and ultimately impair cell growth, replicative capabilities, and/or promote aging. Stress-induced damage to biological macromolecules such as DNA, proteins and membrane components has been intensively studied and characterized [1–4]. However, also cellular metabolites and the flux through metabolic pathways are affected. The impact of stress on metabolism and its small-molecule cellular constituents has attracted notable attention in areas of chemistry and toxicology [5,6], but only recently research has begun to systemically analyze the wide-ranging consequences at the systems level [7,8]. Since metabolism is a basal property of life and a main driving force during evolution, being the closest-to-phenotype cellular attribute, and given the reactive nature of many metabolites, the identification of stress-related metabolic processes appears to be of fundamental importance.

2. Chemical Damage to Metabolites: Oxidants and Antioxidants

One of the most important causes of metabolic damage is unspecific oxidation. Basal levels of oxidizing molecules arise constantly from normal cellular processes involving redox reactions or in the mitochondrial or photosynthetic electron transport chains [9]. Normal levels of oxidants are compensated by cellular metabolism, and being integral part of physiology, are required for biochemical reactions and as signaling molecules [10]. In a range of conditions referred to as "oxidative stress" [11], normal redox homeostasis is however disrupted and the balance between reducing and oxidizing molecules shifts towards the latter. Under oxidative stress, reactive oxygen species (ROS) represent one of the most important groups of cellular oxidants and comprise oxygen-derived molecules such as hydrogen peroxide (H_2O_2), superoxide anion (O_2^-), or hydroxyl radical (•HO). Non-enzymatic chemical reactivity plays a crucial role in this scenario, as for example in the context of the Fe(II)-catalyzed Fenton reaction, that converts H_2O_2 into •HO [12]. Superoxide anion and hydroxyl radical are highly reactive and indiscriminately damage metabolites, lipids, enzymes and nucleic acids via peroxidation [13–15]. This damage causes "errors" in metabolism [16,17], (Figure 1 (i)) such as the propagation of lipid peroxidation and carbohydrate oxidation [18,19]. In addition to the strong potential of superoxide anion to react with transition metals, it also combines with other chemical species, e.g., nitric oxide, and generates highly oxidant species such as peroxynitrite, causing further damage to lipids, altered nucleobases, and nitration of amino acid residues [20–23]. Under oxidative stress, the functionality of cellular systems declines, increasing the risk of cellular damage, apoptosis, aging, neurodegeneration and cancer [24–26].

Common targets to oxidative damage are amino acids, both in their free and protein integrated forms that are both highly abundant. Especially cysteine (which will be discussed later in the text) and methionine are the most prone to oxidation [27]. When the sulphur of methionine reacts with oxidizing molecules, it can form methionine-*S*- or methionine-*R*-sulfoxides, whose accumulation has been associated with various diseases and aging [28,29].

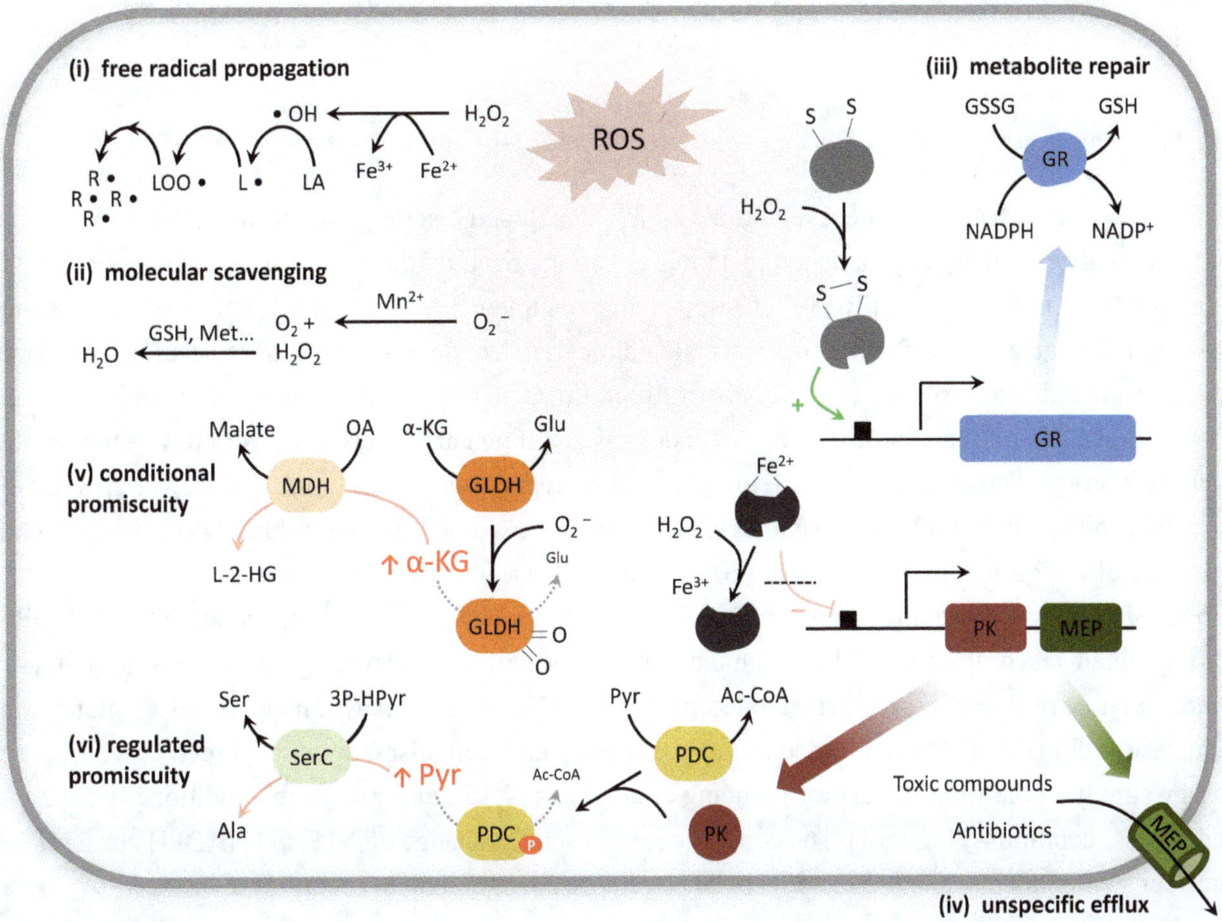

Figure 1. Formation of non-canonical metabolites and repair strategies. Chemical modification of metabolites leads to unwanted reactions, causing damage to both macromolecules and small molecules (**i**). Molecular scavengers cleanse and channel ROS toward less toxic products (**ii**). Stress sensors activate genes involved in specific protein-based repair responses (**iii**); or in unspecific transport processes (**iv**). As a general outcome, the metabolome reconfigures, affecting metabolite levels and fluxes; this in turn affects the specificity of enzymatic reactions. (**v–vi**) Selection of relevant examples from metabolism. Abbreviations: ROS, reactive oxygen species; Met, Methionine; GR, glutathione reductase; GSSG, oxidized glutathione or glutathione disulfide; GSH, reduced glutathione; MEP, multidrug efflux pump; LA, lipid; L•, lipid radical; LOO•, lipid peroxyl radical; R• complex radicals; GLDH, glutamate dehydrogenase; α-KG, α-ketoglutarate; Glu, glutamate; MDH, malate dehydrogenase; OA, oxaloacetate; L-2-HG, L-2-hydroxyglutarate; PK, protein kinase; PDC, pyruvate dehydrogenase complex; Pyr, pyruvate; Ac-CoA, acetyl-coenzyme A; SerC, phosphoserine transaminase; 3P-HPyr, 3-phosphohydroxypyruvate; Ser, serine; Ala, alanine.

Biological systems exploit this property of metabolites and use them as scavengers for ROS. This applies not only to methionine and cysteine, but also to pyruvate, α-tocopherol, ascorbic acid, or other amino acids [12,30–32] (Figure 1 (ii)). While free methionine acts as general redox buffer, protein-coded methionine residues absorb oxidizing agents by positioning on the exposed protein surface and surrounding the active site, thereby preventing more severe oxidative damage on targeted proteins [33,34]. This oxidation of methionine residues at the protein surface can proceed without significant reduction in enzyme activity, as for example in glutamine synthetase, which is shown extremely resistant to H_2O_2 [33].

2.1. Not only Protein and DNA Damage, but also Metabolites Can Be Repaired

In analogy to DNA and protein repair, cells have developed a series of control mechanisms to cope with the oxidized/chemically damaged metabolites, summarized by the term "metabolite repair" [35] or "metabolite-damage control" [36–38]. These involve complementary strategies that either: (i) "repair" the chemical damage to metabolites (e.g., reversion through reduction of an oxidized molecule); (ii) degrade or convert the altered molecules into less harmful products or functional metabolites; or (iii) export the non-canonical metabolite. Some of these mechanisms are of general nature and overlap with the parallel occurring protein repair mechanisms (e.g., glutathione reductase, thioredoxin). Others are specifically targeted against small-molecule derivatives, as for example Glutathione peroxidase 4 (GPx4), which is a protein of the glutathione peroxidase (GPx) family with wide substrate specificity, able to reduce a variety of lipid hydroperoxides, preventing radical propagation through lipid peroxidation [39,40]. GPx4 activity obtains its redox potential through the reduced form of the tripeptide glutathione (GSH) [41,42], a molecular scavenger which is regenerated in a NADPH-dependent reaction catalyzed by glutathione reductase (GR) (Figure 1 (iii)). Indeed, most of the repair mechanisms require, directly or indirectly, energy supply in the form of ATP or reducing equivalents [4,15], and many are conditionally activated under stress conditions (Figure 1). The consequence is a sudden change of ATP and NADPH consumption once cells enter a stress situation and its balance requires metabolism to reconfigure.

An illustrative example for specific oxidative stress-related repair concerns methionine. Repair of oxidized methionine does occur when the amino acid is in its free form as well as when it is part of a protein [43,44] (Figure 2). Three enzyme families of methionine sulfoxide reductases, MsrA, MsrB and MsrC, metabolize the derivatives in a stereospecific manner. Of them, MsrA and MsrB are highly conserved and reduce the *S*-sulfoxide and the *R*-sulfoxide residues, respectively, predominantly within a protein context [45]. MsrC (also named f*R*Msr) is only present in prokaryotes [46] and in some eukaryotes [47], and has its main affinity for free methionine-*R*-sulfoxide. Interestingly, MsrC exhibits no sequence homology with MsrA and MsrB, indicating independent evolutionary origin of these functionally overlapping proteins. In higher plants, where no specific MsrC-like activity exists, the repair function on free methionine-*R*-sulfoxide is efficiently covered by a subtype of MsrB proteins (which would have evolved by segmental duplication) [48], evidencing the evolution of alternative adaptive strategies to cope with a similar type of damage (Figure 2). As a common observation, repair mechanisms appear widely distributed, irrespective of whether metabolic damage proceeds through direct metabolite oxidation, non-enzymatic reactivity, enzyme inactivation, or enzyme promiscuity. Although most of these repair functions are non-essential under normal conditions, the respective mutants commonly show growth defects, lower survival rates or higher mutation frequencies [36,37,47,49,50]. In humans, deficiencies in metabolite detoxification systems are related to encephalopathic disorders [51], cancer predisposition [52], drug intolerance [53], or aging and a decreased lifespan [54]. Evidence has also been presented that neuronal clearance of toxic metabolites could be the explanation of why metazoans require sleep [55].

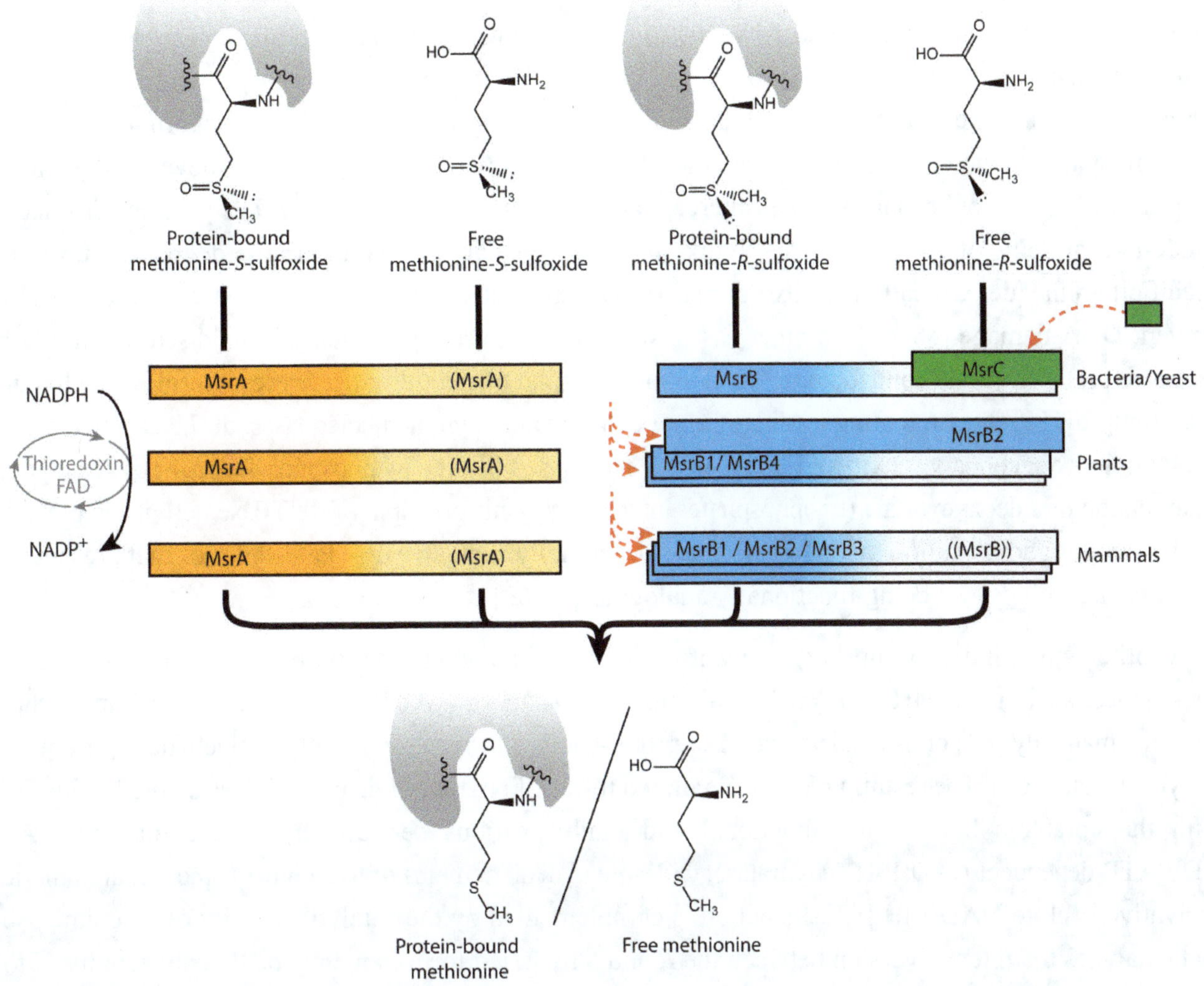

Figure 2. Oxidative damage on methionine and its repair enzymes. Methionine is oxidized to methionine-*S*-sulfoxide and methionine-*R*-sulfoxide derivatives. Three families of methionine sulfoxide reductases (Msr, different colors), revert the process depending on thioredoxin, each being stereo-specific against the *R*- or the *S*-enantiomeric form. While MsrA and MsrB are highly conserved with preference for protein integrated methionine sulfoxides, MsrC (f*R*Msr) is limited to unicellular organisms and reduces free methionine-*R*-sulfoxide. MsrB has evolved by duplication and divergence (red dashed arrows), accounting for coverage of diverse subcellular localizations, and enabling in plants the appearance of a subtype active against the free methionine-*R*-sulfoxide. Color intensities illustrate different strength in catalytic efficiency.

2.2. Other Forms of Metabolite Damage: Examples of Nucleotide and Cofactor Metabolism

Metabolism comprises a large diversity of molecules. The spectrum of possible combinatorial reactions and modifications on metabolites extends to a huge space of chemical reactions. Some modifications occur however more frequent than others. For instance, nucleotide metabolism is by its nature composed of many metabolites possessing high structural similarities. A range of abnormal nucleotides is generated as a consequence of spontaneous oxidation [56]. Others arise from unspecific enzymatic activities, as for example by methylation of nucleobases. Several of these aberrant compounds with only slight structural differences from the standard bases, can subsequently be recognized by downstream enzymes which propagate the error via incorporation. The mischarge of nucleoside diphosphate (NDP) kinases, which can promiscuously accept a broad range of base-differing substrates, releases abnormal NTPs that cause mispairing and mutagenesis if incorporated into DNA [56,57]. Cells have developed various families of NTP pyrophosphatases that intercept and hydrolyze non-canonical NTPs [37]. For instance, MutT-like proteins of the widespread Nudix superfamily catalyze—with variable degree of substrate specificity—the detoxification of oxo-purine dNTPs generated during oxidative stress, and thereby prevent DNA damage [58–60]. In other cases, substrate promiscuity (which will be discussed in more detail below) is the main contributor to the pool of abnormal nucleotides: dUTPases hydrolyze with high specificity dUTP, discriminating against normal cytosine and thymine nucleobases. dUTP can arise from UDP through a secondary activity of ribonucleotide reductase followed by phosphorylation, or by (catalytic) deamination of a deoxycytidine (tri-)phosphate. Interestingly, this principle of dUTPase activity is present in all three kingdoms of life, but is fulfilled by two mutually exclusive dUTPase families that present no sequence homology, yet being functionally analogous [61,62].

Another quantitatively highly relevant case concerns chemical modification of enzymatic coenzymes, such as NADH or FADH. Hydration of NADH or NADPH is common, occurring either non-enzymatically [63] or as a side reaction of promiscuous enzymes such as glyceraldehyde 3-phosphate dehydrogenase [64]. The resulting *R* and *S* hydrated forms of NAD(P)H display high structural similarity with the cofactors but are non-functional and inhibit various dehydrogenase reactions [65]. An ADP/ATP-dependent NAD(P)H dehydratase, widespread in all domains of life, converts the *S* enantiomeric derivative back to NAD(P)H [63]. Its action is complemented by the catalytic activity of an epimerase that catalyzes the interconversion between the *R* and *S* hydrated forms, an enzyme that appears fused to the dehydratase in some species such as *Escherichia coli*. Interestingly, in higher plants [66,67] and also in mammals [68] both genes possess organellar targeting sequences, allowing the proteins to be directed and co-localize within diverse subcellular compartments.

3. Non-Enzymatic Metabolic Reactions Are Affected by Stress Conditions

A second aspect of the susceptibility of metabolism concerns non-enzymatic reactions that occur as part of normal metabolism or through non-specific chemical reactivity but that contribute to canonical metabolic pathways [69,70]. Such non-enzymatic reactions are of high importance for the evolution of metabolism, as exemplified by the metabolism-like non-enzymatic interconversions that can occur between central carbon metabolites, which might have given rise to the early evolution of glycolysis and pentose phosphate pathway [71]. The presence of an enzyme catalyst in modern cells does however not prevent these non-enzymatic chemical reactions to occur in parallel to the enzymatic pathways. Termed Class I non-enzymatic metabolic reactions, a subset of non-enzymatic reactions are of low substrate specificity and can affect multitude of metabolites and modify a series of macromolecules [69]. These include Maillard reactions, unspecific conjugations of amino acids that are accelerated under heat and UV exposure, and also unspecific protein modifications such as protein (poly-)phosphorylation, glycosylation or acylation. The term "underground metabolism" [70] has been suggested to summarize the repertoire of this chemical reactivity, both non-enzymatic and enzyme-promiscuous, occurring in parallel to the flux of functional metabolic pathways. This reactivity is distinguishable from other more specific non-enzymatic reactions that are part of the metabolic network, and occur either exclusively non-enzymatically (Class II non-enzymatic metabolic reactions) or in parallel to existing enzyme function (Class III) [69].

Non-enzymatic reactions are dependent on the chemical environment (metal availability, pH, temperature, ionic strength), and must therefore be particularly sensitive to stress conditions. This is well illustrated for the case of non-enzymatic protein acylation: A number of essential, endogenous thioester metabolites, such as acetyl-CoA or succinyl-CoA, can unspecifically cross-acylate protein lysines [72,73]. Protein acylation appears most prominent in mitochondria, correlates with the levels of these reactive metabolites and with mitochondrial energy-metabolism, and has been linked to a form of "carbon stress" [74]. Non-enzymatic modification of macromolecules can alter protein function and may require quality-control responses. In this case, the sirtuin family of deacetylase enzymes (which has received notable attention by a potential connection to the benefits arising from a calorie-restriction diet) has been proposed to counteract the potential deleterious effects of non-enzymatic protein acylation and thus to play a role in stress resistance [74]. Similarly, for small molecules, intermediates generated in non-enzymatic reactions can, if accumulated, interfere with metabolic pathways by acting as competitive inhibitors of enzymes, or serve as alternative substrates, and thus need to be cleared.

While specific repair mechanisms might have evolved for non-canonical metabolites which are produced in higher quantities or present strong cytotoxic effects, it is reasonable to assume that not all metabolic side products possess specific clearance mechanisms. This becomes illustrative, as the number of potential chemical reaction products from non-enzymatic reactivity exceeds the number of enzymes encoded in a typical genome by several orders of magnitude. A broad range of metabolites, including especially those for which no specific cleaning enzyme exists [75–77], might however be cleared via unspecific cellular export, which is largely mediated by efflux pumps and transmembrane channels such as multidrug transporters (Figure 1 (iv)). In bacteria, numerous studies report the involvement of membrane transporters in multidrug resistance [78–80]. *E. coli*'s resistance to a high number of compounds is mediated by the outer-membrane pore-forming protein TolC [81]. This transporter acts in concert with the inner membrane TolC-dependent efflux pump AcrB and with cognate periplasmic proteins (e.g., AcrA) to form tripartite trans-periplasmic exporters that push xenobiotics out of the cell [82,83]. There is growing evidence suggesting that TolC-mediated extrusion is not limited to xenobiotics [84]: *E. coli* tolC mutants show lower fitness phenotypes in certain stress conditions [85], accumulate cell-synthesized enterotoxins and potentially-deleterious intermediary metabolites [86], and overexpress key stress response regulators, presumably triggered by the abnormal retention of toxic cellular products [87]. Similar transport mechanisms operate in many other bacteria [80].

4. Enzymes That Change Substrate Specificity during Stress Conditions

A third important aspect of metabolism during stress conditions does not concern the direct chemical modification of metabolites, but affects them indirectly by altering enzyme function. Upon damage, several enzymes decrease in substrate specificity and simultaneously increase in promiscuity. When enzymes are modified, novel subsidiary activities and interactions can arise concomitantly from the altered proteins, enabling additional metabolic reactions, some of potential physiological significance [70,88]. It is a general property of metabolism that most metabolic enzymes are not as specific as sometimes depicted in textbooks, and are in fact error-prone or ambiguous in their function [89,90] (Figure 3) (The concept of "enzyme promiscuity" has been massively used recently, but in different contexts and with different meanings [91]. The classification that we make below follows that by Hult and Berglund [91].).

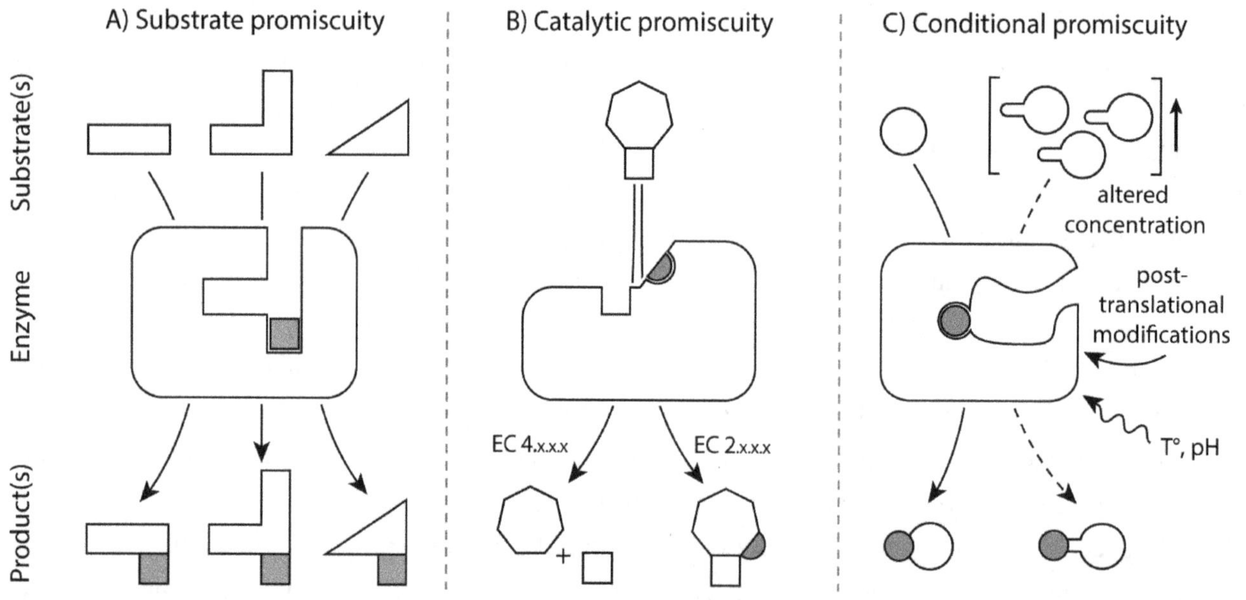

Figure 3. Mechanistic classification of enzyme promiscuity. (**A**) Substrate promiscuity or multispecificity: A certain enzyme can perform the same catalytic reaction on a diverse set of substrates indistinctly; (**B**) Catalytic promiscuity: Different chemical transformations are allowed by the same enzyme, according to which this can be classified with various E.C. numbers. (**C**) Conditional promiscuity: Latent secondary activities of an enzyme might gain activity in response to environmental changes, e.g., due to an increase in the concentration of substrate analogs with lower affinity for the enzyme, or by post-translational signals related with induced conformational changes. The three examples shown are merely illustrative but inspired by transaminase TyrB, cytosine methyltransferase and thymidine kinase, respectively.

Most enzymes accept multiple substrates and possess affinities for a wider range of compounds, a phenomenon referred to as substrate promiscuity (Figure 3A) (Some authors consider that the capacity of an enzyme to catalyze various natural substrates may be more conveniently described by the term "substrate multispecificity" or "cross-reactivity", so that the term "substrate promiscuity" is reserved to adventitious secondary activities different from the one the enzyme has evolved: see [90,92].). A recent example of notable attention is the enzyme TP53-induced glycolysis and apoptosis regulator (TIGAR), an enzyme that possesses an important role in cancer cell metabolism. TIGAR was first identified as fructose-2,6-bisphosphatase (F26BPase) [93], but it accepts multiple substrates. TIGAR shows its highest activity as phosphoglycolate-independent 2,3-bisphosphoglycerate phosphatase [94], acting most efficiently on a metabolite whose function is so far unclear.

Other enzymes are able to catalyze more than one type of chemical reaction on a given set of substrates (catalytic promiscuity, Figure 3B). This is the case of certain cytosine methyltransferases that are able to catalyze cytosine methylation as well as cytosine deamination. Cytosine deamination yields thymine, implying that this catalytic activity causes mutations and could play a potential role in tumourogenesis [95].

A classical example of enzymes that possess promiscuous activities that can lead to toxic by-products are malate dehydrogenase (MDH) and isocitrate dehydrogenase (IDH1 and IDH2). MDH typically catalyzes the interconversion between malate and oxaloacetate, but possesses certain affinity towards α-ketoglutarate. Despite the much higher affinity for the canonical substrate, high intracellular levels of α-ketoglutarate promote that MDH also produces L-2-hydroxyglutarate (see conditional promiscuity, Figure 1 (v), Figure 3C) [35]. A widely distributed FAD-dependent enzyme, the L-2-hydroxyglutarate dehydrogenase, specifically converts this side-product back into α-ketoglutarate [36]. Deleterious mutations in this enzyme in humans are responsible for an inherited metabolic disease called L-2-hydroxyglutaric aciduria, attributed to the accumulation of the aberrant product [51]. An analogous detoxification system exists also for its enantiomer D-2-hydroxyglutarate [96]. This normal metabolic intermediate is overproduced by mutant forms of isocitrate dehydrogenase-1 and -2 (IDH1 and IDH2) [97]. These mutations, preferentially occurring in cancer, alter the enzyme towards an increased production of this metabolite, which is otherwise formed in a side reaction and at a low rate only. D-2-hydroxyglutarate is also found at increased levels in patients suffering from the metabolic syndrome D-2-hydroxyglutaric aciduria [98], a genetically heterogeneous neurometabolic disorder attributed to germline mutations in IDH2 [99] and mutant genotypes of the D-2-hydroxyglutarate dehydrogenase, the enzyme responsible for D-2-hydroxyglutarate repair [98].

Importantly, such secondary activities of enzymes are often latent in normal growth conditions, but can become relevant in response to environmental changes or stresses (conditional promiscuity, Figure 3C), either by: (i) an enhanced bioavailability of a low-affinity substrate analog in comparison with the concentration of the natural substrate; (ii) conformational (allosteric) changes of the enzyme induced by direct exposure to stressors; (iii) conformational changes regulated by specific post-translational modifications exerted by stress-related signaling pathways. Therefore, enzyme promiscuity must be of critical importance for an integral view of metabolic reconfiguration under stress conditions (Figure 1).

An illustrative case is represented by metalloenzymes that exhibit a tendency toward mis-metallation during stress conditions. The activity of most metalloproteins depends on the binding of the correct metal, with a mis-metallation resulting in change of reactivity, or partial to total inactivation. Bacterial ribulose 5-phosphate epimerase (Rpe) and some other enzymes that are canonically iron-dependent, actually accept both iron Fe(II) and manganese Mn(II) as cofactor [100,101]. As pointed out by Cotruvo and Stubbe [102], Fe(II)- and Mn(II)- binding affinities are comparable in these proteins, and thus the discrimination between both metals is roughly determined by the differential bioavailability. While the enzymes are hence bound to Fe(II) under normal cellular growth conditions, they appear charged with Mn(II) under oxidative stress or iron-limiting conditions. This ambiguity is regarded as an adaptive strategy to protect enzymes from irreversible inactivation and severe damage [100] (Table 1): (i) Fe(II) centers are sensitive to ROS and oxidize to Fe(III), causing enzyme inactivation [101]; (ii) free iron pools react with ROS propagating the toxic free radical formation [103]. Interestingly, the same transcriptional control mechanisms that are responsible to ensure iron sequestration and control during oxidative stress [103] are linked to Mn(II) import and the promotion of the Mn metabolism [102]. Noteworthy, Mn(II), when bound to phosphate or carbonate, can act as an efficient non-enzymatic catalyst for the superoxide disproportionation at physiological conditions [104], which can in fact rescue yeast superoxide dismutase-knockouts [105], conferring an additional role to Mn(II) as antioxidant.

Adaptive Promiscuity: When Altered Substrate Specificity Becomes Advantageous

Despite being an energetic burden to the cell, promiscuity and multi-substrate specificity is an integral biological property in metabolic networks, and is essential as it is necessary for the evolution of metabolism [106]. In some instances, the ability to process multiple substrates is beneficial to cells in stress situations (Table 1). Perhaps one of the most illustrative examples is the genetic code "ambiguity". Translational fidelity, and thus an adequate protein biosynthesis, relies on the accurate aminoacylation of the different tRNA molecules by specific aminoacyl-tRNA synthetases and on unambiguous codon-anticodon pairing within the ribosome. Yet a certain degree of permissiveness is known to exist due to nucleobase wobbling, accounting for the degeneracy in the genetic code. Indeed, numerous exceptions and variants to the standard decoding have been reported in different organisms [107–109]. Netzer *et al.* [110] demonstrated how the genetic code accuracy can also change within the lifespan of an individual cell, as a consequence of induced changes in the acylation fidelity of aminoacyl-tRNA synthetases. A small but significant amount of methionine (Met) residues (approximately 1% of all coded Met) misacylate to non-cognate tRNA in normal unstressed mammalian cells. This fraction sharply increases (to approximately 13%) under ROS-related stress conditions, for example when exposing cultured cells to viral infections, bacterial wall factors or physico-chemical stressors. This results in proteins with high content in mistranslated Met. As discussed above, Met is a biological scavenger of ROS, proposing an adaptive role of methionyl-tRNA synthetase (MetRS) promiscuity and controlled Met-misacylation in coping with oxidative damage (Table 1). A recent study [111] reveals that the modulation of MetRS specificity is mediated by particular serine phosphorylations driven by the protein kinase ERK, which is a common mediator in the signaling response to stress. This represents a further indication about the *in vivo* protective role of Met-misincorporation for cell survival and growth under ROS stress.

The antioxidant function of Met residues might be also the cause of AUA codon reassignment in mitochondria, from isoleucine (Ile) to Met, which is widespread in vertebrates and also occurs in several invertebrates with a highly aerobic metabolism (like insects) and in some unicellular eukaryotes, including *Saccharomyces cerevisiae* [112]. With the respiratory chain as continuous source of ROS, one might hypothesize that a permanent redefinition of the genetic code (with one extra codon assigned to Met, and 2 instead of 3 codons for Ile) could have been a more suitable evolutionary adaptation in mitochondria than an inducible mechanism to tune promiscuous methionylation activities on non-methionyl tRNAs. Integral proteins of the mitochondrial inner membrane are overloaded with Met on their exposed surfaces (a classical preferential position for Ile residues), and in this way a significant containment of oxidative damage is achieved [112].

Table 1. Possible physiological benefits of various promiscuous and redundant activities induced in response to stress or environmental changes.

Stress	Enzyme	Induced Activity	Alleged Function	Reference
Oxidative stress	Mammalian MetRS	Mismethionylation of non-cognate tRNAs	Protect enzymes from inactivation	[110]
Biotic interaction	SerRS/LeuRS from *Candida*	Misleucinylation of Ser-tRNA	Increased invasivity, immune invisibility	[113]
Biotic interaction, anaerobiosis, starvation...	Bacterial transaminases	Redundancy in Ala & Lys biosynthesis	Robust peptidoglycan biosynthesis	[114]
Oxidative stress	Bacterial Pentosephosphate epimerase (Rpe)	Mismetallation with Mn(II)	Protect enzymes from inactivation	[102]

Another case in which ambiguous tRNA acylation has been proven beneficial is that of the fungus *Candida albicans* [115] (Table 1). Here the codon CUG can be translated both as a leucine (Leu) and as a serine (Ser) (with 5% and 95% frequency, respectively). This duality arises from the capacity of two aminoacyl-tRNA synthetases (LeuRS and SerRS) to recognize the particular unique serine tRNA (CAG) and allows statistically distributed populations of different mistranslated proteins from the same gene. This increase in phenotypic diversity is exploited in pathogenicity [113]. In fact, genes encoding cell surface adhesins are especially rich in CUG codons: partial CUG mistranslation as Leu results in morphological variations and heterogeneity in the cell wall that modify *Candida* interactions with the host microenvironment. In particular, experimental evidence shows that adherence capacities improve while recognition by the hosts immune system decreases [113], both being related with *Candida* virulence.

5. Changing Activity and Metabolic Switches: Stress Situations That Affect Flux Distribution

Enzyme activities are tunable and the degree of affinity and/or efficiency (K_m, k_{cat}) for different metabolites may change according to conditions [91] (Figure 3C). This also applies during the individual lifetime of an enzyme, e.g., due to age related conformational alterations [116]. As a consequence, aside from the immediate effects of protein damage on the enzymes natural function and its promiscuous activities, the inactivation of a metabolic enzyme commonly involves an adaptation in total metabolic flux and/or results in a redistribution of flux toward other pathways. Due to its redox capacity, cysteine plays a pivotal role in many enzyme catalytic mechanisms, and the redox sensitivity of this amino acid renders several enzymes vulnerable to redox imbalances and ROS accumulation [27,117,118]. Interestingly, this redox sensitivity is exploited by cells to intensify metabolic protection in stress situations. In yeast, two enzymes of glycolysis, glyceraldehyde-3-phosphate dehydrogenase (GAPDH) and pyruvate kinase (PK), have been shown to be sensitive to oxidation, leading to a rapidly induced (in seconds-scale) but temporal glycolytic block [119,120]. Termed the glycolysis/PPP transition, the oxidative part of the pentose phosphate pathway (PPP) that is more redox resistant, increases in activity, resulting in enhanced NADPH production under the conditions where this cofactor is required by the anti-oxidant machinery [119,121]. The inactivation of GAPDH during oxidative stress is not only an adverse side reaction; the enzyme possesses evolved structural features that compete for free H_2O_2 to assure rapid inactivation when levels of this metabolite rise [122]). Increased PPP activity and stress resistance augmentation in yeast are also observed when the activities of triosephosphate isomerase [123] and pyruvate kinase [124,125] are reduced by genetic modification, implying that increased NADPH production in the oxidative PPP is a recurrent consequence to blocks in glycolysis. Recently, time-resolved metabolomics has revealed this mechanism to be conserved in mammalian skin cells, revealing however that the mechanistic trigger of the glycolytic/PPP transition is not yet understood [126]. In this context, it is however important to keep in mind that the contribution of distinct metabolic pathways, including the oxidative PPP, to the total NADPH pool differs between yeast and mammalian cell types, and even between tissues [127], so that the individual cells and organisms may distribute NADPH production in a different manner.

Another oxidant-induced change in metabolic flux distribution is connected to the polyamine metabolites spermine, spermidine and putrescine. These polycations are implicated in a multitude of cellular functions, including the stress response [128]. Treatment with peroxides affects the intracellular concentrations of spermidine and putrescine in correlation with the expression of the polyamine transporter *TPO1* [129]. Yeast cells overexpressing this transporter are sensitive to oxidants and are unable to induce canonical stress response proteins, such as HSP70 or HSP104. Effects of the transporter overexpression on metabolite levels are only detected upon application of the oxidant; stress-induced transport of these metabolites thus appears critical for the interaction of the metabolome and the proteome in stress situations [129]. In this respect, polyamine catabolism has attracted attention as a potential contributor to cytotoxicity, with H_2O_2 being a product of polyamine oxidation [130]. Studies in human breast cancer cell lines demonstrate the involvement of the polyamine degradative enzyme spermine oxidase SMO (PAOh1) in the production of H_2O_2 upon induction by a spermine analogue [131]. Overexpressing SMO leads to sub-lethal DNA damage and this confers sensitivity to radiation [132,133].

6. Evolution and the Impact of Metabolic Inaccuracy on the Genotype to Phenotype Relationship

Carbonell *et al.* [134] implies that promiscuous enzymes are widespread in the phylogenetic tree of life and appear statistically more shared across species (allegedly more ancient) than specific ones, consistent with the hypothesis of an evolutionary trend toward the specialization of enzymes [135,136]. The different levels of specificity of current metabolic enzymes would have evolved by segmental duplication of previously promiscuous protein-coding genes. Enzymes would diverge at different rates depending on trade-offs and selective pressures over their particular function and the environmental context [106]. Interestingly, the content in promiscuous and unspecific enzymes differs between organisms and correlates with their ecology; taxonomic groups facing stronger environmental changes possess more promiscuous enzymes [134]. Recently, evidence has been provided that a fraction of latent enzymatic side activities maps to functional products of the existing metabolic network, and confers advantage under novel growth environments [137].

Also metabolite repair—potentially in the form of much more rudimentary mechanisms—might date back to the very early stages of evolution and could already have been important for the development of metabolism during the origin of life [138], as first metabolic networks likely required mechanisms to achieve a minimally robust functioning [139]. During subsequent evolution, non-enzymatic reactions and enzyme promiscuity did contribute to the increase in the phenotypic variability of a cell. Persisting in modern cells, metabolic ambiguity is to be added to the sources of variation at the transcriptional and translational levels, required to understand the genotype-to-phenotype relationship (Figure 4): During the flow of information from gene over protein to metabolite, errors and variability are propagated, increasing the spectrum of phenotypes to be reached from the same genotype. Whereas most of these variants may involve a fitness cost for a well-adapted organism, flexibility facilitates adaptation when conditions change (Figure 4).

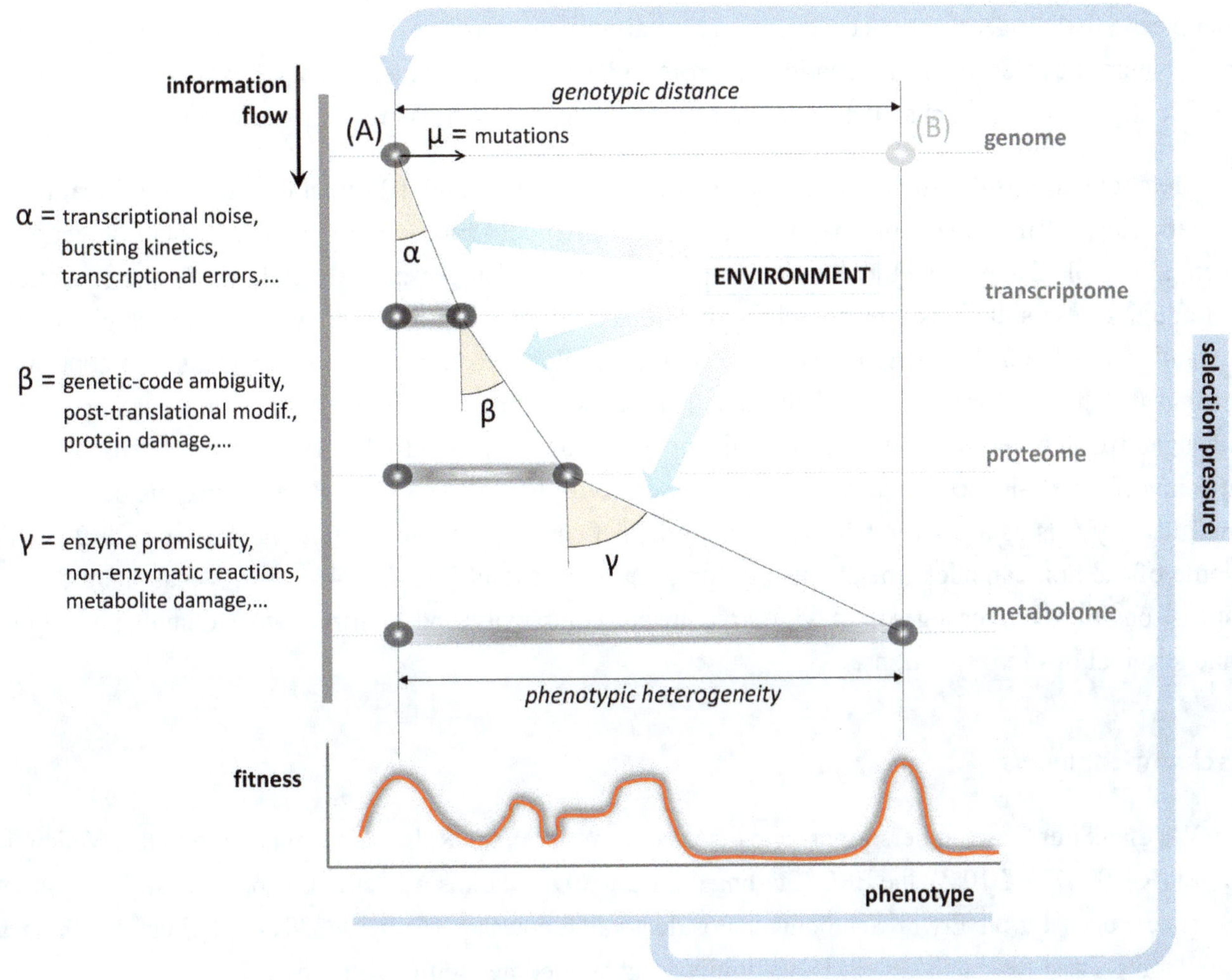

Figure 4. Impact of metabolic inaccuracy on the genotype-to-phenotype relationship. Biological processes, including transcription, translation and metabolism are not totally accurate. During the gene-protein-metabolite information flow, errors generated at each operational layer that are not compensated (depicted as angles α, β, γ) propagate, overall shaping the phenotypic space to be achieved from the same genotype (variation in the x-axis). Phenotypic variability arising due to error propagation can be enhanced by environmental stress conditions and enables a wide space of positions in the fitness landscape (red curve), all of which contribute to define the selection pressure over the given genotype (**A**). Promiscuity, noise, and unspecificity may thus facilitate the physiological adaptation to changing conditions, which would otherwise require the evolution to a different genotype (**B**).

7. Conclusions

With the advent of systems biology and analytical techniques that allow analyzing multiple cellular metabolites in parallel, it has become clear that metabolism is constituted by a highly dynamic network. In this short review, we discuss examples that reveal the impact of stress conditions on metabolites, metabolic reactions and metabolic flux. Non-enzymatic reactions, chemical reactivity between metabolites, and enzyme promiscuity are inevitable consequence of the chemical properties of essential metabolites, and mechanistic and structural constraints of the participating enzymes. In stress conditions, these factors play synergetic roles in the propagation of errors and damage within metabolism, but also form the basis to reconfigure metabolism so that cells can adapt to changing or stress-inducing environments [90,140].

Metabolism in stress situations appears to be mainly affected by: (i) direct chemical modification of metabolites; or (ii) damage on enzymes that change their substrate specificity and/or catalytic activity. Temporary blockage of metabolic pathways translates as changes in the metabolic flux distribution. Although stress situations appear to be deleterious in the first instance, they have been exploited by evolution to extend the metabolic network, and to develop dynamic response mechanism to cope with changes in the environment. This includes either the use of promiscuous metabolites in signaling, so that repair and damage responses are initiated and function as stress sensors [10,27,141], or direct defense systems that involve stress-induced redistribution of the metabolic flux [70,119,120]. Finally, enzyme conditional promiscuity acts as a reservoir of new functions that can evolve into tunable mechanisms [142], with some of the non-canonical metabolites turning out to be useful [70,106,143–145]. Metabolism during stress conditions thus changes in activity, structure and complexity; and contributes to evolution, phenotype, and survival in stress situations.

Acknowledgments

We thank our lab-members for critical discussion. Work in the Ralser lab is funded from the Wellcome Trust (RG 093735/Z/10/Z), the ERC (Starting grant 260809). Markus A. Keller is supported by the Austrian Science Funds by an Erwin Schrödinger postdoctoral fellowship (FWF, J 3341). Markus Ralser is a Wellcome Trust Research Career Development and Wellcome-Beit Prize fellow.

Author Contributions

Gabriel Piedrafita, Markus A Keller and Markus Ralser have written this Review article.

References

1. Van Houten, B. Nucleotide excision repair in *Escherichia coli*. *Microbiol. Rev.* **1990**, *54*, 18–51.
2. Mary, J.; Vougier, S.; Picot, C.R.; Perichon, M.; Petropoulos, I.; Friguet, B. Enzymatic reactions involved in the repair of oxidized proteins. *Exp. Gerontol.* **2004**, *39*, 1117–1123.
3. Perrone, G.G.; Tan, S.-X.; Dawes, I.W. Reactive oxygen species and yeast apoptosis. *Biochim. Biophys. Acta* **2008**, *1783*, 1354–1368.

4. Morano, K.A.; Grant, C.M.; Moye-Rowley, W.S. The response to heat shock and oxidative stress in *Saccharomyces cerevisiae*. *Genetics* **2012**, *190*, 1157–1195.
5. Harman, D. Aging: A theory based on free radical and radiation chemistry. *J. Gerontol.* **1956**, *11*, 298–300.
6. Sies, H. Biochemistry of oxidative stress. *Angew. Chem. Int. Ed. Engl.* **1986**, *25*, 1058–1071.
7. Liu, J.; Litt, L.; Segal, M.R.; Kelly, M.J.S.; Pelton, J.G.; Kim, M. Metabolomics of oxidative stress in recent studies of endogenous and exogenously administered intermediate metabolites. *Int. J. Mol. Sci.* **2011**, *12*, 6469–6501.
8. Noctor, G.; Lelarge-Trouverie, C.; Mhamdi, A. The metabolomics of oxidative stress. *Phytochemistry* **2014**, *112*, 33–53.
9. Ott, M.; Gogvadze, V.; Orrenius, S.; Zhivotovsky, B. Mitochondria, oxidative stress and cell death. *Apoptosis* **2007**, *12*, 913–922.
10. D'Autréaux, B.; Toledano, M.B. ROS as signalling molecules: Mechanisms that generate specificity in ROS homeostasis. *Nat. Rev. Mol. Cell Biol.* **2007**, *8*, 813–824.
11. Sies, H. Oxidative stress: A concept in redox biology and medicine. *Redox Biol.* **2015**, *4C*, 180–183.
12. Valko, M.; Morris, H.; Cronin, M.T.D. Metals, toxicity and oxidative stress. *Curr. Med. Chem.* **2005**, *12*, 1161–1208.
13. Du, J.; Gebicki, J.M. Proteins are major initial cell targets of hydroxyl free radicals. *Int. J. Biochem. Cell Biol.* **2004**, *36*, 2334–2343.
14. Liu, C.-C.; Gebicki, J.M. Intracellular GSH and ascorbate inhibit radical-induced protein chain peroxidation in HL-60 cells. *Free Radic. Biol. Med.* **2012**, *52*, 420–426.
15. Davies, K.J. Oxidative stress, antioxidant defenses, and damage removal, repair, and replacement systems. *IUBMB Life* **2000**, *50*, 279–289.
16. Halliwell, B.; Gutteridge, J. *Free Radicals in Biology and Medicine*; 4th ed.; Oxford University Press: Oxford, OX, USA, 2007.
17. Breen, A.P.; Murphy, J.A. Reactions of oxyl radicals with DNA. *Free Radic. Biol. Med.* **1995**, *18*, 1033–1077.
18. Gilbert, B.C.; King, D.M.; Thomas, C.B. Radical reactions of carbohydrates. Part 2. An electron spin resonance study of the oxidation of D-glucose and related compounds with the hydroxyl radical. *J. Chem. Soc. Perkin Trans.* **1981**, *2*, 1186–1199.
19. Manini, P.; la Pietra, P.; Panzella, L.; Napolitano, A.; d'Ischia, M. Glyoxal formation by Fenton-induced degradation of carbohydrates and related compounds. *Carbohydr. Res.* **2006**, *341*, 1828–1833.
20. Arteel, G.E.; Briviba, K.; Sies, H. Protection against peroxynitrite. *FEBS Lett.* **1999**, *445*, 226–230.
21. Sawa, T.; Akaike, T.; Maeda, H. Tyrosine nitration by peroxynitrite formed from nitric oxide and superoxide generated by xanthine oxidase. *J. Biol. Chem.* **2000**, *275*, 32467–32474.
22. Klotz, L.-O.; Schroeder, P.; Sies, H. Peroxynitrite signaling: Receptor tyrosine kinases and activation of stress-responsive pathways. *Free Radic. Biol. Med.* **2002**, *33*, 737–743.
23. Klotz, L.-O.; Sies, H. Defenses against peroxynitrite: Selenocompounds and flavonoids. *Toxicol. Lett.* **2003**, *140–141*, 125–132.
24. Williams, T.I.; Lynn, B.C.; Markesbery, W.R.; Lovell, M.A. Increased levels of 4-hydroxynonenal and acrolein, neurotoxic markers of lipid peroxidation, in the brain in Mild Cognitive Impairment and early Alzheimer's disease. *Neurobiol. Aging* **2006**, *27*, 1094–1099.

25. Pamplona, R. Membrane phospholipids, lipoxidative damage and molecular integrity: A causal role in aging and longevity. *Biochim. Biophys. Acta* **2008**, *1777*, 1249–1262.
26. Thanan, R.; Oikawa, S.; Hiraku, Y.; Ohnishi, S.; Ma, N.; Pinlaor, S.; Yongvanit, P.; Kawanishi, S.; Murata, M. Oxidative stress and its significant roles in neurodegenerative diseases and cancer. *Int. J. Mol. Sci.* **2015**, *16*, 193–217.
27. Cremers, C.M.; Jakob, U. Oxidant sensing by reversible disulfide bond formation. *J. Biol. Chem.* **2013**, *288*, 26489–26496.
28. Moskovitz, J. Methionine sulfoxide reductases: Ubiquitous enzymes involved in antioxidant defense, protein regulation, and prevention of aging-associated diseases. *Biochim. Biophys. Acta* **2005**, *1703*, 213–219.
29. Stadtman, E.R.; van Remmen, H.; Richardson, A.; Wehr, N.B.; Levine, R.L. Methionine oxidation and aging. *Biochim. Biophys. Acta* **2005**, *1703*, 135–140.
30. O'Donnell-Tormey, J.; Nathan, C.F.; Lanks, K.; DeBoer, C.J.; de la Harpe, J. Secretion of pyruvate. An antioxidant defense of mammalian cells. *J. Exp. Med.* **1987**, *165*, 500–514.
31. Troxell, B.; Zhang, J.-J.; Bourret, T.J.; Zeng, M.Y.; Blum, J.; Gherardini, F.; Hassan, H.M.; Yang, X.F. Pyruvate protects pathogenic spirochetes from H_2O_2 killing. *PLoS ONE* **2014**, *9*, e84625.
32. Rose, R.C.; Bode, A.M. Biology of free radical scavengers: An evaluation of ascorbate. *FASEB J.* **1993**, *7*, 1135–1142.
33. Levine, R.L.; Mosoni, L.; Berlett, B.S.; Stadtman, E.R. Methionine residues as endogenous antioxidants in proteins. *Proc. Natl. Acad. Sci. USA* **1996**, *93*, 15036–15040.
34. Luo, S.; Levine, R.L. Methionine in proteins defends against oxidative stress. *FASEB J.* **2009**, *23*, 464–472.
35. Rzem, R.; Vincent, M.-F.; van Schaftingen, E.; Veiga-da-Cunha, M. L-2-hydroxyglutaric aciduria, a defect of metabolite repair. *J. Inherit. Metab. Dis.* **2007**, *30*, 681–689.
36. Linster, C.L.; van Schaftingen, E.; Hanson, A.D. Metabolite damage and its repair or pre-emption. *Nat. Chem. Biol.* **2013**, *9*, 72–80.
37. Galperin, M.Y.; Moroz, O.V.; Wilson, K.S.; Murzin, A.G. House cleaning, a part of good housekeeping. *Mol. Microbiol.* **2006**, *59*, 5–19.
38. Van Schaftingen, E.; Rzem, R.; Marbaix, A.; Collard, F.; Veiga-da-Cunha, M.; Linster, C.L. Metabolite proofreading, a neglected aspect of intermediary metabolism. *J. Inherit. Metab. Dis.* **2013**, *36*, 427–434.
39. Ursini, F.; Maiorino, M.; Brigelius-Flohé, R.; Aumann, K.D.; Roveri, A.; Schomburg, D.; Flohé, L. Diversity of glutathione peroxidases. *Methods Enzymol.* **1995**, *252*, 38–53.
40. Labunskyy, V.M.; Hatfield, D.L.; Gladyshev, V.N. Selenoproteins: Molecular pathways and physiological roles. *Physiol. Rev.* **2014**, *94*, 739–777.
41. Galano, A.; Alvarez-Idaboy, J.R. Glutathione: Mechanism and kinetics of its non-enzymatic defense action against free radicals. *RSC Adv.* **2011**, *1*, 1763–1771.
42. Haenen, G.R.M.M.; Bast, A. Glutathione revisited: A better scavenger than previously thought. *Front. Pharmacol.* **2014**, doi:10.3389/fphar.2014.00260.
43. Stadtman, E.R. Oxidation of free amino acids and amino acid residues in proteins by radiolysis and by metal-catalyzed reactions. *Annu. Rev. Biochem.* **1993**, *62*, 797–821.
44. Lee, B.C.; Gladyshev, V.N. The biological significance of methionine sulfoxide stereochemistry. *Free Radic. Biol. Med.* **2011**, *50*, 221–227.

45. Boschi-Muller, S.; Olry, A.; Antoine, M.; Branlant, G. The enzymology and biochemistry of methionine sulfoxide reductases. *Biochim. Biophys. Acta* **2005**, *1703*, 231–238.
46. Lin, Z.; Johnson, L.C.; Weissbach, H.; Brot, N.; Lively, M.O.; Lowther, W.T. Free methionine-(*R*)-sulfoxide reductase from *Escherichia coli* reveals a new GAF domain function. *Proc. Natl. Acad. Sci. USA* **2007**, *104*, 9597–9602.
47. Le, D.T.; Lee, B.C.; Marino, S.M.; Zhang, Y.; Fomenko, D.E.; Kaya, A.; Hacioglu, E.; Kwak, G.-H.; Koc, A.; Kim, H.-Y.; *et al.* Functional analysis of free methionine-*R*-sulfoxide reductase from *Saccharomyces cerevisiae*. *J. Biol. Chem.* **2009**, *284*, 4354–4364.
48. Le, D.T.; Tarrago, L.; Watanabe, Y.; Kaya, A.; Lee, B.C.; Tran, U.; Nishiyama, R.; Fomenko, D.E.; Gladyshev, V.N.; Tran, L.-S.P. Diversity of plant methionine sulfoxide reductases B and evolution of a form specific for free methionine sulfoxide. *PLoS ONE* **2013**, *8*, e65637.
49. Heurlier, K.; Vendeville, A.; Halliday, N.; Green, A.; Winzer, K.; Tang, C.M.; Hardie, K.R. Growth deficiencies of *Neisseria meningitidis pfs* and *luxS* mutants are not due to inactivation of quorum sensing. *J. Bacteriol.* **2009**, *191*, 1293–1302.
50. Lyu, L.-D.; Tang, B.-K.; Fan, X.-Y.; Ma, H.; Zhao, G.-P. Mycobacterial MazG safeguards genetic stability via housecleaning of 5-OH-dCTP. *PLoS Pathog.* **2013**, *9*, e1003814.
51. Van Schaftingen, E.; Rzem, R.; Veiga-da-Cunha, M. L-2-Hydroxyglutaric aciduria, a disorder of metabolite repair. *J. Inherit. Metab. Dis.* **2009**, *32*, 135–142.
52. Nakabeppu, Y. Cellular levels of 8-oxoguanine in either DNA or the nucleotide pool play pivotal roles in carcinogenesis and survival of cancer cells. *Int. J. Mol. Sci.* **2014**, *15*, 12543–12557.
53. Marinaki, A.M.; Duley, J.A.; Arenas, M.; Ansari, A.; Sumi, S.; Lewis, C.M.; Shobowale-Bakre, M.; Fairbanks, L.D.; Sanderson, J. Mutation in the ITPA gene predicts intolerance to azathioprine. *Nucleosides. Nucleotides Nucleic Acids* **2004**, *23*, 1393–1397.
54. Sakamoto, T.; Maebayashi, K.; Nakagawa, Y.; Imai, H. Deletion of the four phospholipid hydroperoxide glutathione peroxidase genes accelerates aging in *Caenorhabditis elegans*. *Genes Cells* **2014**, *19*, 778–792.
55. Inoué, S.; Honda, K.; Komoda, Y. Sleep as neuronal detoxification and restitution. *Behav. Brain Res.* **1995**, *69*, 91–96.
56. Delaney, S.; Jarem, D.A.; Volle, C.B.; Yennie, C.J. Chemical and biological consequences of oxidatively damaged guanine in DNA. *Free Radic. Res.* **2012**, *46*, 420–441.
57. Kamiya, H. Mutagenic potentials of damaged nucleic acids produced by reactive oxygen/nitrogen species: Approaches using synthetic oligonucleotides and nucleotides: Survey and summary. *Nucleic Acids Res.* **2003**, *31*, 517–531.
58. Bessman, M.J.; Frick, D.N.; O'Handley, S.F. The MutT proteins or "Nudix" hydrolases, a family of versatile, widely distributed, "housecleaning" enzymes. *J. Biol. Chem.* **1996**, *271*, 25059–25062.
59. Sekiguchi, M.; Tsuzuki, T. Oxidative nucleotide damage: Consequences and prevention. *Oncogene* **2002**, *21*, 8895–8904.
60. McLennan, A.G. The Nudix hydrolase superfamily. *Cell. Mol. Life Sci.* **2006**, *63*, 123–143.
61. Persson, R.; Cedergren-Zeppezauer, E.S.; Wilson, K.S. Homotrimeric dUTPases; structural solutions for specific recognition and hydrolysis of dUTP. *Curr. Protein Pept. Sci.* **2001**, *2*, 287–300.
62. Hemsworth, G.R.; González-Pacanowska, D.; Wilson, K.S. On the catalytic mechanism of dimeric dUTPases. *Biochem. J.* **2013**, *456*, 81–88.

63. Marbaix, A.Y.; Noël, G.; Detroux, A.M.; Vertommen, D.; van Schaftingen, E.; Linster, C.L. Extremely conserved ATP- or ADP-dependent enzymatic system for nicotinamide nucleotide repair. *J. Biol. Chem.* **2011**, *286*, 41246–41252.
64. Oppenheimer, N.J.; Kaplan, N.O. Glyceraldehyde-3-phosphate dehydrogenase catalyzed hydration of the 5–6 double bond of reduced β-nicotinamide adenine dinucleotide (βNADH). Formation of β-6-hydroxy-1,4,5,6-tetrahydronicotinamide adenine dinucleotide. *Biochemistry* **1974**, *13*, 4685–4694.
65. Yoshida, A.; Dave, V. Inhibition of NADP-dependent dehydrogenases by modified products of NADPH. *Arch. Biochem. Biophys.* **1975**, *169*, 298–303.
66. Niehaus, T.D.; Richardson, L.G.L.; Gidda, S.K.; ElBadawi-Sidhu, M.; Meissen, J.K.; Mullen, R.T.; Fiehn, O.; Hanson, A.D. Plants utilize a highly conserved system for repair of NADH and NADPH hydrates. *Plant Physiol.* **2014**, *165*, 52–61.
67. Colinas, M.; Shaw, H.V.; Loubéry, S.; Kaufmann, M.; Moulin, M.; Fitzpatrick, T.B. A pathway for repair of NAD(P)H in plants. *J. Biol. Chem.* **2014**, *289*, 14692–14706.
68. Marbaix, A.Y.; Tyteca, D.; Niehaus, T.D.; Hanson, A.D.; Linster, C.L.; van Schaftingen, E. Occurrence and subcellular distribution of the NADPHX repair system in mammals. *Biochem. J.* **2014**, *460*, 49–58.
69. Keller, M.A.; Piedrafita, G.; Ralser, M. The widespread role of non-enzymatic reactions in cellular metabolism. *Curr. Opin. Biotechnol.* **2015**, *34*, 153–161.
70. D'Ari, R.; Casadesús, J. Underground metabolism. *Bioessays* **1998**, *20*, 181–186.
71. Keller, M.A.; Turchyn, A.V.; Ralser, M. Non-enzymatic glycolysis and pentose phosphate pathway-like reactions in a plausible Archean ocean. *Mol. Syst. Biol.* **2014**, doi:10.1002/msb.20145228.
72. Weinert, B.T.; Iesmantavicius, V.; Moustafa, T.; Schölz, C.; Wagner, S.A.; Magnes, C.; Zechner, R.; Choudhary, C. Acetylation dynamics and stoichiometry in *Saccharomyces cerevisiae*. *Mol. Syst. Biol.* **2014**, doi:10.1002/msb.134766.
73. Wagner, G.R.; Payne, R.M. Widespread and enzyme-independent Nε-acetylation and Nε-succinylation of proteins in the chemical conditions of the mitochondrial matrix. *J. Biol. Chem.* **2013**, *288*, 29036–29045.
74. Wagner, G.R.; Hirschey, M.D. Nonenzymatic protein acylation as a carbon stress regulated by sirtuin deacylases. *Mol. Cell* **2014**, *54*, 5–16.
75. Zhang, Y.; Xiao, M.; Horiyama, T.; Zhang, Y.; Li, X.; Nishino, K.; Yan, A. The multidrug efflux pump MdtEF protects against nitrosative damage during the anaerobic respiration in *Escherichia coli*. *J. Biol. Chem.* **2011**, *286*, 26576–26584.
76. Zhou, K.; Zou, R.; Stephanopoulos, G.; Too, H.-P. Metabolite profiling identified methylerythritol cyclodiphosphate efflux as a limiting step in microbial isoprenoid production. *PLoS ONE* **2012**, *7*, e47513.
77. Fung, D.K.C.; Lau, W.Y.; Chan, W.T.; Yan, A. Copper efflux is induced during anaerobic amino acid limitation in *Escherichia coli* to protect iron-sulfur cluster enzymes and biogenesis. *J. Bacteriol.* **2013**, *195*, 4556–4568.
78. Piddock, L.J.V. Clinically relevant chromosomally encoded multidrug resistance efflux pumps in bacteria. *Clin. Microbiol. Rev.* **2006**, *19*, 382–402.
79. Nikaido, H.; Zgurskaya, H.I. AcrAB and related multidrug efflux pumps of *Escherichia coli*. *J. Mol. Microbiol. Biotechnol.* **2001**, *3*, 215–218.
80. Li, X.-Z.; Nikaido, H. Efflux-mediated drug resistance in bacteria: An update. *Drugs* **2009**, *69*, 1555–1623.

81. Bleuel, C.; Grosse, C.; Taudte, N.; Scherer, J.; Wesenberg, D.; Krauss, G.J.; Nies, D.H.; Grass, G. TolC is involved in enterobactin efflux across the outer membrane of *Escherichia coli*. *J. Bacteriol.* **2005**, *187*, 6701–6707.
82. Du, D.; Wang, Z.; James, N.R.; Voss, J.E.; Klimont, E.; Ohene-Agyei, T.; Venter, H.; Chiu, W.; Luisi, B.F. Structure of the AcrAB-TolC multidrug efflux pump. *Nature* **2014**, *509*, 512–515.
83. Takatsuka, Y.; Chen, C.; Nikaido, H. Mechanism of recognition of compounds of diverse structures by the multidrug efflux pump AcrB of *Escherichia coli*. *Proc. Natl. Acad. Sci. USA* **2010**, *107*, 6559–6565.
84. Poole, K. Bacterial multidrug efflux pumps serve other functions. *Microbe* **2008**, *3*, 179–185.
85. Deininger, K.N.W.; Horikawa, A.; Kitko, R.D.; Tatsumi, R.; Rosner, J.L.; Wachi, M.; Slonczewski, J.L. A requirement of TolC and MDR efflux pumps for acid adaptation and GadAB induction in *Escherichia coli*. *PLoS ONE* **2011**, *6*, e18960.
86. Tatsumi, R.; Wachi, M. TolC-dependent exclusion of porphyrins in *Escherichia coli*. *J. Bacteriol.* **2008**, *190*, 6228–6233.
87. Rosner, J.L.; Martin, R.G. Reduction of cellular stress by TolC-dependent efflux pumps in *Escherichia coli* indicated by BaeSR and CpxARP activation of spy in efflux mutants. *J. Bacteriol.* **2013**, *195*, 1042–1050.
88. Nagy, P.; Lechte, T.P.; Das, A.B.; Winterbourn, C.C. Conjugation of glutathione to oxidized tyrosine residues in peptides and proteins. *J. Biol. Chem.* **2012**, *287*, 26068–26076.
89. Bar-Even, A.; Noor, E.; Savir, Y.; Liebermeister, W.; Davidi, D.; Tawfik, D.S.; Milo, R. The moderately efficient enzyme: Evolutionary and physicochemical trends shaping enzyme parameters. *Biochemistry* **2011**, *50*, 4402–4410.
90. Khersonsky, O.; Tawfik, D.S. Enzyme promiscuity: A mechanistic and evolutionary perspective. *Annu. Rev. Biochem.* **2010**, *79*, 471–505.
91. Hult, K.; Berglund, P. Enzyme promiscuity: Mechanism and applications. *Trends Biotechnol.* **2007**, *25*, 231–238.
92. Copley, S.D. Enzymes with extra talents: Moonlighting functions and catalytic promiscuity. *Curr. Opin. Chem. Biol.* **2003**, *7*, 265–272.
93. Bensaad, K.; Tsuruta, A.; Selak, M.A.; Vidal, M.N.C.; Nakano, K.; Bartrons, R.; Gottlieb, E.; Vousden, K.H. TIGAR, a p53-inducible regulator of glycolysis and apoptosis. *Cell* **2006**, *126*, 107–120.
94. Gerin, I.; Noël, G.; Bolsée, J.; Haumont, O.; van Schaftingen, E.; Bommer, G.T. Identification of TP53-induced glycolysis and apoptosis regulator (TIGAR) as the phosphoglycolate-independent 2,3-bisphosphoglycerate phosphatase. *Biochem. J.* **2014**, *458*, 439–448.
95. Yebra, M.J.; Bhagwat, A.S. A cytosine methyltransferase converts 5-methylcytosine in DNA to thymine. *Biochemistry* **1995**, *34*, 14752–14757.
96. Engqvist, M.K.M.; Eßer, C.; Maier, A.; Lercher, M.J.; Maurino, V.G. Mitochondrial 2-hydroxyglutarate metabolism. *Mitochondrion* **2014**, *19*, 275–281.
97. Ward, P.S.; Patel, J.; Wise, D.R.; Abdel-Wahab, O.; Bennett, B.D.; Coller, H.A.; Cross, J.R.; Fantin, V.R.; Hedvat, C.V.; Perl, A.E.; *et al.* The common feature of leukemia-associated IDH1 and IDH2 mutations is a neomorphic enzyme activity converting alpha-ketoglutarate to 2-hydroxyglutarate. *Cancer Cell* **2010**, *17*, 225–234.

98. Struys, E.A.; Salomons, G.S.; Achouri, Y.; van Schaftingen, E.; Grosso, S.; Craigen, W.J.; Verhoeven, N.M.; Jakobs, C. Mutations in the D-2-hydroxyglutarate dehydrogenase gene cause D-2-hydroxyglutaric aciduria. *Am. J. Hum. Genet.* **2005**, *76*, 358–360.
99. Kranendijk, M.; Struys, E.A.; van Schaftingen, E.; Gibson, K.M.; Kanhai, W.A.; van der Knaap, M.S.; Amiel, J.; Buist, N.R.; Das, A.M.; de Klerk, J.B.; *et al.* IDH2 mutations in patients with D-2-hydroxyglutaric aciduria. *Science* **2010**, doi:10.1126/science.1192632.
100. Sobota, J.M.; Imlay, J.A. Iron enzyme ribulose-5-phosphate 3-epimerase in *Escherichia coli* is rapidly damaged by hydrogen peroxide but can be protected by manganese. *Proc. Natl. Acad. Sci. USA* **2011**, *108*, 5402–5407.
101. Anjem, A.; Imlay, J.A. Mononuclear iron enzymes are primary targets of hydrogen peroxide stress. *J. Biol. Chem.* **2012**, *287*, 15544–15556.
102. Cotruvo, J.A.; Stubbe, J. Metallation and mismetallation of iron and manganese proteins *in vitro* and *in vivo*: The class I ribonucleotide reductases as a case study. *Metallomics* **2012**, *4*, 1020–1036.
103. Frawley, E.R.; Fang, F.C. The ins and outs of bacterial iron metabolism. *Mol. Microbiol.* **2014**, *93*, 609–616.
104. Barnese, K.; Gralla, E.B.; Valentine, J.S.; Cabelli, D.E. Biologically relevant mechanism for catalytic superoxide removal by simple manganese compounds. *Proc. Natl. Acad. Sci. USA* **2012**, *109*, 6892–6897.
105. Sanchez, R.J.; Srinivasan, C.; Munroe, W.H.; Wallace, M.A.; Martins, J.; Kao, T.Y.; Le, K.; Gralla, E.B.; Valentine, J.S. Exogenous manganous ion at millimolar levels rescues all known dioxygen-sensitive phenotypes of yeast lacking CuZnSOD. *J. Biol. Inorg. Chem.* **2005**, *10*, 913–923.
106. Copley, S.D. Toward a systems biology perspective on enzyme evolution. *J. Biol. Chem.* **2012**, *287*, 3–10.
107. Söll, D. Enter a new amino acid. *Nature* **1988**, *331*, 662–663.
108. Prat, L.; Heinemann, I.U.; Aerni, H.R.; Rinehart, J.; O'Donoghue, P.; Söll, D. Carbon source-dependent expansion of the genetic code in bacteria. *Proc. Natl. Acad. Sci. USA* **2012**, *109*, 21070–21075.
109. Li, L.; Boniecki, M.T.; Jaffe, J.D.; Imai, B.S.; Yau, P.M.; Luthey-Schulten, Z.A.; Martinis, S.A. Naturally occurring aminoacyl-tRNA synthetases editing-domain mutations that cause mistranslation in *Mycoplasma parasites*. *Proc. Natl. Acad. Sci. USA* **2011**, *108*, 9378–9383.
110. Netzer, N.; Goodenbour, J.M.; David, A.; Dittmar, K.A.; Jones, R.B.; Schneider, J.R.; Boone, D.; Eves, E.M.; Rosner, M.R.; Gibbs, J.S.; *et al.* Innate immune and chemically triggered oxidative stress modifies translational fidelity. *Nature* **2009**, *462*, 522–526.
111. Lee, J.Y.; Kim, D.G.; Kim, B.-G.; Yang, W.S.; Hong, J.; Kang, T.; Oh, Y.S.; Kim, K.R.; Han, B.W.; Hwang, B.J.; *et al.* Promiscuous methionyl-tRNA synthetase mediates adaptive mistranslation to protect cells against oxidative stress. *J. Cell Sci.* **2014**, *127*, 4234–4245.
112. Bender, A.; Hajieva, P.; Moosmann, B. Adaptive antioxidant methionine accumulation in respiratory chain complexes explains the use of a deviant genetic code in mitochondria. *Proc. Natl. Acad. Sci. USA* **2008**, *105*, 16496–16501.
113. Miranda, I.; Silva-Dias, A.; Rocha, R.; Teixeira-Santos, R.; Coelho, C.; Gonçalves, T.; Santos, M.A.S.; Pina-Vaz, C.; Solis, N.V.; Filler, S.G.; *et al. Candida albicans* CUG mistranslation is a mechanism to create cell surface variation. *MBio* **2013**, *4*, doi:10.1128/mBio.00285-13.

114. Lal, P.B.; Schneider, B.L.; Vu, K.; Reitzer, L. The redundant aminotransferases in lysine and arginine synthesis and the extent of aminotransferase redundancy in *Escherichia coli*. *Mol. Microbiol.* **2014**, *94*, 843–856.
115. Miranda, I.; Rocha, R.; Santos, M.C.; Mateus, D.D.; Moura, G.R.; Carreto, L.; Santos, M.A.S. A genetic code alteration is a phenotype diversity generator in the human pathogen *Candida albicans*. *PLoS ONE* **2007**, *2*, e996.
116. Rothstein, M. Aging and the alteration of enzymes: A review. *Mech. Ageing Dev.* **1975**, *4*, 325–338.
117. Van Montfort, R.L.M.; Congreve, M.; Tisi, D.; Carr, R.; Jhoti, H. Oxidation state of the active-site cysteine in protein tyrosine phosphatase 1B. *Nature* **2003**, *423*, 773–777.
118. Lee, J.-G.; Baek, K.; Soetandyo, N.; Ye, Y. Reversible inactivation of deubiquitinases by reactive oxygen species *in vitro* and in cells. *Nat. Commun.* **2013**, doi:10.1038/ncomms2532.
119. Ralser, M.; Wamelink, M.M.; Kowald, A.; Gerisch, B.; Heeren, G.; Struys, E.A.; Klipp, E.; Jakobs, C.; Breitenbach, M.; Lehrach, H.; *et al.* Dynamic rerouting of the carbohydrate flux is key to counteracting oxidative stress. *J. Biol.* **2007**, doi:10.1186/jbiol61.
120. Ralser, M.; Wamelink, M.M.C.; Latkolik, S.; Jansen, E.E.W.; Lehrach, H.; Jakobs, C. Metabolic reconfiguration precedes transcriptional regulation in the antioxidant response. *Nat. Biotechnol.* **2009**, *27*, 604–605.
121. Stincone, A.; Prigione, A.; Cramer, T.; Wamelink, M.M.C.; Campbell, K.; Cheung, E.; Olin-Sandoval, V.; Gruening, N.-M.; Krueger, A.; Tauqeer Alam, M.; *et al.* The return of metabolism: Biochemistry and physiology of the pentose phosphate pathway. *Biol. Rev. Camb. Philos. Soc.* **2015**, *90*, 927–963.
122. Peralta, D.; Bronowska, A.K.; Morgan, B.; Dóka, É.; van Laer, K.; Nagy, P.; Gräter, F.; Dick, T.P. A proton relay enhances H_2O_2 sensitivity of GAPDH to facilitate metabolic adaptation. *Nat. Chem. Biol.* **2015**, *11*, 156–163.
123. Ralser, M.; Heeren, G.; Breitenbach, M.; Lehrach, H.; Krobitsch, S. Triose phosphate isomerase deficiency is caused by altered dimerization—Not catalytic inactivity—Of the mutant enzymes. *PLoS ONE* **2006**, *1*, e30.
124. Grüning, N.-M.; Du, D.; Keller, M.A.; Luisi, B.F.; Ralser, M. Inhibition of triosephosphate isomerase by phosphoenolpyruvate in the feedback-regulation of glycolysis. *Open Biol.* **2014**, doi:10.1098/rsob.130232.
125. Grüning, N.-M.; Rinnerthaler, M.; Bluemlein, K.; Mülleder, M.; Wamelink, M.M.C.; Lehrach, H.; Jakobs, C.; Breitenbach, M.; Ralser, M. Pyruvate kinase triggers a metabolic feedback loop that controls redox metabolism in respiring cells. *Cell Metab.* **2011**, *14*, 415–427.
126. Kuehne, A.; Emmert, H.; Soehle, J.; Winnefeld, M.; Fischer, F.; Wenck, H.; Gallinat, S.; Terstegen, L.; Lucius, R.; Hildebrand, J.; *et al.* Acute activation of oxidative pentose phosphate pathway as first-line response to oxidative stress in human skin cells. *Mol. Cell* **2015**, doi:10.1016/j.molcel.2015.06.017.
127. Fan, J.; Ye, J.; Kamphorst, J.J.; Shlomi, T.; Thompson, C.B.; Rabinowitz, J.D. Quantitative flux analysis reveals folate-dependent NADPH production. *Nature* **2014**, *510*, 298–302.
128. Miller-Fleming, L.; Olin-Sandoval, V.; Campbell, K.; Ralser M. Remaining mysteries of molecular biology: The role of polyamines in the cell. *J. Mol. Biol.* **2015**, doi:10.1016/j.jmb.2015.06.020.

129. Krüger, A.; Vowinckel, J.; Mülleder, M.; Grote, P.; Capuano, F.; Bluemlein, K.; Ralser, M. Tpo1-mediated spermine and spermidine export controls cell cycle delay and times antioxidant protein expression during the oxidative stress response. *EMBO Rep.* **2013**, *14*, 1113–1119.
130. Casero, R.A.; Pegg, A.E. Polyamine catabolism and disease. *Biochem. J.* **2009**, *421*, 323–338.
131. Pledgie, A.; Huang, Y.; Hacker, A.; Zhang, Z.; Woster, P.M.; Davidson, N.E.; Casero, R.A. Spermine oxidase SMO(PAOh1), not N1-acetylpolyamine oxidase PAO, is the primary source of cytotoxic H_2O_2 in polyamine analogue-treated human breast cancer cell lines. *J. Biol. Chem.* **2005**, *280*, 39843–39851.
132. Bianchi, M.; Bellini, A.; Cervelli, M.; Degan, P.; Marcocci, L.; Martini, F.; Scatteia, M.; Mariottini, P.; Amendola, R. Chronic sub-lethal oxidative stress by spermine oxidase overactivity induces continuous DNA repair and hypersensitivity to radiation exposure. *Biochim. Biophys. Acta* **2007**, *1773*, 774–783.
133. Amendola, R.; Cervelli, M.; Fratini, E.; Sallustio, D.E.; Tempera, G.; Ueshima, T.; Mariottini, P.; Agostinelli, E. Reactive oxygen species spermine metabolites generated from amine oxidases and radiation represent a therapeutic gain in cancer treatments. *Int. J. Oncol.* **2013**, *43*, 813–820.
134. Carbonell, P.; Lecointre, G.; Faulon, J.-L. Origins of specificity and promiscuity in metabolic networks. *J. Biol. Chem.* **2011**, *286*, 43994–44004.
135. Jensen, R.A. Enzyme recruitment in evolution of new function. *Annu. Rev. Microbiol.* **1976**, *30*, 409–425.
136. Voordeckers, K.; Brown, C.A.; Vanneste, K.; van der Zande, E.; Voet, A.; Maere, S.; Verstrepen, K.J. Reconstruction of ancestral metabolic enzymes reveals molecular mechanisms underlying evolutionary innovation through gene duplication. *PLoS Biol.* **2012**, *10*, e1001446.
137. Notebaart, R.A.; Szappanos, B.; Kintses, B.; Pal, F.; Gyorkei, A.; Bogos, B.; Lazar, V.; Spohn, R.; Csorg, B.; Wagner, A.; *et al.* Network-level architecture and the evolutionary potential of underground metabolism. *Proc. Natl. Acad. Sci. USA* **2014**, *111*, 11762–11767.
138. Cornish-Bowden, A.; Cardenas, M.L. Self-organization at the origin of life. *J. Theor. Biol.* **2008**, *252*, 411–418.
139. Piedrafita, G.; Montero, F.; Morán, F.; Cárdenas, M.L.; Cornish-Bowden, A. A simple self-maintaining metabolic system: Robustness, autocatalysis, bistability. *PLoS Comput. Biol.* **2010**, doi:10.1371/journal.pcbi.1000872.
140. Tawfik, D.S. Messy biology and the origins of evolutionary innovations. *Nat. Chem. Biol.* **2010**, *6*, 692–696.
141. Benoit, R.; Auer, M. A direct way of redox sensing. *RNA Biol.* **2011**, *8*, 18–23.
142. Andersson, D.I. Evolving promiscuously. *Proc. Natl. Acad. Sci. USA* **2011**, *108*, 1199–1200.
143. Miller, B.G.; Raines, R.T. Identifying latent enzyme activities: Substrate ambiguity within modern bacterial sugar kinases. *Biochemistry* **2004**, *43*, 6387–6392.
144. Patrick, W.M.; Quandt, E.M.; Swartzlander, D.B.; Matsumura, I. Multicopy suppression underpins metabolic evolvability. *Mol. Biol. Evol.* **2007**, *24*, 2716–2722.
145. Patrick, W.M.; Matsumura, I. A study in molecular contingency: Glutamine phosphoribosylpyrophosphate amidotransferase is a promiscuous and evolvable phosphoribosylanthranilate isomerase. *J. Mol. Biol.* **2008**, *377*, 323–336.

7

Biological Activities of Reactive Oxygen and Nitrogen Species: Oxidative Stress *vs.* Signal Transduction

Adelheid Weidinger and Andrey V. Kozlov *

Ludwig Boltzmann Institute for Experimental and Clinical Traumatology, AUVA Research Center, Donaueschingenstraße 13, 1200 Vienna, Austria; E-Mail: adelheid.weidinger@trauma.lbg.ac.at

* Author to whom correspondence should be addressed; E-Mail: Andrey.Kozlov@trauma.lbg.ac.at;

Academic Editors: Peter Eckl and Michael Breitenbach

Abstract: In the past, reactive oxygen and nitrogen species (RONS) were shown to cause oxidative damage to biomolecules, contributing to the development of a variety of diseases. However, recent evidence has suggested that intracellular RONS are an important component of intracellular signaling cascades. The aim of this review was to consolidate old and new ideas on the chemical, physiological and pathological role of RONS for a better understanding of their properties and specific activities. Critical consideration of the literature reveals that deleterious effects do not appear if only one primary species (superoxide radical, nitric oxide) is present in a biological system, even at high concentrations. The prerequisite of deleterious effects is the formation of highly reactive secondary species (hydroxyl radical, peroxynitrite), emerging exclusively upon reaction with another primary species or a transition metal. The secondary species are toxic, not well controlled, causing irreversible damage to all classes of biomolecules. In contrast, primary RONS are well controlled (superoxide dismutase, catalase), and their reactions with biomolecules are reversible, making them ideal for physiological/pathophysiological intracellular signaling. We assume that whether RONS have a signal transducing or damaging effect is primarily defined by their quality, being primary or secondary RONS, and only secondly by their quantity.

Keywords: superoxide radical; hydrogen peroxide; hydroxyl radical; nitric oxide; mitochondria

---

1. Introduction

Reactive oxygen and nitrogen species (RONS) include two classes of chemically-reactive molecules containing oxygen (reactive oxygen species, ROS) and nitrogen (reactive nitrogen species, RNS). Both classes are referred to as RONS. The majority of RONS carries unpaired electrons and is called free radicals. In mammalians, a major function of specialized enzymes, such as NADPH-oxidase, myeloperoxidase and nitric oxide synthase (NOS), is the generation of RONS. The controlled generation of RONS in the extracellular space by these enzymes was developed evolutionarily as part of the innate immune system to kill bacteria. However, an overwhelming release of RONS may also induce deleterious effects, causing damage to host biological structures. Another group of enzymes release RONS intracellularly as a byproduct of metabolic processes. For instance, superoxide ($O_2^{\bullet-}$) is released as a byproduct of mitochondrial respiration and monooxygenase activity of cytochrome p450. Intracellular RONS, as well as excessive release of extracellular RONS were thought to induce deleterious effects, causing oxidative damage to different kinds of biomolecules. These processes are referred to as “oxidative stress”.

Oxidative stress is believed to significantly contribute to the development of a number of diseases, particularly age-related diseases. However, more and more evidence suggest that intracellular generation of RONS is an important component of intracellular signaling cascades regulating several physiological functions, such as regulation of vascular tone, insulin synthesis, activation of hypoxia-inducible factor (HIF), cell proliferation, differentiation and migration. It took over 50 years for a clear understanding of the chemical basis of free radical/RONS biology to emerge. In the following 50 years, studies on the biological effects of free radicals with biological targets were undertaken. The aim of this review is to put together old and recent ideas on the chemical and pathophysiological role of RONS for a better understanding of their properties and specific activities. This review is predominantly based on selected reviews, elaborating on different aspects of RONS activity and thought to be a guide through a large body of literature existing on this topic.

2. Chemical Basics

The current knowledge on RONS biology is based on studies of free radical reactions conducted more than 100 years ago. Free radicals are defined as molecules having an unpaired electron. Their physical properties are similar to those of free electrons, giving a signal at $g = 2.0023$ in the electron paramagnetic resonance spectrum [1]. The chemical mechanisms underlying the formation and toxicity of free radicals were proposed by the British chemist Henry J. H. Fenton in 1894 and later developed by the Austrian chemist Joseph Weiss and the German chemist and Nobel Prize winner Fritz Haber in 1934. Henry J. H. Fenton showed that the formation of toxic hydroxyl radicals ($^{\bullet}OH$) from hydrogen peroxide (H_2O_2) is catalyzed by iron ions, called the “Fenton reaction” ([2], reviewed in [3]). He pointed out that iron ions are necessary to form toxic $^{\bullet}OH$ radicals. Joseph Weiss and Fritz Haber discovered that $O_2^{\bullet-}$ can be converted into H_2O_2 and further to $^{\bullet}OH$, called the Haber–Weiss reaction ([4], reviewed in [5,6]). This reaction shows that one free radical can give rise to another secondary radical. Already in those days, the transformation of one ROS ($O_2^{\bullet-}$) to another ($^{\bullet}OH$)

was associated with the presence of iron ions as a catalyst (reviewed in [7]). Later, other transition metals, such as copper ions, were shown to generate toxic RONS.

Another important step in understanding the biological function of RONS was the discovery of free radical chain reactions. This was done in 1935 by the Russian chemist and Nobel Prize winner Nikolai Semenov. He described four types of free radical reactions, namely initiation, propagation, branching and termination [8]. The same reactions occur in biological membranes upon pathological conditions and are termed lipid peroxidation, the major mechanism of oxidative damage to biological membranes. Importantly, the branching chain reaction of lipid peroxidation, the cleavage of lipid peroxides, is catalyzed by ferrous ions similar to the Fenton reaction, which is a cleavage of H_2O_2 to $^{\bullet}OH$ by ferrous ions. The branching chain reaction between lipid peroxides and iron ions accelerates lipid peroxidation [9], again suggesting that iron ions are the prerequisite for the toxic effects of lipid peroxidation.

In the 1950s, researchers started to associate free radical chemistry with biomedical questions. It has been suggested that most of the damaging effects of oxygen in living systems are due to the formation of free radicals (reviewed in [10]). This assumption promoted the application of the knowledge of free radical chemistry to biological systems.

3. Oxidative Stress

However, in 1968, a major breakthrough in the field of free radical biology was done by Irvin Fridovich who discovered superoxide dismutase (SOD), a specific enzyme catalyzing the transition of $O_2^{\bullet-}$ into H_2O_2 ([11,12], reviewed in [13]). A few years later, Chance and coauthors reported that mitochondria are the key generator of $O_2^{\bullet-}$ in cells ([14], reviewed in [15]). These two findings are crucial, as they show that free radicals, on the one hand, are produced in biological systems, and on the other hand, there is an enzymatic mechanism regulating their concentration. This clearly suggests that free radicals occur in biological systems and probably have a specific function. Since then, numerous studies have been performed to understand the biological function of free radicals.

Until the mid-1970s, the literature almost exclusively refers to free radicals. Later, it became evident that not only free radicals, but also non-radical products, such as H_2O_2 or hypochlorous acid (HOCl), which are also powerful oxidizing agents, participate in free radical reactions (reviewed in [16]). To take into account both the radical and the non-radical species, the more general term "reactive oxygen species" (ROS) was introduced. Later, nitric oxide ($NO^{\bullet}$) and peroxynitrite ($ONOO^-$) were also shown to interact with ROS, and all of these species were termed RONS. Primarily, the toxic properties of RONS were of interest. It was shown that an excessive generation of RONS damaged almost all classes of biomolecules, such as lipids [17], proteins [18] and DNA (reviewed in [19,20]). In the 1970s and 1980s, the term "oxidative stress" was used for these deleterious processes. Later, "oxidative stress" was defined by the German biochemist Helmut Sies as an imbalance between oxidants and antioxidants in favor of the oxidants, potentially leading to damage ([21], reviewed in [22]). Evolutionarily, the induction of oxidative stress was possibly developed as an important part of the innate immune system as a defense mechanism against bacteria [23]. However, it was also shown that RONS, produced by the immune system, can also damage host cells [24].

Careful observation of oxidative damage reactions of biomolecules shows that primary RONS, such as $O_2^{\bullet -}$, H_2O_2 or $NO^\bullet$, in most cases reversibly react with the target molecules. $NO^\bullet$, for instance, reversibly binds to heme proteins, whereas $O_2^{\bullet -}$ reacts with proteins, changing their redox state without damage to their structure. For instance, the reaction between $O_2^{\bullet -}$ and cytochrome c results in the reduction of heme, which is used to detect $O_2^{\bullet -}$ [25]. The damage is predominantly associated with secondary RONS, such as $^\bullet OH$, $ONOO^-$ and HOCl [26–28]. All of these toxic species are formed if more than one reactive species is present. Two major reactions leading to the formation of toxic RONS are: (i) the Fenton reaction between ferrous ions and H_2O_2 yielding $^\bullet OH$; and (ii) the reaction of $O_2^{\bullet -}$ with $NO^\bullet$ yielding $ONOO^-$. Furthermore, the formation of HOCl, formed in an enzymatic reaction from H_2O_2 and Cl^-, is associated with damage to host tissues [28]. In addition, iron ions can directly react with organic peroxides, inducing lipid peroxidation. Moreover, the presence of iron and copper ions induces double-strand breaks of DNA, a DNA damage that is difficult to repair (reviewed in [29]). Some anticancer drugs are based on mechanisms causing double-strand breaks of DNA catalyzed by transition metals [30]. Oxidative damage to proteins is often associated with the reaction between amino acids and $ONOO^-$, resulting in the formation of nitrated amino acids, such as nitrotyrosine. $ONOO^-$ is a secondary RONS formed in the reaction between $NO^\bullet$ and $O_2^{\bullet -}$. $ONOO^-$ and $NO^\bullet$ may have quite opposite biological effects. For instance, $NO^\bullet$ has an inhibitory effect on lipid peroxidation, while $ONOO^-$ activates this process (reviewed in [31]). Another important regulatory mechanism based on the interaction of $NO^\bullet$ and $O_2^{\bullet -}$ is driven by decreased $NO^\bullet$ levels, rather than by increased $ONOO^-$ levels (reviewed in [32,33]). The interaction between $NO^\bullet$ and $O_2^{\bullet -}$ might also result in vasoconstriction by inactivation of prostacyclin synthase (reviewed in [34]).

The data described above suggest that primary ROS ($O_2^{\bullet -}$, $NO^\bullet$, H_2O_2) only have a weak damaging potential, whereas secondary RONS are more toxic. Primary species are well controlled by SOD, catalase and NO synthases, while secondary species are less controllable, since there is no specific enzymatic system controlling their levels.

Interestingly, $O_2^{\bullet -}$-controlling systems are different inside the cells and in extracellular fluids. SOD, the intracellular enzyme for removing $O_2^{\bullet -}$, and the extracellular SOD (ecSOD) produce potentially dangerous H_2O_2. H_2O_2 itself is relatively inactive, but can lead to the formation of toxic $^\bullet OH$. In contrast, ceruloplasmin in the blood inactivates $O_2^{\bullet -}$, yielding H_2O [35]. However, extracellular SOD, which mainly occurs in tissue ([36,37]; reviewed in [38,39]), can also be found in plasma under specific pathological conditions [40] and contribute to the elimination of $O_2^{\bullet -}$. The fact that SOD, not ceruloplasmin, occurs in cells indirectly suggests that H_2O_2 may have a physiological function inside cells, but not in extracellular fluids. The data gathered in the last few decades suggest that primary RONS formed in mitochondria ($O_2^{\bullet -}$ and H_2O_2) and $NO^\bullet$ are associated with intracellular signaling cascades. Since NO-mediated signaling pathways have already been extensively reviewed ([41–47]), in this review, we focus on non-NO-mediated signaling pathways.

4. Signaling

There is a solid body of literature supporting the essential role of mitochondrial ROS in intracellular signaling. The data on the involvement of mitochondrial ROS in intracellular signaling pathways related to inflammation have been summarized in recent reviews [48–50]. In the last few years, the role of ROS in positive and negative regulation of insulin signaling has also been intensively studied and reviewed [51]. Furthermore, the role of mitochondrial ROS in activation of HIF has been intensively studied and debated (reviewed in [52,53]). In addition, the role of ROS has been demonstrated for NF-κB-dependent gene transcription and a number of other signaling cascades (reviewed in [54]). Notably, most of the publications on oxidative stress referred to specific types of RONS involved in oxidative damage, whereas data on signaling are predominantly addressed to ROS and RONS in general. This led to a large knowledge gap on the mechanisms of intracellular signaling concerning RONS, since is not clear whether all or only specific RONS contribute to these signaling pathways. In the following section, we will focus on studies where specific types of RONS contributing to intracellular signaling cascades were determined.

5. Superoxide Radical

A number of reports suggest $O_2^{\bullet-}$ as part of intracellular signaling cascades. This species is predominantly produced by mitochondrial complexes I and III (reviewed in [55]). Evidence of the involvement of mitochondrial $O_2^{\bullet-}$ in intracellular signaling cascades can be shown by:

1. Alteration of mitochondrial function, particularly of complexes I and III, by pharmacological or genetic modulation, which has an effect on signaling pathways.
2. The correlation of a certain $O_2^{\bullet-}$ level with an effective signaling cascade.
3. Application of mitochondria-targeted antioxidants (mtAOX) or radical scavengers has an effect on signaling pathways.
4. Genetic manipulation of mitochondrial SOD and cytoplasmic SOD decreases the efficiency of specific signaling cascades.

Mitochondrial $O_2^{\bullet-}$, the primary mitochondrial ROS, was often associated with the regulation of inflammatory pathways, such as activation of the inflammasome, regulation of inflammatory cytokines synthesis and mechanisms of innate immunity. The involvement of mitochondrial $O_2^{\bullet-}$ in the activation of the inflammasome was suggested by Zhou *et al.* [56]. The authors showed that specific inhibition of mitochondrial complexes I and III, the major sources of ROS in mitochondria, significantly diminished the activation of the "nucleotide-binding domain, leucine-rich family and pyrin domain containing 3" (NLRP3) inflammasome, suggesting that mitochondrial $O_2^{\bullet-}$ is involved in this signaling cascade. Another important feature of inflammation is the release of cytokines. Bulua *et al.* [57] showed that LPS-stimulated IL-6 production could be reduced by treatment with MitoQ, a mitochondria-targeted radical scavenger. This effect was coincident with increased levels of mitochondrial $O_2^{\bullet-}$, suggesting a key role of $O_2^{\bullet-}$ in this signaling pathway. Weidinger *et al.* demonstrated that mitochondria-targeted antioxidants reduce the expression of IL-6 and iNOS in a model of systemic inflammatory response induced by LPS [58]. In leukocytes, Kröller-Schön [59] showed that elevated mtROS formation activated NADPH-oxidase at the posttranslational level. Inhibition of the mitochondrial permeability transition pore, which is supposed to facilitate the transport of $O_2^{\bullet-}$ from mitochondria to the cytoplasm, prevented activation of NADPH-oxidase. In contrast, the deficiency of mitochondrial SOD intensified the stimulation of NADPH-oxidase, suggesting that this process is mediated by $O_2^{\bullet-}$, rather than by H_2O_2. Applying specific mitochondrial inhibitors and direct detection of mtROS, Dikalov *et al.*

suggested that mtROS, presumably $O_2^{\bullet-}$, are able to activate NADPH-oxidase via activation of protein kinase C (PKC) [49]. These data suggest that mitochondrial $O_2^{\bullet-}$ orchestrate cellular ROS production upon inflammation. The same group has shown that stimulation of endothelial cells with angiotensin II elevates mitochondrial $O_2^{\bullet-}$ levels and simultaneously the activity of NADPH-oxidase in this non-immune cell type [60]. However, treatment with mitoTEMPO, a mitochondria-targeted antioxidant, or overexpression of mitochondrial SOD captured $O_2^{\bullet-}$ and decreased the activation of vascular NADPH oxidases [60]. NADPH oxidases, in turn, may regulate important cellular processes, such as cell migration [61], differentiation [62] and proliferation [63] (reviewed in [64]). These data again suggest that $O_2^{\bullet-}$ rather than H_2O_2 is involved in this signaling cascade. However, in the past, it was believed that $O_2^{\bullet-}$ does not participate in signaling, as it cannot exit mitochondria due to its polarity. Consequently, H_2O_2 formed from $O_2^{\bullet-}$ was considered a necessary intermediate of $O_2^{\bullet-}$-mediated actions. H_2O_2 is a nonpolar molecule and can easily diffuse through the membranes. Recently, however, the situation has shifted. $O_2^{\bullet-}$ has been shown to leave mitochondria via the mitochondrial permeability transition pore [65] and anion channels [66]. We also showed that $O_2^{\bullet-}$ can be released from mitochondria by using $O_2^{\bullet-}$-sensitive spin probes and electron spin resonance spectroscopy [67]. These data strongly support the postulation that $O_2^{\bullet-}$ can directly contribute to intracellular signal transduction pathways.

Nevertheless, other groups propose H_2O_2 being the RONS-based messenger in intracellular signaling cascades, as well. In contrast to $O_2^{\bullet-}$, H_2O_2 is a neutral molecule and relatively inactive. Thus, H_2O_2 is able to cover relatively large distances, up to several cell diameters, before it reacts with its target or is catabolized [68]. Therefore, it is considered as a suitable ROS-dependent signaling component.

6. Hydrogen Peroxide

In the literature, the impact of H_2O_2 on intracellular signaling is supported by the following evidence:

1. Exogenous H_2O_2 has a direct effect on signaling cascades.
2. The H_2O_2 level correlates with the effectiveness of intracellular signal transduction.
3. Genetic manipulation of catalase has an effect on signaling.
4. Upregulation of MnSOD and/or Cu/ZnSOD activates signaling.

Treatment with H_2O_2 increased the proliferation of endothelial cells in a study of Chen *et al.*, suggesting that H_2O_2 directly interferes with pathways regulating proliferation [69]. Wang *et al.* [70] showed that overexpression of the mitochondria-targeted catalase construct suppressed vascular endothelial growth factor (VEGF)-induced cell migration, suggesting the involvement of H_2O_2 in the regulation of cell migration. Schmidt *et al.* [71] demonstrated that overexpression of catalase in cell lines caused a deficiency in the activation of NF-κB in response to tumor necrosis factor (TNF), while the catalase inhibitor, aminotriazole, restored the induction of NF-κB. Overexpression of Cu/Zn-dependent SOD elevated NF-κB activation. These data suggest H_2O_2 rather than $O_2^{\bullet-}$ as the mediator of NF-kB pathway activation. Brunelle *et al.* [72] demonstrated that overexpressing glutathione peroxidase or catalase, but not SOD, stabilized HIF-1 in cells, suggesting that H_2O_2 acts as signaling molecule in the process of HIF-1 regulation. West *et al.* [73] showed that overexpressing catalase in mitochondria results in impaired bacterial killing by leukocytes, suggesting the predominant role of H_2O_2. Hoarau *et al.* [74] demonstrated that H_2O_2 plays an essential role in the development of β-cells, as it activates the ERK1/2 pathway. Other studies on H_2O_2-mediated signaling are summarized in several reviews [75–77].

7. Secondary RONS

The majority of papers on RONS-mediated intracellular signaling suggest either $O_2^{\bullet -}$ or H_2O_2 as the major signaling molecule. Only a few studies suggest that signaling molecules may be derivatives of H_2O_2. Garlid *et al.* [78] studied ROS-mediated opening of mitochondrial ATP-sensitive potassium channels and suggested an unknown derivative of H_2O_2 as a contributor to this pathway. Others suggested that HOCl may participate in extracellular, but not in intracellular signaling, for instance by interaction with TGF-ß1 [79]. A few more publications can be found on peroxynitrite-mediated signaling (reviewed in [80,81]). It has been assumed that peroxynitrite has an impact on pathways, which, under physiological conditions, are regulated by tyrosine phosphorylation and dephosphorylation. $ONOO^-$ causes tyrosine nitration, which blocks the respective signaling cascades. Tyrosine nitration seems to have a significant impact on a number of pathways, such as MAP kinase, STAT3, ERK and PKC-mediated pathways (reviewed in [80,81]). The fact that $ONOO^-$ irreversibly binds to proteins has a pathological impact on cellular function, rather than contributing to physiological intracellular signaling. This suggests that the biological impact of primary and secondary RONS is different. Primary RONS are predominantly associated with signaling, whereas secondary RONS are associated with oxidative stress. Primary RONS are regulated by SOD, catalase and peroxidases and have a specific physiological function for the regulation of intracellular signaling. The secondary RONS were evolutionarily developed for extracellular actions, predominantly as part of the innate immune system for killing bacteria. The intracellular release of such secondary RONS leads to deleterious consequences, as these are catalytically highly active, without a reliable control system for intracellular levels. We assume that, in evolution, the primary species were developed for intracellular physiological signaling and the secondary species for extracellular actions, such as killing of bacteria. However, at the same time, these species are able to cause damage to the cell.

8. Conclusions

We assume that whether RONS have a beneficial or deleterious effect is primarily defined by their quality, being primary or secondary RONS, and only secondly by their quantity (Figure 1). Therefore, we think that the common statement that at low concentrations ROS regulate physiological processes and at high concentrations are deleterious is not completely correct. Critical consideration of the existing literature shows that deleterious effects, termed as oxidative stress, do not appear if only one primary species is present in a biological system, even at high concentrations. To develop deleterious effects, a primary species reacts with another or a transition metal, yielding highly reactive secondary species, such as $ONOO^-$ or $^\bullet OH$. The secondary RONS are catalytically very active, not tightly controlled and consequently may not act as signal transducers. In contrast, primary RONS are well controlled; their reactions with targets are reversible; and they do not damage target molecules. This makes them ideal for intracellular signaling processes. Unfortunately, the majority of papers on signaling refers to RONS without specifying their types. In this review, we highlight an approach allowing one to distinguish the contribution of different RONS. This can be used to define the origin of RONS contributing to intracellular signaling cascades in future studies.

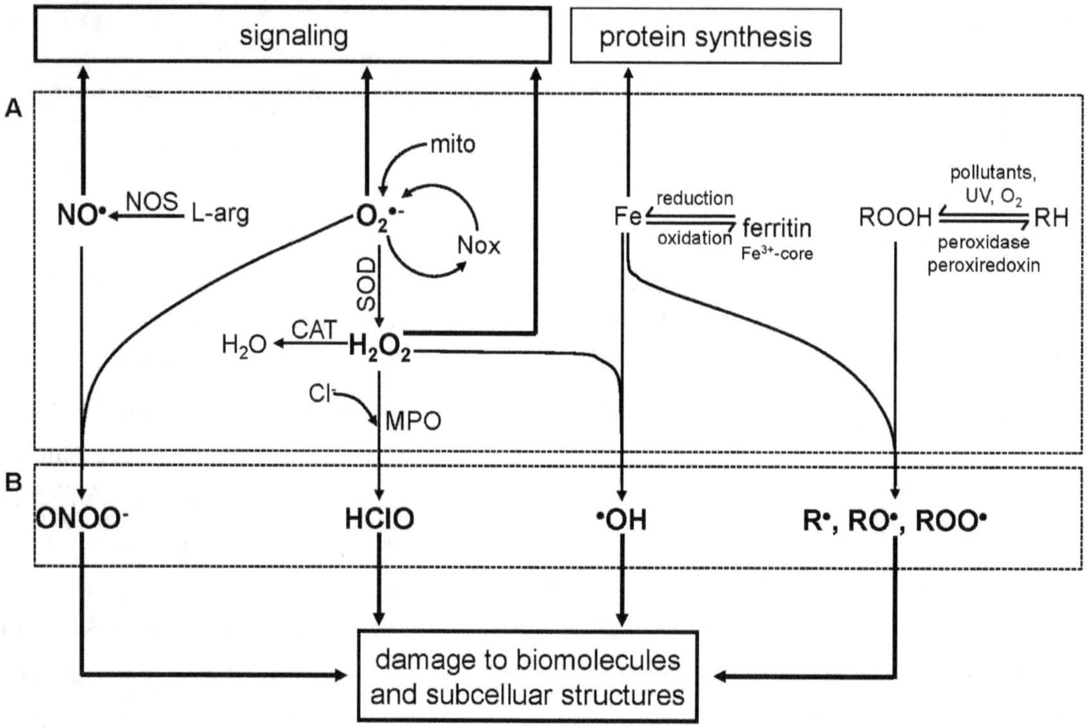

Figure 1. Scheme illustrating physiological and pathophysiological reactions of different reactive species. A, primary reactive species (NO•, $O_2^{•-}$, Fe, ROOH) and the products of the interaction of two identical reactive species (dismutation of O_2 to H_2O_2) and transition metals (reactive oxygen, nitrogen and metal species = RONMS). B, secondary products of reactions between two different RONMS. Primary products predominantly contribute to physiological processes (e.g., signaling, protein synthesis); secondary products exert deleterious effects on diverse cell functions. Abbreviations: NO, nitric oxide; $O_2^{•-}$, superoxide; Fe, iron; ROOH, lipid peroxide; H_2O_2 hydrogen peroxide; RH, non-oxidized lipid; R•, RO•, ROO•, lipid radicals; NOS, nitric oxide synthase; L-arg, L-arginine; $ONOO^-$, peroxynitrite; NOX, NADPH oxidase; mito, mitochondria; SOD, superoxide dismutase; CAT, catalase; H_2O, water; Cl^-, chloride ion; MPO, myeloperoxidase; HClO, hypochlorous acid; •OH, hydroxyl radical; UV, ultraviolet radiation.

Acknowledgments

The authors thank Asmita Banerjee for assistance in the preparation of this manuscript. Adelheid Weidinger was supported by FFG (Austria, COMET K-Project 825329, BioPersMed) and by FWF (Austria, Project P211221-B11).

References

1. Wertz, J.E.; Bolton, J.R. *Electron Spin Resonance: Elementary Theory and Practical Applications*; John Wiley & Sons: New York, NY, USA, 1972.
2. Fenton, H.J.H. Oxidation of tartaric acid in presence of iron. *J. Chem. Soc. Trans.* **1894**, *65*, 899–910.
3. Koppenol, W.H. The centennial of the Fenton reaction. *Free Radic. Biol. Med.* **1993**, *15*, 645–651.
4. Haber, F.; Weiss, J. The catalytic decomposition of hydrogen peroxide. *Proc. R. Soc.* **1934**, *147*, 332–351.
5. Koppenol, W.H. The Haber-Weiss cycle—70 Years later. *Redox Rep.* **1994**, *6*, 229–234.
6. Kehrer, J.P. The Haber-Weiss reaction and mechanisms of toxicity. *Toxicology* **2000**, *149*, 43–50.
7. Ullrich, V.; Kissner, R. Redox signaling: Bioinorganic chemistry at its best. *J. Inorg. Biochem.* **2006**, *100*, 2079–2086.
8. Semenov, N.N. *Some Problems in Chemical Kinetics and Reactivity*; Princeton University Press: Princeton, NJ, USA, 1959.
9. Vladimirov, Y.A.; Olenev, V.; Suslova, T.; Cheremisina, Z. Lipid peroxidation in mitochondrial membrane. *Adv. Lipid Res.* **1980**, *17*, 173–249.
10. Gilbert, D.L. *Oxygen and Living Processes: An Interdisciplinary Approach*; Springer: New York, NY, USA, 1981.
11. McCord, J.M.; Fridovich, I. The reduction of cytochrome c by hypoxanthine and xanthine oxidase. *Biochem. J.* **1968**, *243*, 5753–5760.
12. Mccord, J.M.; Fridovich, I. Superoxide dismutase: An enzymic function for erythrocuprein (hemocuprein). *J. Biol. Chem.* **1969**, *244*, 6049–6055.
13. Fridovich, I. Superoxide anion radical, superoxide dismutases, and related matters. *J. Biol. Chem.* **1997**, *272*, 18515–18517.
14. Boveris, B.A.; Oshino, N.; Chance, B. The cellular production of hydrogen peroxide. *Biochem. J.* **1972**, *128*, 617–630.
15. Chance, B.; Sies, H.; Boveris, A. Hydroperoxide metabolism in mammalian organs. *Physiol. Rev.* **1979**, *59*, 527–605.
16. Halliwell, B. Free radicals, reactive oxygen species and human disease: A critical evaluation with special reference to atherosclerosis. *Br. J. Exp. Pathol.* **1989**, *70*, 737–757.
17. Azizova, O.A.; Panasenko, O.; Volnov, T.; Vladimirov, Y.A. Free radical lipid oxidation affects cholesterol transfer between lipoproteins and erythrocytes. *Free Radic. Biol. Med.* **1989**, *7*, 251–257.
18. Lyras, L.; Perry, R.H.; Perry, E.K.; Ince, P.G.; Jenner, A.; Jenner, P.; Halliwell, B. Oxidative damage to proteins, lipids, and DNA in cortical brain regions from patients with dementia with Lewy bodies. *J. Neurochem.* **1998**, *71*, 302–312.
19. Halliwell, B.; Aruoma, O.I. DNA damage by oxygen-derived species. *FEBS Lett.* **1991**, *281*, 9–19.
20. Cadet, J.; Douki, T.; Ravanat, J.-L. Oxidatively generated base damage to cellular DNA. *Free Radic. Biol. Med.* **2010**, *49*, 9–21.
21. Sies, H. Oxidative stress: Introductory remarks. In *Oxidative Stress*; Academic Press: London, UK, 1985; pp. 1–8.
22. Sies, H. Oxidative stress: Oxidants and antioxidants. *Exp. Physiol.* **1997**, *82*, 291–295.

23. Rosen, H.; Klebanoff, S.J.; Wang, Y.; Brot, N.; Heinecke, J.W.; Fu, X. Methionine oxidation contributes to bacterial killing by the myeloperoxidase system of neutrophils. *Proc. Natl. Acad. Sci. USA* **2009**, *106*, 18686–18691.
24. Schwab, L.; Goroncy, L.; Palaniyandi, S.; Gautam, S.; Triantafyllopoulou, A.; Mocsai, A.; Reichardt, W.; Karlsson, F.J.; Radhakrishnan, S.V.; Hanke, K.; *et al.* Neutrophil granulocytes recruited upon translocation of intestinal bacteria enhance graft-versus-host disease via tissue damage. *Nat. Med.* **2014**, *20*, 648–654.
25. Arthur, M.; Kowalski-Saunders, P.; Gurney, S.; Tolcher, R.; Bull, F.; Wright, R. Reduction of ferricytochrome C may underestimate superoxide production by monocytes. *J. Immunol. Methods* **1987**, *98*, 63–69.
26. Cadet, J.; Wagner, J.R. Oxidatively generated base damage to cellular DNA by hydroxyl radical and one-electron oxidants: Similarities and differences. *Arch. Biochem. Biophys.* **2014**, *557*, 47–54.
27. Salgo, M.G.; Squadrito, G.L.; Pryor, W.A. Peroxynitrite causes apoptosis in rat thymocytes. *Biochem. Biophys. Res. Commun.* **1995**, *215*, 1111–1118.
28. Panasenko, O.M.; Evgina, S.A.; Driomina, E.S.; Sharov, V.S.; Sergienko, V.I.; Vladimirov, Y.A. Hypochlorite induces lipoproteins and lipid peroxidation in blood phospholipid liposomes. *Free Radic. Biol. Med.* **1995**, *19*, 133–140.
29. Chevion, M. A site-specific mechanism for free radical induced biological damage: The essential role of redox-active transition metals. *Free Radic. Biol. Med.* **1988**, *5*, 27–37.
30. Roy, B.; Hecht, S.M. Hairpin DNA sequences bound strongly by bleomycin exhibit enhanced double-strand cleavage. *J. Am. Chem. Soc.* **2014**, *136*, 4382–4393.
31. Violi, F.; Milite, M.T.; Medica, I.C.; La, Á. Nitric oxide and its role in lipid peroxidation. *Diabetes Metab. Res. Rev.* **1999**, *15*, 283–288.
32. Cai, H.; Harrison, D.G. Endothelial dysfunction in cardiovascular disease. The role of oxidant stress. *Circ. Res.* **2000**, *87*, 840–844.
33. Yung, L.M.; Leung, F.P.; Yao, X.; Chen, Z.-Y.; Huang, Y. Reactive oxygen species in vascular wall. *Cardiovasc. Hematol. Disord. Drug Targets* **2006**, *6*, 1–19.
34. Schildknecht, S.; Ullrich, V. Peroxynitrite as regulator of vascular prostanoid synthesis. *Arch. Biochem. Biophys.* **2009**, *484*, 183–189.
35. Samokyszyns, V.M.; Miller, D.M.; Reif, D.W.; Austq, S.D. Inhibition of superoxide and ferritin-dependent lipid peroxidation by ceruloplasmin. *J. Biol. Chem.* **1989**, *264*, 21–26.
36. Marklund, S.L. Extracellular superoxide dismutase and other superoxide dismutase isoenzymes in tissues from nine mammalian species. *Biochem. J.* **1984**, *222*, 649–655.
37. Marklund, S.L. Extracellular superoxide dismutase in human tissues and human cell lines. *J. Clin. Invest.* **1984**, *74*, 1398–1403.
38. Fattman, C.L.; Schaefer, L.M.; Oury, T.D. Extracellular superoxide dismutase in biology and medicine. *Free Radic. Biol. Med.* **2003**, *35*, 236–256.
39. Fukai, T.; Ushio-Fukai, M. Superoxide dismutases: Role in redox signaling, vascular function, and diseases. *Antioxid. Redox Signal.* **2011**, *15*, 1583–1606.
40. Sandstrom, J.; Nilsson, P.; Karlsson, K.; Marklund, S.L. 10-Fold increase in human plasma extracellular superoxide dismutase content caused by a mutation in heparin-binding domain. *J. Biol. Chem.* **1994**, *269*, 19163–19166.

41. Stamler, J.S.; Carolina, N. Redox signaling: Nitrosylation and related target interactions of nitric oxide. *Cell* **1994**, *76*, 931–936.
42. Foster, M.W.; McMahon, T.J.; Stamler, J.S. S-nitrosylation in health and disease. *Trends Mol. Med.* **2003**, *9*, 160–168.
43. Guikema, B.; Lu, Q.I.; Heuil, D.J. Chemical considerations and biological selectivity of protein nitrosation: Implications for NO-mediated signal transduction. *Antioxid. Redox Signal.* **2005**, *7*, 593–606.
44. Ahmed, K.A.; Sawa, T.; Akaike, T. Protein cysteine S-guanylation and electrophilic signal transduction by endogenous nitro-nucleotides. *Amino Acids* **2011**, *41*, 123–130.
45. Piantadosi, C.A. Regulation of mitochondrial processes by protein S-nitrosylation. *Biochim. Biophys. Acta* **2012**, *1820*, 712–721.
46. Sha, Y.; Marshall, H.E. S-nitrosylation in the regulation of gene transcription. *Biochim. Biophys. Acta* **2012**, *1820*, 701–711.
47. Sawa, T.; Zaki, M.H.; Okamoto, T.; Akuta, T.; Tokutomi, Y.; Kim-Mitsuyama, S.; Ihara, H.; Kobayashi, A.; Yamamoto, M.; Fujii, S.; *et al.* Protein S-guanylation by the biological signal 8-nitroguanosine 3',5'-cyclic monophosphate. *Nat. Chem. Biol.* **2007**, *3*, 727–735.
48. Daiber, A. Redox signaling (cross-talk) from and to mitochondria involves mitochondrial pores and reactive oxygen species. *Biochim. Biophys. Acta* **2010**, *1797*, 897–906.
49. Dikalov, S.I.; Nazarewicz, R.R. Angiotensin II-induced production of mitochondrial reactive oxygen species: Potential mechanisms and relevance for cardiovascular disease. *Antioxid. Redox Signal.* **2013**, *19*, 1085–1094.
50. Schulz, E.; Wenzel, P.; Münzel, T.; Daiber, A. Mitochondrial redox signaling: Interaction of mitochondrial reactive oxygen species with other sources of oxidative stress. *Antioxid. Redox Signal.* **2014**, *20*, 308–324.
51. Bashan, N.; Kovsan, J.; Kachko, I.; Ovadia, H.; Rudich, A. Positive and negative regulation of insulin signaling by reactive oxygen and nitrogen species. *Physiol. Rev.* **2009**, *89*, 27–71.
52. Cash, T.P.; Pan, Y.; Simon, M.C. Reactive oxygen species and cellular oxygen sensing. *Free Radic. Biol. Med.* **2007**, *43*, 1219–1225.
53. Klimova, T.; Chandel, N.S. Mitochondrial complex III regulates hypoxic activation of HIF. *Cell Death Differ.* **2008**, *15*, 660–666.
54. Thannickal, V.J.; Fanburg, B.L. Reactive oxygen species in cell signaling. *Am. J. Physiol. Lung Cell. Mol. Physiol.* **2000**, *279*, L1005–L1028.
55. Murphy, M.P. How mitochondria produce reactive oxygen species. *Biochem. J.* **2009**, *417*, 1–13.
56. Zhou, R.; Yazdi, A.S.; Menu, P.; Tschopp, J. A role for mitochondria in NLRP3 inflammasome activation. *Nature* **2011**, *469*, 221–225.
57. Bulua, A.C.; Simon, A.; Maddipati, R.; Pelletier, M.; Park, H.; Kim, K.-Y.; Sack, M.N.; Kastner, D.L.; Siegel, R.M. Mitochondrial reactive oxygen species promote production of proinflammatory cytokines and are elevated in TNFR1-associated periodic syndrome (TRAPS). *Exp. Med.* **2011**, *208*, 519–533.
58. Weidinger, A.; Müllebner, A.; Paier-Pourani, J.; Banerjee, A.; Miller, I.; Lauterböck, L.; Duvigneau, J.C.; Skulachev, V.P.; Redl, H.; Kozlov, A.V.; *et al.* Vicious inducible nitric oxide

synthase-mitochondrial reactive oxygen species cycle accelerates inflammatory response and causes liver injury in rats. *Antioxid. Redox Signal.* **2015**, *22*, 572–586.

59. Kröller-Schön, S.; Steven, S.; Kossmann, S.; Scholz, A.; Daub, S.; Oelze, M.; Xia, N.; Hausding, M.; Mikhed, Y.; Zinssius, E.; *et al.* Molecular mechanisms of the crosstalk between mitochondria and NADPH oxidase through reactive oxygen species-studies in white blood cells and in animal models. *Antioxid. Redox Signal.* **2014**, *20*, 247–266.
60. Dikalova, A.E.; Bikineyeva, A.T.; Budzyn, K.; Nazarewicz, R.R.; McCann, L.; Lewis, W.; Harrison, D.G.; Dikalov, S.I. Therapeutic targeting of mitochondrial superoxide in hypertension. *Circ. Res.* **2010**, *107*, 106–116.
61. Ullevig, S.; Zhao, Q.; Lee, C.F.; Seok Kim, H.; Zamora, D.; Asmis, R. NADPH oxidase 4 mediates monocyte priming and accelerated chemotaxis induced by metabolic stress. *Arterioscler. Thromb. Vasc. Biol.* **2012**, *32*, 415–426.
62. Li, J.; Stouffs, M.; Serrander, L.; Banfi, B.; Bettiol, E.; Charnay, Y.; Steger, K.; Krause, K.-H.; Jaconi, M.E. The NADPH oxidase NOX4 drives cardiac differentiation: Role in regulating cardiac transcription factors and map kinase activation. *Mol. Biol. Cell* **2006**, *17*, 3978–3988.
63. Crosas-Molist, E.; Bertran, E.; Sancho, P.; López-Luque, J.; Fernando, J.; Sánchez, A.; Fernández, M.; Navarro, E.; Fabregat, I. The NADPH oxidase NOX4 inhibits hepatocyte proliferation and liver cancer progression. *Free Radic. Biol. Med.* **2014**, *69*, 338–347.
64. Crestani, B.; Besnard, V.; Boczkowski, J. Signalling pathways from NADPH oxidase-4 to idiopathic pulmonary fibrosis. *Int. J. Biochem. Cell Biol.* **2011**, *43*, 1086–1089.
65. Hou, Y.; Ghosh, P.; Wan, R.; Ouyang, X.; Cheng, H.; Mattson, M.; Cheng, A. Permeability transition pore-mediated mitochondrial superoxide flashes. *Neurobiol. Aging* **2014**, *35*, 975–989.
66. Lustgarten, M.S.; Bhattacharya, A.; Muller, F.L.; Jang, Y.C.; Shimizu, T.; Shirasawa, T.; Richardson, A.; van Remmen, H. Complex I generated, mitochondrial matrix-directed superoxide is released from the mitochondria through voltage dependent anion channels. *Biochem. Biophys. Res. Commun.* **2012**, *422*, 515–521.
67. Piskernik, C.; Haindl, S.; Behling, T.; Gerald, Z.; Kehrer, I.; Redl, H.; Kozlov, A.V. Antimycin A and lipopolysaccharide cause the leakage of superoxide radicals from rat liver mitochondria. *Biochim. Biophys. Acta* **2008**, *1782*, 280–285.
68. Chen, K. Beyond LDL oxidation: ROS in vascular signal transduction. *Free Radic. Biol. Med.* **2003**, *35*, 117–132.
69. Chen, K.; Thomas, S.R.; Albano, A.; Murphy, M.P.; Keaney, J.F. Mitochondrial function is required for hydrogen peroxide-induced growth factor receptor transactivation and downstream signaling. *J. Biol. Chem.* **2004**, *279*, 35079–35086.
70. Wang, Y.; Zang, Q.S.; Liu, Z.; Wu, Q.; Maass, D.; Dulan, G.; Shaul, P.W.; Melito, L.; Frantz, D.E.; Kilgore, J.A.; *et al.* Regulation of VEGF-induced endothelial cell migration by mitochondrial reactive oxygen species. *Am. J. Physiol.* **2011**, *301*, C695–C704.
71. Schmidt, K.N.; Amstad, P.; Cerutti, P.; Baeuerle, P.A. The roles of hydrogen peroxide and superoxide as messengers in the activation of transcription factor NF-κB. *Chem. Biol.* **1995**, *2*, 13–22.
72. Brunelle, J.K.; Bell, E.L.; Quesada, N.M.; Vercauteren, K.; Tiranti, V.; Zeviani, M.; Scarpulla, R.C.; Chandel, N.S. Oxygen sensing requires mitochondrial ROS but not oxidative phosphorylation. *Cell Metab.* **2005**, *1*, 409–414.

73. West, A.P.; Brodsky, I.E.; Rahner, C.; Woo, D.K.; Erdjument-Bromage, H.; Tempst, P.; Walsh, M.C.; Choi, Y.; Shadel, G.S.; Ghosh, S.; *et al.* TLR signalling augments macrophage bactericidal activity through mitochondrial ROS. *Nature* **2011**, *472*, 476–480.
74. Hoarau, E.; Chandra, V.; Rustin, P.; Scharfmann, R.; Duvillie, B. Pro-oxidant/antioxidant balance controls pancreatic β-cell differentiation through the ERK1/2 pathway. *Cell Death Dis.* **2014**, doi:10.1038/cddis.2014.441.
75. Henriksen, E.J. Effects of H_2O_2 on insulin signaling the glucose transport system in mammalian skeletal muscle. *Methods Enzymol.* **2013**, *528*, 269–278.
76. Marinho, H.S.; Real, C.; Cyrne, L.; Soares, H.; Antunes, F. Hydrogen peroxide sensing, signaling and regulation of transcription factors. *Redox Biol.* **2014**, *2*, 535–562.
77. Bretón-Romero, R.; Lamas, S. Hydrogen peroxide signaling in vascular endothelial cells. *Redox Biol.* **2014**, *2*, 529–534.
78. Garlid, A.O.; Jaburek, M.; Jacobs, J.P.; Garlid, K.D. Mitochondrial reactive oxygen species: Which ROS signals cardioprotection? *Am. J. Physiol. Heart Circ. Physiol.* **2013**, *305*, H960–H968.
79. Feng, X.-X.; Liu, M.; Yan, W.; Zhou, Z.-Z.; Xia, Y.-J.; Tu, W.; Li, P.-Y.; Tian, D.-A. β3 Integrin promotes TGF-β1/H_2O_2/HOCl-mediated induction of metastatic phenotype of hepatocellular carcinoma cells by enhancing TGF-β1 signaling. *PLOS ONE* **2013**, *8*, e79857.
80. Klotz, L.-O.; Schroeder, P.; Sies, H. Peroxynitrite signaling: Receptor tyrosine kinases and activation of stress-responsive pathways. *Free Radic. Biol. Med.* **2002**, *33*, 737–743.
81. Szabó, C.; Ischiropoulos, H.; Radi, R. Peroxynitrite: Biochemistry, pathophysiology and development of therapeutics. *Nat. Rev.* **2007**, *6*, 662–680.

Permissions

All chapters in this book were first published by MDPI; hereby published with permission under the Creative Commons Attribution License or equivalent. Every chapter published in this book has been scrutinized by our experts. Their significance has been extensively debated. The topics covered herein carry significant findings which will fuel the growth of the discipline. They may even be implemented as practical applications or may be referred to as a beginning point for another development.

The contributors of this book come from diverse backgrounds, making this book a truly international effort. This book will bring forth new frontiers with its revolutionizing research information and detailed analysis of the nascent developments around the world.

We would like to thank all the contributing authors for lending their expertise to make the book truly unique. They have played a crucial role in the development of this book. Without their invaluable contributions this book wouldn't have been possible. They have made vital efforts to compile up to date information on the varied aspects of this subject to make this book a valuable addition to the collection of many professionals and students.

This book was conceptualized with the vision of imparting up-to-date information and advanced data in this field. To ensure the same, a matchless editorial board was set up. Every individual on the board went through rigorous rounds of assessment to prove their worth. After which they invested a large part of their time researching and compiling the most relevant data for our readers.

The editorial board has been involved in producing this book since its inception. They have spent rigorous hours researching and exploring the diverse topics which have resulted in the successful publishing of this book. They have passed on their knowledge of decades through this book. To expedite this challenging task, the publisher supported the team at every step. A small team of assistant editors was also appointed to further simplify the editing procedure and attain best results for the readers.

Apart from the editorial board, the designing team has also invested a significant amount of their time in understanding the subject and creating the most relevant covers. They scrutinized every image to scout for the most suitable representation of the subject and create an appropriate cover for the book.

The publishing team has been an ardent support to the editorial, designing and production team. Their endless efforts to recruit the best for this project, has resulted in the accomplishment of this book. They are a veteran in the field of academics and their pool of knowledge is as vast as their experience in printing. Their expertise and guidance has proved useful at every step. Their uncompromising quality standards have made this book an exceptional effort. Their encouragement from time to time has been an inspiration for everyone.

The publisher and the editorial board hope that this book will prove to be a valuable piece of knowledge for researchers, students, practitioners and scholars across the globe.

List of Contributors

Anton Hermann and Thomas M. Weiger
Department of Cell Biology, Division of Cellular and Molecular Neurobiology, University of Salzburg, Salzburg 5020, Austria

Guzel F. Sitdikova
Department of Physiology of Man and Animals, Kazan Federal University, Kazan 420008, Russia

Kerstin Nowotny, Tobias Jung, Annika Höhn, Daniela Weber and Tilman Grune
Department of Molecular Toxicology, German Institute of Human Nutrition Potsdam-Rehbruecke, Arthur-Scheunert-Allee 114-116, 14558 Nuthetal, Germany

Rudolf J. Schaur
Institute of Molecular Biosciences, University of Graz, Heinrichstrasse 33a, 8010 Graz, Austria

Werner Siems
Institute for Medical Education, KortexMed GmbH, Hindenburgring 12a, 38667 Bad Harzburg, Germany

Nikolaus Bresgen
Division of Genetics, Department of Cell Biology, University of Salzburg, Hellbrunnerstasse 34, 5020 Salzburg, Austria

Peter M. Eckl
Division of Genetics, Department of Cell Biology, University of Salzburg, Hellbrunnerstasse 34, 5020 Salzburg, Austria

Alessandra da Silva Dantas
Departamento de Biologia Celular e Genética, Instituto de Biologia Roberto Alcantara Gomes, Universidade do Estado do Rio de Janeiro (UERJ), Rio de Janeiro 20550-013, Brazil

Alison Day, Mélanie Ikeh, Iaroslava Kos, Beatrice Achan and Janet Quinn
Institute for Cell and Molecular Biosciences, Faculty of Medical Sciences, Newcastle University, Newcastle upon Tyne NE2 4HH, UK

H. P. Vasantha Rupasinghe, Mannfred M. A. Boehm, Satvir Sekhon-Loodu and Indu Parmar
Department of Environmental Sciences, Faculty of Agriculture, Dalhousie University, Truro, NS B2N 5E3, Canada

Bob Bors
Department of Plant Sciences, University of Saskatchewan, Saskatoon, SK S7N 5A8, Canada

Andrew R. Jamieson
Agriculture and Agri-Food Canada, Kentville, NS B4N 1J5, Canada

Gabriel Piedrafita and Markus A Keller
Department of Biochemistry, University of Cambridge, 80 Tennis Court Rd, Cambridge CB2 1GA, UK

Markus Ralser
Department of Biochemistry, University of Cambridge, 80 Tennis Court Rd, Cambridge CB2 1GA, UK
The Francis Crick Institute, Mill Hill Laboratory, The Ridgeway, London NW1 7AA, UK

Adelheid Weidinger and Andrey V. Kozlov
Ludwig Boltzmann Institute for Experimental and Clinical Traumatology, AUVA Research Center, Donaueschingenstraße 13, 1200 Vienna, Austria

Index

www.ingramcontent.com/pod-product-compliance
Lightning Source LLC
LaVergne TN
LVHW081014210926
848485LV00004B/860